PHYSIOLOGICAL CONTROLS
and
REGULATIONS

WILLIAM S. YAMAMOTO and JOHN R. BROBECK

Editors

Bicentennial Volume

Department of Physiology, School of Medicine

University of Pennsylvania

W. B. SAUNDERS COMPANY

Philadelphia and London, 1965

Physiological Controls and Regulations

In memory of

HENRY CUTHBERT BAZETT, C.B.E.

1885–1950

Professor of Physiology, University of Pennsylvania, 1921–1950,
whose interest in physiological controls and regulations remains
very much alive in our Department

Contributors

JOHN R. BROBECK, Ph.D., M.D.

> Professor and Chairman of the Department of Physiology, School of Medicine, University of Pennsylvania.

FRANK P. BROOKS, M.D.

> Associate Professor of Physiology and of Medicine, School of Medicine, University of Pennsylvania; Chief of the Gastrointestinal Section, Hospital of the University of Pennsylvania.

RICHARD A. DAVIS, M.D.

> Assistant Professor of Neurosurgery, School of Medicine, University of Pennsylvania.

RALPH B. DELL, M.D.

> Visiting Scholar, College of Physicians and Surgeons, Columbia University; Visiting Fellow in Pediatrics, Babies Hospital, Columbia-Presbyterian Medical Center, New York, N.Y.

McIVER W. EDWARDS, JR., M.D.

> Associate in Physiology and Scholar of the Pennsylvania Plan, School of Medicine, University of Pennsylvania.

C. L. HAMILTON, Ph.D.

> Research Psychologist, Veterans Administration Hospital, Coatesville, Pennsylvania; Assistant Professor of Physiology, School of Medicine, University of Pennsylvania.

H. T. HAMMEL, Ph.D.

Fellow, John B. Pierce Foundation Laboratory, New Haven, Connecticut.

JAMES D. HARDY, Ph.D.

Fellow, John B. Pierce Foundation Laboratory, New Haven, Connecticut; Professor of Physiology, School of Medicine, Yale University.

S. M. McCANN, M.D.

Professor and Chairman of the Department of Physiology, Southwestern Medical School, The University of Texas, Dallas, Texas.

LYSLE H. PETERSON, M.D.

Professor of Physiology, School of Medicine, University of Pennsylvania; Director, Bockus Research Institute, Graduate Hospital of the University of Pennsylvania.

GERARD P. SMITH, M.D.

Assistant Professor of Physiology, School of Medicine, University of Pennsylvania.

JAMES A. F. STEVENSON, M.D.

Professor and Head of the Department of Physiology, University of Western Ontario, London, Ontario, Canada.

SAUL WINEGRAD, M.D.

Assistant Professor of Physiology, School of Medicine, University of Pennsylvania.

ROBERT W. WINTERS, M.D.

Professor of Pediatrics, College of Physicians and Surgeons, Columbia University; Attending Pediatrician, Babies Hospital and Vanderbilt Clinic, Columbia-Presbyterian Medical Center, New York, N.Y.

WILLIAM S. YAMAMOTO, M.D.

Associate Professor of Physiology, School of Medicine, University of Pennsylvania.

Preface

A few years ago when most of us were members of the faculty of the same Department of Physiology, we decided to write a book about controls and regulations because they are the common focus of our several research interests. The Bicentennial Year of the School of Medicine of the University of Pennsylvania was selected as the date of publication. Happily, it is also the hundredth anniversary of the publication of Claude Bernard's *L'Introduction à l'Étude de la Médicine Expérimentale*.

It is obvious that our group is not large enough or diverse enough in talent to review all possible examples of biological control systems. We have written about those we are studying, at the level of complexity at which we work. To help us make the volume more comprehensive, however, Professor James A. F. Stevenson of the University of Western Ontario has contributed the chapter on water exchange; we are grateful for his aid. We also wish to thank the staff of the W. B. Saunders Co. for their encouragement and generous help.

One can see in these essays the intention of the several authors to survey their special areas of study from the point of view that regular behavior in physiological systems is neither fortuitous nor unsystematic. Moreover, we hope that these pages reflect the excitement and ferment that have infused many areas of physiological study as a result of the collision of traditional physiology with traditional control systems engineering. Third, we want to try to show that apparently unorganized "facts" in physiology can be articulated into logically pleasing patterns of relationships. Because the authors differ in their realization of these three goals, the volume displays some of the range of interpretation which may be given to the words "regulation" and "control." To mathematicians or control systems engineers, the level of formal abstraction in these chapters may seem disappointing and "behind" the "state of the art." Conversely, to some of us, a "state of the art" engineering exposition of biological control systems may seem disappointing and "behind" contemporary experimental physiology. Between the converging disciplines of physiology and control systems engineering there is still a substantial gap.

We offer this volume mainly for the interest of medical and graduate students looking for places to begin their own exploration of these relationships. We hope that this record of our experience and conclusions will be of value, at least as a means of encouraging their endeavor.

W. S. Y.

J. R. B.

Contents

Chapter 1

EXCHANGE,
CONTROL, AND REGULATION

JOHN R. BROBECK

To be alive has different meanings in different fields of biology. It may mean respiring, reproducing, or reacting to stimuli; or the occurrence of typical biochemical reactions; or the existence of biophysical states or processes having certain specified properties. But it also means the phenomena which can be identified as "exchanging," since living animals and plants carry on almost continuous exchanges of energy and substance between their own economy and their environment. Physiology is in part the study of these exchanges. It can be expected to answer three questions, namely, what happens, how fast does it happen, and what determines how fast? For any given exchange these questions may be phrased as follows: (1) What is the nature of the exchange from a purely descriptive viewpoint? (2) What is the rate of exchange under varied physiological and environmental conditions? (3) What physiological mechanisms govern the rate of exchange? The last of these, in turn, may be stated as two subsidiary questions: What mechanisms govern rate of gain? What governs rate of loss?

Physiologists of earlier generations have thought about these questions and given us a few principles which should be reviewed. For example, Lavoisier, the French chemist, biochemist, and physiologist of the eighteenth century, provided a basic generalization for study of exchanges, and introduced the concept of their regulation, in a statement which Adolph (1961) has translated as follows: "Regulation may consist in governed exchanges of substance." Lavoisier knew that living systems must be able to carry on these exchanges, and in order to preserve and perpetuate themselves they must be able to control the rate.

A second name is equally familiar, that of Claude Bernard, the distinguished French physiologist of the middle of the nineteenth century. Bernard wrote of the stability which the body maintains, notably in the composition of its fluids, the plasma and interstitial fluids, which he called the internal environment. He discovered by chemical analysis that their composition is practically constant from time to time. Many years later the American physiologist, Walter Bradford Cannon (1929), invented a name for this constancy, the term "homeostasis." His word has come to have two different meanings, either to designate the constancy, or the processes by which the constancy is preserved. Cannon proposed it for the former, as a synonym for "steady state"; but other authors have applied it to reactions for attaining the steady state, possibly because true equilibrium is never established, or possibly because the means of approaching constancy may be more interesting than the steady state, per se.

Fourth on the list of prominent names is Rudolf Schoenheimer (1942). Some thirty years ago he was one of several investigators who showed, using compounds identified by isotopic methods, that the atoms of the body do not have a static constancy but a dynamic one. Schoenheimer gave a series of lectures published in a small yet fascinating volume, *The Dynamic State of the Body Constituents*. The book was at once recognized as a landmark in the history of biological exchanges— and in the history of biology, for that matter. Perhaps a dynamic state could have been predicted from observations like those of Lavoisier. Yet it was only when molecules could be labeled by isotopic tags that it became possible to measure the rate of turnover of the materials in the body and thus to disclose the surprising speed and apparent ease with which atoms and molecules are replaced.

The discoveries of these four men, Lavoisier, Bernard, Cannon, and Schoenheimer, are only a part of the history of physiological exchanges, of course. Other names could be added, including E. F. Adolph because of the influence of his monograph on *Physiological Regulations* (1943) and his studies of water exchange and other regulations. Another important figure was the late André Mayer, father of Jean Mayer of Harvard (whose work has popularized the "glucostatic theory" of control of food intake). André Mayer was an outstanding French investigator in physiology, biochemistry, and nutrition. Just before the outbreak of the Second World War, he and Gasnier wrote an important series of papers on regulation of exchange of body substance, including food intake and the compounds the body derives from its food (Gasnier and Mayer, 1939). The exchanges they described are, in fact, subdivisions of the overall energy exchange of the body.

TYPICAL EXCHANGES

Energy is one of the most important of the exchanges of living systems. Plants utilize it in radiant form for growth, for movement of fluids, and for reproduction. Some of the energy they absorb from the sun is, in turn, radiated as heat. Animals are practically unable to absorb radiant energy for any synthetic purpose or for the accomplishment of any kind of work. Their energy must come from compounds previously synthesized either by plants or by other animals. As a general principle one can say that these compounds must have a chemical composition similar to the constituents of the body itself. The energy is present in chemical bondings, which upon rearrangement may release energy for growth, motion, secretion, reabsorption, or electrochemical events. With food as the energy source, heat, work, and the storage of compounds within the body are the forms in which energy is disposed of.

Heat is especially interesting among these variables, because the body may be said to have a heat exchange as well as an energy exchange. These two types of exchange are interrelated in that heat is produced during the course of energy exchange. But at the same time they are distinct because once energy has been converted into heat it cannot re-enter the economy of the body for conversion into any other type of energy. Perhaps this explains why the study of "temperature regulation," i.e., control of heat exchange, has developed as a relatively independent subject. It is comparatively easy to measure the rate of heat production within the body as well as the rate of gain or loss to the environment. Mechanisms which alter rate of gain or loss are apparent since they are on the surface of the body. Moreover, their control by the nervous system can be explored fairly readily. This kind of research employs techniques based upon well-established physical principles, and it has gone ahead perhaps more successfully than study of any other exchange. Complications do arise, however, when subtle changes in rate of heat production must be taken into account. At this point a study of thermal exchange encounters the problems of the intermediary metabolism, or biochemistry and biophysics of organs and tissues, and immediately becomes more difficult.

Other exchanges of the whole animal are equally well known; some of them, too, have been studied since the time of Lavoisier. The respiratory gases, oxygen and carbon dioxide, and the water gained and lost by the body might well serve as prototypes in any discussion of physiological exchange. Rates of their gain and loss are variable and are controlled by physiological mechanisms. This may be true, also, of the exchange of other materials, such as the individual cations including sodium, potassium, calcium, and magnesium, or the anions of chloride and phosphate. For these electrolytes, however, only the mechanisms of loss have been studied, and not the mechanisms of gain. Most discussions of their exchange imply that the body receives with no specific control a certain amount of these ions as components of its food, and then regulates the amount within the body by adjustments in the rate of loss. We do not know that this is true; there is evidence that animals will seek for an electrolyte—salt, for example—in a fashion which suggests a behavioral control of intake; more data are needed on these questions. When it comes to less common elements such as iron, sulfur, and others, almost nothing is known about control of either gain or loss. As a matter of fact, for one of the most plentiful elements in the body, its nitrogen, the mechanisms controlling gain and loss are practically unknown.

A few moments of attention to these overall exchanges makes it clear that what happens for the body as a whole happens also in its several organs and parts. The arm or leg has its own characteristic energy exchange, with foodstuffs transported by the blood as its energy sources, and work, heat, and storage (of muscle protein, for example, or muscle glycogen) as destination of the energy it receives. In a similar fashion one can characterize the exchanges which take place within organs such as the liver or kidney. The blood stream brings energy, oxygen, glucose, amino acids, and a multitude of other compounds at a given temperature, and carries away at some other temperature much of what it brought plus whatever may have been added by the organ, or less whatever may have been taken for excretion into the bile or the urine. Even a relatively simple physiological system like the gallbladder has its own kinds of exchange—to and from the blood, and by way of the bile it receives from the hepatic ducts and discharges via the cystic duct into the duodenum.

The concept of exchange is so fundamental in physiology that one is tempted to try to apply it to every system. What about the glands? Do they have exchanges? They do for energy, for respiratory gases, and for materials they pass from the blood into their own secretions. They do, also, for compounds they may synthesize from other materials. In many instances one can identify processes of synthesis (gain), storage, and release (loss) of a product (a hormone like thyroxin is a good example), and can study the mechanisms controlling the rate of synthesis and those governing the rate of release. The quantity stored is determined by these rates. An analogous situation is found in the circulatory system, where the great vessels of the arterial system gain blood at a rate determined by mechanisms controlling the output of the heart, and lose blood through arterioles and capillaries at a rate determined by the vasomotor system controlling the size of these small vessels. The amount of blood within the large arteries depends upon how rapidly it is put in and how quickly it flows out. What is true of blood is true also of mechanical energy. The heart is the source of input of energy into the vascular system, the vessels are the site of energy dissipation by friction as the blood flows along. How much energy is stored within the aorta and larger arteries is measured under most conditions by determining the force the blood exerts against their walls. This force is known as the blood pressure; it has received much attention in studies of control or regulation.

CONTROL AND REGULATION

Blood pressure is one of the physiological quantities to which the term, regulation, is most commonly applied. As Lavoisier used the term, it means a governed exchange, of energy in this case. Blood pressure is said to be regulated for one reason, because it remains relatively constant under so many different conditions. For example, if a wound produces hemorrhage, bodily changes occur which tend to prevent a fall in pressure; the cardiac output will be enhanced if possible, vasoconstriction will occur in other parts of the body, and vasoconstriction locally will tend to limit further blood loss through the wound. Some of these changes occur in reverse when extra blood is introduced via transfusion into the arterial tree. So pressure is said to be regulated because physiological mechanisms tend to keep it constant. It is also said to be regulated for another reason, namely, that the arterial system contains specialized cells which respond to changes in blood pressure and by their activity tend to preserve the constant level of pressure. (Later we shall see that they are not stimulated by pressure per se, but by stretching which the pressure induces.) We shall combine both of these meanings and shall use the word, regulation, to denote the preservation of a relatively constant value by means of physiological mechanisms which include a specialized detector for the value or some function of it.

The word *control* is sometimes used interchangeably with regulation, but there are good reasons for distinguishing between the two terms. Control describes management. For a physiological exchange it is management of a rate of functioning. With reference to blood pressure, a control of heart rate and stroke volume and a control of peripheral blood flow are required for regulation of the amount of blood within the arterial system and thus for regulation of blood pressure. Partial pressure of respiratory gases within body fluids is regulated; rate of oxygen consumption and of pulmonary ventilation are controlled. Body temperature is regulated—or, at least, the temperature of some idealized and possibly theoretical mechanism in the body is regulated; rate of heat production in muscle and rate of heat loss from skin and respiratory passages are controlled. Food intake is controlled; loss of heat is controlled; work output is controlled; but energy exchange may be regulated in that the energy content of the body tends to remain constant. But perhaps this is not a regulation after all, because we do not know of the existence of any specialized cells capable of responding to changes in energy content or to any variable proportional to this content. These examples are given to illustrate the differences in usage of the words *regulation* and *control;* it is apparent that controls are required to achieve regulation.

CHARACTERISTICS OF REGULATING SYSTEMS

With these ideals as an introduction, it is possible to review more systematically some of the characteristics of physiological exchanges, controls, and regulations. An outline of their interrelation illustrates that

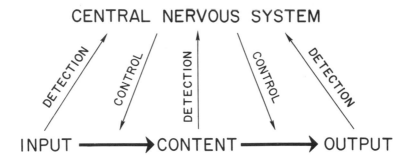

FIGURE 1-1. Relationships of input, content, and output, with outline of how physiological mechanisms detect and control rates of gain and loss.

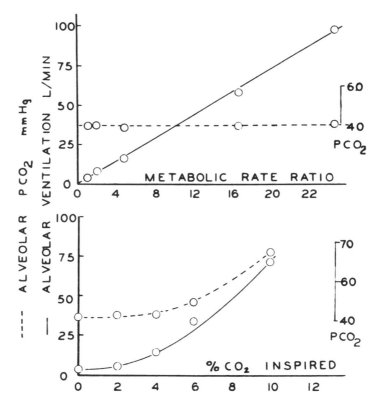

FIGURE 1-2. Alveolar P_{CO_2} and alveolar ventilation when metabolic rate is increased (upper graph) or when CO_2 is added to inspired gas mixture. (Yamamoto, 1960.)

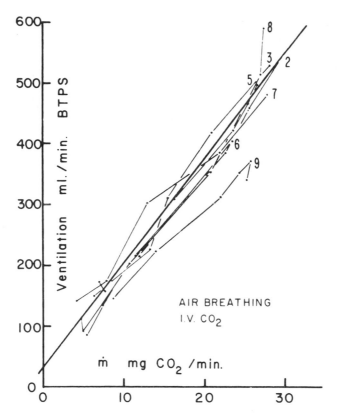

FIGURE 1-3. Regression of ventilation (ml/min) upon CO₂ output (mg/min) for anesthetized rats given an infusion of blood artificially loaded with CO_2. (Yamamoto and Edwards, 1960.)

in every case* there are (1) an input, gain, or intake; (2) an output, loss, or expenditure; (3) a content; (4) a mechanism for detecting content and/or changes in content; and (5) mechanisms for controlling intake and output, respectively (Fig. 1-1). In addition to these five categories, there may or may not be also (6) mechanisms for detecting rate of intake, rate of output, and/or changes in these rates. Homeostasis is possible because the body employs mechanisms of these several types in a manner which maintains a relatively constant content.

INTAKE, OUTPUT, AND CONTENT

Examples of the relationships between these three variables could be taken from almost any of the body's exchanges. Carbon dioxide is well suited for this purpose because it has been studied quantitatively.

* Higher animals only.

Yamamoto and Edwards have summarized the nature of the exchange when the carbon dioxide is produced only within the body (Fig. 1-2), and when it is introduced artificially into the blood via an apparatus for loading blood with carbon dioxide (Fig. 1-3). In Figure 1-2 the data show how, as the metabolic rate is increased some fivefold, the rate of pulmonary ventilation rises proportionately. The extra carbon dioxide is eliminated from the lungs as rapidly as it is produced, and the pressure of carbon dioxide in arterial blood does not rise. Here on the upper graph the dashes illustrate the constancy of content, while the solid line illustrates the variable rate of exchange by which the constancy is attained. Figure 1-3 shows this same variability for the other situation, in which the added carbon dioxide did not come from working muscles but from a gas exchanger perfused with the rat's blood. On this graph the abscissa is the rate of addition of carbon dioxide, and the ordinate is the rate of pulmonary ventilation. Yamamoto and Edwards showed that there was a straight-line relation between the amount of carbon dioxide added and the rate of ventilation. The slope of this line defines what may be called an "input-output" relation for exchange of carbon dioxide. This relation is important for every type of exchange. If there is known to be a regulation in which the value of a given content remains practically constant, then one can be certain that the input-output relation is like that of Figure 1-3. Conversely, if the relation is like the one described, there must be a regulation. Data of this nature demonstrate the quantitative interaction of intake, output, and content as a basis for more fundamental study of their control and regulation.

DETECTION

The fourth of the typical components of an exchanging system, the detector, has already been mentioned (in discussing blood flow and pressure) as a mechanism containing specialized cells responding to changes in pressure or in stretch. Cells having this function are known as receptors, as sensory cells, or as detectors. In the circulatory system they lie within the walls of the arteries and other vessels, and are especially prominent in the carotid sinus in the neck at the division of the carotid artery into internal and external branches. When they are stretched they send nerve impulses via the carotid sinus nerve into the brainstem, where they eventually make reflex connections with motor neurons which alter heart rate and blood flow through vascular beds. The study of detector elements requires first of all that their presence within the body be established, that their location be identified, and finally that their sensitivity be determined. The stretch receptors of the carotid sinus have been known for many years; their sensitivity to stretch, and to rate of change of stretch, have been measured much more recently.

Other detectors have received similiar attention, including those for the respiratory gases. For carbon dioxide they seem to be found in at least three locations (Masland and Yamamoto, 1962), and when all three are removed or denervated, animals no longer increase their ventilation when carbon dioxide accumulates in body fluids. Still other types of detectors are known to lie within the brain, and their responses and sensitivity have been investigated. In the rostral part of the hypothalamus near the optic chiasm and optic tracts are neurons which seem to be stimulated by changes in osmolarity—that is, in water concentration. Their activity has been recorded as the rate of discharge of single units under conditions where osmolarity was altered. The rostral hypothalamus and the part of the brain just ahead of it also contain elements which respond to temperature. When this part of the brain is warmed or cooled, regulation of body temperature is affected so that heat loss from the body is increased or decreased, respectively. Injury to this region prevents this control of rate of loss. Furthermore, Hardy and his colleagues (1964; Nakayama et al., 1961) have studied the sensitivity of single units to measured temperature changes, and have established their role in detection of temperature in very much the same fashion that other investigators have studied the properties of sensory cells in other parts of the body.

A further word may be added about what these units are detecting. Neither for carbon dioxide, nor for water, nor for heat does the specific detector actually respond to the overall content of the variable it senses. That is, there is no meter for volume of all the water in the body. Rather, the detectors respond to a function of content which is either the partial pressure of carbon dioxide, or the osmolarity, or the temperature. Each of these may be spoken of as a "potential," a driving force or a measure of the tendency of the variable to be transferred. Thus, carbon dioxide moves from a region of high partial pressure to one where it is lower; water moves from high water concentration to low; while heat moves from regions at high temperatures to regions at lower temperature.

Still another consideration is interesting. For these three quantities the detectors not only are bathed in solutions having a given potential of each, but they also produce the variable in their internal reactions. Carbon dioxide, water, and heat are what school children know as "waste" products of cellular life. Like all other cells of the body, the detector elements must be able to cast off into their environment these products of their own chemistry. This is but another example of a physiological exchange, at a cellular level. One does not need to say that these detector cells function in order to regulate an exchange for the body as a whole. More exactly, one says that they respond to variations in their own exchange and utilize the mechanisms of the rest of the body as appliances for controlling their own rate of gain or loss. In a sense, the skin is employed as a heat exchanger for cells within the hypothalamus.

Control of Gain or Loss

Mechanisms through which the exchanges of the body ultimately take place are so well known that they do not need review. They include, of course, the skin, lungs, gastrointestinal tract, and kidneys. If one considers not the whole organism but its parts, the mechanisms of exchange include the vascular system, ducts of glands, and other specialized tissues or avenues of intake or output. The control of these mechanisms of exchange is more pertinent to this discussion than are the mechanisms themselves. To generalize about the control, it is accomplished usually by way of a neural or an endocrine system, or some combination of these two.

Energy exchange may be chosen as an example. The rate of intake of energy as food is controlled entirely by the nervous system through the sensory and motor systems which bring about feeding. The rate of expenditure of energy is controlled partly by the nervous system, which manages the muscles of the body via motor nerves, and partly by endocrine glands, which determine the rate of cellular metabolism. Rate of heat production as one component of energy expenditure is similarly controlled by activity of skeletal muscles and by endocrine mechanisms controlling metabolic rate. In many discussions of these controls the autonomic or involuntary nervous system is given a prominent place. This may be justified, but equally well it may be challenged. The tissue with largest mass in the body is skeletal muscle. It also has the largest exchange of energy, heat, respiratory gases, and chemical sources of energy when the body is at rest, and has by far the greatest capacity for increasing the rate of all of these exchanges when the body is working to maximum output. Consequently, the controls of skeletal muscle are bound to be quantitatively the most significant determinants of the rate of any exchange in which muscles participate. These controls, naturally, do not belong to the autonomic, but to the somatic or voluntary division of the nervous system. Similarly, in heat production by shivering, in ventilation of the lungs, in changes in posture or location, or in any control in which behavior is a component—in all of these the contribution of the autonomic is less important than that of the somatic system. It is only by considering all divisions of the nervous system and most of the endocrine glands that one sees the scope of mechanisms controlling the whole spectrum of cells, tissues, organs, and parts of the body where exchanges are in progress.

Systematic study of a regulation often begins with experiments to identify and describe the mechanisms of control of a given process. Thus, in research upon the endocrine glands, the adenohypophysis, for example, the problem is to find out what stimulates what, and when. Because there are variations in the rate of endocrine secretion it is obvious that there must be physiological mechanisms of control; yet no one knows whether there is regulation in the strict sense as already defined (constancy as a result of detection). An outline of the mechanisms controlling the several functions of the digestive tract likewise illustrates the fact that in areas of study which belong to what is known as "classic" physiology there still remain familiar problems to be solved concerning how control systems operate.

DETECTION OF GAIN OR LOSS

Finally, a few comments may be given about the possibility of detection of values other than content or some potential related to it. Many physical systems are controlled on the basis of not only where the system is, but of where it is going. One corrects the position of the steering wheel either because the car is displaced on the highway or because it is becoming displaced. Bank accounts are replenished because the balance is low or is becoming low. These are analogous to a physiological system where, e.g., not only is heat content measured as temperature, but changes in heat content or changes in temperature are also detected. This type of response is very common in physiology.

Yet a further refinement is also possible, the detection of gain or loss without reference to content. A bank account can be kept solvent (if it was originally so) by balancing deposits against withdrawals without any knowledge of the total amount of money held by the bank. Similarly, a water supply system can be managed by measuring rate of inflow and rate of outflow, the rate of gain or loss of water from the system. This procedure may have certain advantages, in that sometimes flow is easier to monitor than content. If the control is based only upon flow, however, it may encounter systematic errors which in the long run may become devastating. For example, if the inflow and outflow gauges were to register the same value but one was in error by 0.1 percent, eventually the amount of water in the reservoir would show a considerable change. Consequently, the measurement of water level from week to week provides a necessary check upon the accuracy of flow measurements, even though water level cannot be measured very accurately from minute to minute.

Physiological detectors of rate of intake or output are not plentiful for illustration of this principle, but they do exist. The sensory elements responding to heat or cold in the skin may be mentioned. Before the body has gained any large amount of heat from a heat lamp, the cutaneous receptors will signal to the nervous system that heat gain is occurring. The converse is true during exposure to cold. For water balance there is a similar mechanism, the receptors in the mouth and oropharynx which are stimulated by changes in hydration. (One of the older theories of the genesis of thirst was the dry mouth theory, which attributed to these receptors the sensation of thirst and the motivation for drinking to satisfy the thirst.) While the central or hypothalamic detectors respond to changes in water concentration in bodily fluids, the oral detectors signal the availability of water at the top level of the digestive system.

A more elaborate version of a similar mechanism was described by Adolph and his associates as a metering system for water intake. Adolph (1943) noted the precision with which a dog drinks. If the animal's water deficit is measured precisely by laboratory techniques, and the animal is then given water to drink, he will take at one drinking almost exactly the amount the laboratory tests had shown to be needed. Drinking stops at a time when the water can be only in the stomach and not distributed to the body as a whole. For these and other reasons, Adolph concluded that there is within the mouth and pharynx a quantitative measuring of water intake, which is signalled to the central nervous system in some way so as to affect the amount of water taken by a thirsty dog.

Systems Analysis and Communications Theory

All the topics included in this chapter are discussed more fully elsewhere in this volume. Some of them are presented, or analyzed, from a point of view different from the traditional one which we have been using, and thus a word about this newer presentation is appropriate here. If we turn again to regulation of the blood pressure, where the autonomic nervous system controls heart rate and peripheral blood flow, we see that the word *control* can be applied at many other loci in the regulating system. The pressure of the arterial blood upon the vessel walls "controls" the degree of stretching of the sensory or detector elements in the carotid sinus; this stretching controls generation of impulses in the sinus nerve; the nerve impulses control reflex centers in the medulla oblongata; reflex centers control motor systems into the spinal cord which, in turn, control the preganglionic neurons lying within the cord; the preganglionic neurons control postganglionic neurons; the post-ganglionic neurons control the pacemaker of the heart; the pacemaker controls wave of excitation, excitation controls contraction, contraction controls cardiac output, and output controls blood storage in the aorta and thence the arterial pressure. So there is a circuit of control; and if there is a circuit, we must suppose that something moves around the circuit to carry the control from one element to another. Upon first inspection the nerve impulses seem to have this function—indeed, in examples considered earlier in this chapter the control was always carried by either neural or endocrine mechanisms. In regulation of blood pressure, however, and probably in every regulation, there are links in the chain of command which are neither neural nor endocrine. The trans-mission of nerve impulses and the hormones cannot possibly be a complete record of the flow of control throughout the circuit.

Some concept or theory of how the control may be transmitted is essential if we are to have any quantitative description of the inter-actions. How does one compare the control of carotid sinus mechanisms by stretching with control of force of contraction by excitation processes? For such a comparison no one physical, chemical, or biological variable will do. Nevertheless, one senses a possibility and even a necessity for some generalized relationship. To meet this need physiologists are begin-ning to take advantage of techniques developed by engineers for analysis of mechanical, electrical, and electronic control systems. In "systems analysis" the flow of control is regarded as a flow of information, in the sense that the flow of military command is based upon transmission of information. The "codes" in which the information is transcribed are being studied, and methods for measuring the content of information are being invented, so that the different parts of the system can be given the same type of mathematical treatment. In biology this means that the control of stretch by blood pressure can be discussed in the same conceptual framework as the control of generation of nerve impulses by stretch. When systems analysis is attempted upon a biological problem the components of the system continue to be the familiar parts of the body, its cells, organs, and tissues. Their function is identified, however, in a somewhat unfamiliar fashion by analogy with standard abstractions of control systems engineering. A simple and well-known example of these terms is the word *loop,* which designates the pathways over which the control flows as in the circuit for regulation of blood pressure. The circuit which we considered is a closed loop, because when pressure has altered stretching of carotid sinus detectors, the detectors by way of the rest of the circuit eventually alter pressure. This is the meaning of the word *closed* in this context.

The applicability of systems analysis to biological problems was most incisively argued by Norbert Wiener. He created a discipline to which he gave the name of cybernetics—from a Greek word meaning helmsman. The terminology of cybernetics has not replaced anything in more conventional physiological language. On the contrary, it permits us to talk about phenomena for which we previously had no vocabulary. It offers new ways of analyzing biological systems and provides a different kind of insight into their interrelations. One can now say that physiology has its foundations in three—not just in two—fundamental sciences. The first is physics and the second is chemistry, both of which began to be applied to biological problems in the nineteenth century. The third is systems analysis based upon communications theory; its usefulness in biology is only just beginning to be appreciated.

Research in the author's laboratory is supported by Grant G-15930 from the National Science Foundation.

Chapter 2

HOMEOSTASIS, CONTINUITY, AND FEEDBACK

WILLIAM S. YAMAMOTO

In the previous chapter there has been developed by history and example the idea that processes of exchange are pivotal in the discussion of regulatory mechanisms in animals. In fact, the observable regularities in life processes may be separated into two classes, those in which a clearly defined transaction in materials or energy occurs, and the ones in which no such transaction is apparent. Now we would like to develop some of the details of the phenomena involving material transactions in terms of the three words that entitle this chapter. These words offer some hope that there may be a unique and useful way of organizing large segments of the unsegregated wealth of observational detail that characterizes physiological knowledge today.

All the physiological processes which deal with exchange of materials seem bound together and inextricably linked to the idea of homeostasis. This widely respected generalization is attributable to Walter B. Cannon, but the lineage of the concept is ancient and is inseparably linked with almost every great name in physiology. The cluster of ideas which center about this word has a strong claim to being one of the few truly general and basic principles of physiology. It is our immediate purpose to inquire whether the ideas are sufficiently rigorous to be considered mathematically. If one attempts to do so, does it turn out to be too general to be useful? Although the statement of homeostasis represents the culmination of a rather extensive inductive inference, can this inference be used as a premise in a deductive system for the physiology of control and regulation? This preoccupation with the construction of a deductive system is justified on the basis that the ability to devise such a system is one of the major forms of understanding.

To begin this venture, it is not possible to do better than to quote the trenchant and melodious sentence attributed to Claude Bernard, "The fixity of the internal milieu is the necessary condition for free existence" (Bernard, 1878). But for greater detail it is necessary to quote Cannon (1929), since it is he who coined the word homeostasis and identified its context.

"The highly developed living being is an open system having many relations to its surroundings—in the respiratory and alimentary tracts and through surface receptors, neuromuscular organs and bony levers. Changes in the surroundings excite reactions in the system, or affect it directly, so that internal disturbances of the system are produced. Such disturbances are normally kept within narrow limits, because automatic adjustments within the system are brought into action, and thereby wide oscillations are prevented and the internal conditions are held fairly constant. The term 'equilibrium' might be used to designate these constant conditions. The term, however, has come to have exact meaning as applied to relatively simple physico-chemical states in closed systems where known forces are balanced. . . . The present discussion is concerned with the physiological rather than the physical arrangements for attaining constancy. The coordinated physiological reactions which maintain most of the steady states of the body are so complex, and are so peculiar to the living organism, that it has been suggested that a specific designation for these states be employed—homeostasis."

The nub of both statements is that if a particular part of the interior of organisms is examined with respect to stability of composition, the magnitude of changes observed in time is very small. Bernard identified the locus of the stable composition and associated it with freedom of existence—from the tyranny of environment or the vagaries of activity. Cannon emphasized that an organism is a system in open exchange with it surroundings, and therefore any stability that is achieved must be a dynamic one. These ideas are a particularly apt starting point for generalizations about our problems of the first class, those dealing with material exchanges between the environment and the organism. What we are concerned with is the active concentration of substances within a closed space, the envelope of the organism. Knowledge of the concentration may be obtained by a process of simple bookkeeping, keeping track of where particular molecules are located and their rate of accumulation. Basically the unit of exchange is a molecule of a particular identity, and the rate of accumulation of a particular molecular kind is the net difference between the sum of all processes which introduce these molecules from the outside and the sum of all processes that cause them to be extruded to the outside or to lose their identity by catabolism. The formal

expression of this statement has been called either a conservation principle or a continuity expression. Perhaps it is not essential to dignify it
with so sonorous a name, since after all it merely says that profit is the
net difference between income and expenditure. Since animals live in
time, the expression has the following general form:

$$\frac{d\ C(t)\ V(t)}{dt} = I(t) - O(t) \qquad\qquad 2.1$$

where $C(t)$ is the concentration of the molecule in question, $V(t)$ is the
body size or volume of dilution, $I(t)$ is the rate of production or ingestion,
the sum of the activity of all sources, and $O(t)$ is the rate of excretion or
catabolism, the sum of the activity of all sinks.

In this statement, therefore, the functions on the right are all non-
negative. To recapitulate, the equation states that for a given material
the time rate of change of the amount of material in the interior of the
organism is equal to the rate of appearance from all sources minus the
rate of disappearance to all sinks. Unlike the more general conservation
laws stated in the physical sciences, there is a relationship of this sort
for each substance that is a biologically identifiable molecule. Thus glucose
and glycogen, although they are intimately related in the economy of
the organism and are quite freely interconvertible, have separate continuity expressions. In some rare instances the mechanisms that keep
track of the molecules for some important function fail to identify or
distinguish between species of chemicals, and the system seems to go awry,
but this will not invalidate the rule.

Writing the equation in this way may at first seem unsatisfactory,
inasmuch as the term "all sources" includes such distinctive physiological
processes as eating and metabolic synthesis. The one is obviously a physical transfer from outside to inside the envelope, while the latter process
takes place entirely inside the envelope, and failure to distinguish between
ingestion and anabolism may seem an inauspicious start, indeed. Likewise,
sinks include both the catabolic and excretory processes, as well as inevitable loss such as the insensible water loss. Nevertheless, if we look for the
most general expression of the gist of transactions, we must realize that
this equation describes the systems which include animals. The expression
relates an internal variable, the content within the organism, to the rates
of exchange.

Material content is conveniently expressed in terms of two factors,
one of which is the total size, the physical volume occupied by the organism; the second, an intensive factor, is the concentration of material per
unit volume of the space, neglecting for the present any inhomogeneities
in the interior. The formal expansion of the left expression for amount
in terms of its two factors is $C\ dV/dt + V\ dC/dt$. In the first term,
change in size is a factor. The term is comparable in sense to increases
due to growth, albeit the accumulation of edema or the shrinkage of body

size in senescence is equally well represented by these symbols. The second term visualizes the size held constant and looks at the rate of change of concentration. Since all of the statements of stability in organisms deal with phenomena as observed in terms of interior concentrations, this is the point of focus. If the organism is not changing in size, then the rate of change of concentration reflects the balance of the exchanges. If, regardless of what happens to the sources or the sinks or both, the rate of change is zero, the organism will be stable in its history. But if the rate of change of concentration is not zero for any appreciable length of time, an animal could become monumental and isotropic like Lot's wife, or else vanish like the well-known cat, leaving less than a vestigial smile behind. It is probably useful to take the continuity principle as applied to an organism as a definition of error in the total process of exchange.

There is an expression of this sort for each separate substance which composes the animal; the state of the animal as regards its composition can therefore be reduced to an n-dimensional vector, each of whose components is a continuity expression. The intake and output of animals for any single substance are not usually embodied in single mechanisms; for example, to maintain thermal balance, loss to sinks may be represented by alteration of tissue conductivity, change in water loss, creation of artifacts, ingestion of food, and muscular thermogenesis. Therefore, we look for a method of recasting the continuity statement to allow for multiple processes, and at the same time to relate the processes to control systems theory.

The Error Concept: Parallels Between Animals and Machines

We proceed now by analogy. To do so, a short digression is necessary. In considering man-made devices, namely, filters, predictors, and control systems, what ideas are used? These devices deal with sets of situations or signals which may or may not be corrupted by extraneous phenomena such as noise. Their purpose is to perform some operation on the received signals and to produce a response which is as close as possible to a desired output. The output may be, for example, the reproduction of the original uncorrupted signal, or a prediction of what will happen next, or the execution of some action based upon a prescribed contingency. In any case there is a defined response which can be called the ideal or desired response, $F_d(t)$. One way to evaluate how well or how poorly the mechanism operates is to look at the discrepancy or error between the desired response and the actual response. This must be done, of course, for all the possible kinds of input that might actually occur. To start to do this, a first step is the definition of the error of performance. The error equals

orities involving exchanges of other molecules. The sources, as are the sinks, are several and may be quite varied in manner of operation. In a relatively few cases there are single and well-defined avenues of exchange. Thus carbon dioxide comes primarily from oxidative metabolism and exits primarily via the process of pulmonary ventilation. Temperature regulation, on the other hand, involves sources which are metabolic as well as entry through the integument from the environment; the sinks in turn are represented by at least three mechanisms of heat flow. Finally, if one selects the transactions involving water, there are respectively the insensible water loss and the environmental availability of water as uncontrolled phenomena, and the obligatory renal excretion of water as one depending upon competing priority; the mechanisms of thirst and water ingestion and the facultative urinary excretion are processes which the organism can modify to suit needs. The lists may be extended greatly, but in each case it is possible to distinguish between processes over which the organism exercises some control, which we shall call facultative; and a second class which is subject to all the uncontrolled vagaries of existence and which shall be called random.

This division applies equally well to sources and to sinks. The physiological facts of water exchange in mammals may be viewed in either of two ways. We may speak of water balance as the adding up in good bookkeeping fashion of water intake and water loss, or we may look upon it as a contest in which the physiologically dramatic mechanisms, drinking, eating, and the production of urine, are pitted against the unpredictable and uncontrollable challenges of evaporation, breathing, and water supply. The exchanges represented as the balance of both types of process are nevertheless strictly additive, since the continuity expression is in terms of material flow. Conversely, if the controlled mechanisms are to be successful with regard to composition, they must match and oppose the uncontrolled processes in at least the sense of making zero average error.

We can rewrite Equation 2.1 for the j-th substance with this orientation. As before, the rate of change is the net difference between gain and loss rates. The sum of all sources (I_j) is the arithmetical sum of the facultative sources (IF_j) and the random sources (IR_j),

$$I_j = IF_j + IR_j \qquad\qquad 2.5$$

and similarly for the sinks,

$$O_j = OF_j + OR_j \qquad\qquad 2.6$$

It is important to remember that such brief notations as IF_j and OR_j of Equations 2.5 and 2.6 represent arithmetic summation of the several sources or sinks in each class, all of which are functions of time. As long as the processes selected concern molecules of substance in exchange the arithmetic property is retained. Equation 2.1 becomes

$$\frac{dV(t)C_j(t)}{dt} = (IF_j - OF_j) - (OR_j - IR_j) \qquad 2.7$$

Now let us make a conjecture which will take us down one avenue of thought but not invalidate anything we have said thus far. Suppose that the facultative processes, whatever they are, all represent some operation upon the internal concentration of the organism, $C_j(t)$. In this we include the possibility that any or all of the past history of the animal may be used in performing the operation. Let us represent these operations by the symbol, H, as before. Whatever the specific nature of the process that is thus represented, it can be asserted that if there are two or more such, (say, for example, a corrective action taken at one facultative source, H_{ij}, and another taken at a facultative sink, H_{oj},) the net effect of these operations is their additive sum in the arithmetic sense. For each mechanism there will be one such operator, and the sum of the operations will have the distributive property. Thus, define

$$IF_j - OF_j = (H_{ij} + H_{oj}) \, C_j(t) \qquad 2.8$$

where again explicit symbols for the summation over all the mechanisms have been omitted. As a final expression the continuity equation, which stated that materials which accumulate, accumulate at a rate which is the difference between the gain rate and the loss rate (Eq. 2.1), now takes the form

$$\frac{dV(t)C_j(t)}{dt} = (H_{ij} + H_{oj}) \, C_j(t) - (OR_j - IR_j) \qquad 2.9$$

The analogous Equation 2.4 was taken as the defining expression for the error in performance. Hence Equation 2.9 may be taken to say that the error rate of material accumulation inside the organism is the difference between the actions of all controllable processes in the organism as against all the uncontrollable processes, where $E(t)$ of Equation 2.4 corresponds to the term on the left of Equation 2.9. An input function has been selected which is internal to the organism; the possibility of animal reaction is allowed for without specification; and in the place of the desired-output function, which an engineer might select arbitrarily, $F_d(t)$ of Equation 2.4 has been replaced by terms denoting random loss and gain. In an animal which is not increasing in size the equation can be written even more simply and explicitly for the ensuing arguments:

$$dC_j(t)/\, dt = (1/V) \, (H_{ij} + H_{oj}) \, C_j(t) - (1/V) \, (OR_j - IR_j) \qquad 2.10$$

It is the set of such expressions, each describing a material in exchange, that adequately and abstractly describes the state of the internal environment. In more esoteric terms, this is an n-dimensional vector in the state space of the organism. Here is clearly delineated a relationship between internal concentration, material balance, and physiological mechanism. A formulation of this type is our objective, since the principal subject of the statement of both Claude Bernard and Walter Cannon is the internal composition of organisms.

HOMEOSTASIS DEFINED AS AN ERROR-SIGNAL CRITERION

We have arrived at an interesting state of affairs in the scheme of supposing. The expressions written look at once like expressions written by mathematical engineers when they are talking about the properties of filters and predictors with respect to establishing some criterion regarding the quality of performance relative to a desired output, and yet the statements are also deliberately embedded in the scheme of an open system exchanging with environment. Finally, from all biological processes are separated those representing controllable processes which depend upon the composition of the milieu interne.

The similarity of ideas in engineering and homeostasis may be extended even further by inquiries of the following sort: Given the restriction that only the past and present are available and that any device having continuing existence must be physically realizable, is it possible to have zero error at every instant? Given a particular class of filtering systems, is there one which will be optimal, that is, will make the error smaller than any other? What is optimal? The selection of the kind of relationship involving that error expression which is to be the criterion of evaluation is obviously of central importance, because it is the criterion by which the performance of systems will be rated and by which, at least abstractly, an optimum class may be selected. Engineers have only the boundaries of their imagination in the selection of criteria appropriate to their tasks. In physiology the case may be different, since there may be some criterion which nature is using in the selection of the species. Homeostasis is a statement about successful species and must be related in some way to the error expression generated by the continuity equation. First of all we know that we are dealing with material exchange and that negative collections of material do not exist. Hence the cumulative history of all error which is equal to the accumulation of material, j, in the animal throughout the infinite time interval is zero or positive:

$$\int_{-\infty}^{+\infty} (dV(t)C(t)/dt)\ dt \geqq 0$$

2.11

Actually since the life span is finite, the only period that need be considered is the life span; however, the use of mathematically infinite limits makes the expressions more unequivocal and tractable. The cumulative sum of all errors up to the present instant is the amount of the j-th material present now; and if this is divided appropriately by the present volume, the present concentration is a measure of the cumulative errors up to the present. This concentration must also be a positive number. This statement can be symbolized in the following equation:

$$C_j(t) = \int_{-\infty}^{t} (dC(T)/dT)\ dT$$

2.12

This expression is not as general as possible because the occurrence of size change was discarded in Equation 2.10.

Since the life span is finite, then over the life span there is an average value of concentration. One way of relating the formal statement of homeostasis to the symbols is to say that homeostasis is the assertion that the current value of the concentration is never very different from the average value. Now there are several choices regarding the definition of the average. In the first place it cannot be over the infinite interval since this value would be zero for every j and every animal. However, one can select the average over the life span of the animal, in which case one obtains some one number for a given individual. Or, it may seem desirable to pick some arbitrary number such that the deviation of the concentration (any $C(t)$) from the number is small, and further to define this number as the set point. This latter selection, however, presents no advantage over the former, since, if the mechanism is even modestly successful in attaining its set point, the lifetime average must be extremely close to if not identical with the value of the set point. A third choice, which is intriguing since it promises to allow of acclimatization, is to take a moving average terminating at the present instant and beginning at some time in the past. The disadvantage of this choice is that it requires the selection of a certain span of time, which is not today an object of physiological investigation. Let us set $Q(t)$ to represent a function which expresses a measure of the quality of performance of a homeostatic system. The comparisons discussed above between current value and average value can be expressed symbolically as follows, where L is the life span and T_0 is the beginning of the life,

$$Q_1(t) = \int_{-\infty}^{t} \frac{dC_j(T)}{dT}\ dT - \frac{1}{L} \int_{T_0}^{T_0 + L} C_j(T)\ dT$$

2.13

for the comparison over the life span, and

$$Q_2(t) = \int_{-\infty}^{t} \frac{dC_j(T)}{dT}\, dT - \frac{1}{B} \int_{t-B}^{t} C_j(T)\, dT \qquad\qquad 2.14$$

for averaging over a finite past interval B.

A small value for $Q_2(t)$ is the assertion that homeostatic mechanisms are those for which the cumulative errors over the infinite past deviate from the average of a finite, recent set of such errors by a very small amount. It leads to at least one mathematical problem which may be novel enough to interest mathematicians. Since in either equation the measure of quality may be positive or negative, it is desirable to select some more complicated expression which will preclude the possibility that a low net error results simply from the balancing off of positive and negative ones. Recalling that the deviations Q are deviations from the mean, if we select as an error-signal criterion the mean square of the deviations $Q(t)$, this is obviously the variance of the internal composition. The theory can now be related directly to observations in the laboratory on the one hand, and to the Wiener development (see p. 19) on the other, if one so chooses. Nevertheless, to move in this latter direction is fruitless because it seems improbable that any homeostatic mechanism, by this definition, would turn out to be linear or optimal.

Another aspect of the deviations, Q, arises from the fact that concentrations are always positive and hence the maximum negative deviation possible is the same as the negative of the selected mean. In the most general case, the positive concentrations can be considered unlimited; but since only the class of living organisms is included in the objects under consideration, the deviations observable are without doubt bounded positively. In either case, the error criterion deals with the properties of the central moments of positively skewed distributions.

One reason for preferring $Q_2(t)$ rather than $Q_1(t)$ of performance, as the proper definition even though the averaging interval (which in some ways may be termed the memory span) is not abstractly definable, is that it allows of acclimatization. $C_j(t)$ has a variance over the life span and also over selected intervals shorter than the life span. The concept of acclimatization requires that the variance over the smaller sample be less than the variance over the total population of values. We are not certain what kind of a class of functions these general properties describe, but considering the fact of bounded variation along with those we have just decided upon gives a description that is quite unfamiliar mathematically.

To summarize, by insisting upon the application of the continuity expression to the material exchanges of organisms, quite general and unequivocal definitions may be arrived at which can form the basis of further mathematical analysis. The particular statement of homeostasis (Eq. 2.13

or 2.14) can be set into apposition with the error-signal criterion and thereby into the contemporary conceptual machinery of control systems engineering.

STABILITY VS. FREEDOM

From here the trail becomes less clear and more difficult. On the one hand it leads into the realm of the properties of systems whose error distributions are asymmetrical curves limited to the half line, and on the other hand into speculations in which the biological definition is nearly exhausted. One would still like to know whether the term homeostasis can be expected to set apart some animals in some respects from other animals. Exactly how does one discriminate between an animal with the homeostatic property and one without this property? Or is that an empty question since the only animals of the second class are dead ones? I believe that this is not an empty question, and that a discrimination is possible. For example, although homeostasis is a term applied to a continuum of possible behaviors, physiologists agree that with regard to heat a man has homeostasis and a frog does not. The list could be extended. Wherein does this distinction lie?

Now that we have been able to formalize the idea of stability, the other half of homeostasis should be consideration of the phrase "la vie libre," since this is what internal stability purchases. Freedom must refer to the range of worlds and activities which an organism may enter and survive. The experimental data for revealing homeostasis consist of random samples taken from the interior of organisms. These may be in some sequence from a single specimen of the species, or may be taken randomly but in a similar fashion from many individuals of the same kind. Homeostasis attempts to relate the uniformity of the samples to the other major variables in the state of the animal such as activity or location. It is probably essential to note that the samples must be taken over the full range of possible associated conditions. If one samples body temperature in lizards in their natural habitat, the spread of observations is likely to be much narrower than the spread in an assortment of the same species systematically exposed to a full range of artificially induced environmental conditions and then studied for variability of body temperature. This shows that one way to achieve stability is by giving up freedom to invade certain areas or by curtailing the range of activity. An extreme instance is hibernation; one could say that here all activity is traded for the minimal stability needed to continue existence. Another means of attaining stability is the creation or selection of microenvironments; this is a widespread phenomenon used by all species, whether it be schooling of paramecia or the construction of air-conditioned, concrete skyscrapers.

It is possible to distinguish between the phylogenetically more widespread phenomena of behavioral selection of environment and the intrinsic mechanisms an organism may possess internally to render it stable in the face of hostile situations. If we restrict the definition of homeostasis only to the stability of the organism's composition, then the data will very likely lead to widely disparate descriptions of performance, depending upon whether one samples the organism in its natural state with all the behavioral mechanisms acting to keep the milieu interne stable, or whether one deliberately stresses the organism to see what the range of scatter is when it is so stressed. The restricted definition is patently too limiting to be of value to the experimental physiologist.

The problem is even more severe than this because there is no formal agreement as to the boundaries of the meaning of the words. In the case of body temperature we generally agree not to call animals homeotherms if they rely solely upon behavioral regulation for body temperature stability. Yet every species including man avails itself of behavioral adjustments of posture and location to improve internal thermal stability. In the case of food intake all the intake responses are in a sense behavioral since the quest, the act of ingestion, and the refusal on satiety are descriptions of behavioral conditions. Perhaps the key to the dilemma is to decide, and it should be admitted that the decision is arbitrary, that the stable processes of organisms fall into three categories.

1. Those that are uncontrolled, that is, whose range of stability is determined only inasmuch as a certain range is found in living animals; when this is exceeded the individuals perish, and hence any collection of statistics does not include them. The object or substances which display stability of this sort would be outside the ken of sensory mechanisms in the organism and arouse no response other than disease and disability.

2. Those that are controlled by activity of the whole organism, and which may be very specialized as in drinking or eating, or quite nonspecialized behavior like evasion or random motion. We can contend that in a sense each such regulatory action requires that the organism be able to sense the state of affairs anent the material and to respond appropriately. Appropriateness is a key characteristic of this response, and we should exclude consequences of a beneficial nature that occur fortuitously as a result of random responses. It is important to notice that in the strictest sense the organism loses some of its freedom of existence in making these responses. Thus, when thirsty and responding to this thirst the animal is not free to seek an even more arid environment or to indulge in any of several other activities that it may be capable of doing.

3. The third and final category for which the homeostatic idea is apt in the most particular sense is the maintenance of stability of the organism by mechanisms contained entirely within the organism but not requiring what would otherwise be considered as overt behavior. These are the processes most familiar to the systematic physiologist, and concern the arrangement of specialized organs, internal sensing of state, and internalized communication via the nervous system and hormonal system.

There may be some question whether the least precise of some of these controls is significantly less precise than stabilization achievable by behavioral mechanisms. On the other hand, what these processes accomplish is done with very little burden upon the overt external processes of life. Perhaps we misinterpret the famous phrase of Claude Bernard, ". . . the condition of free and independent life." Although the burden of staying alive is substantial at every phylogenetic level, as organisms increase in complexity, there appears to be more and more time during a life span which need not be devoted to survival needs.

The principal argument we are attempting to evolve is further to subdivide mechanisms of stability according to some clear conceptual line, even at the risk that this subdivision may be in substantial disagreement with some of the ideas that have been expressed in the past. The argument revolves around the thought that freedom and stability are to an extent exchangeable, and that in many physiological responses the trading of freedom for stability is striking. This admittedly is a conclusion which may arouse some antipathy, since some would not even agree that freedom and stability are separate concepts.

A particular point emphasized by Cannon is the fact that the internal and external environments are independent of each other. In homeostatic systems a variation in the environment affects minimally or not at all the interior state of the animal. This situation can be described using the experimental observations. If successive and independent measurements of the external and internal environments are made, the requirement of homeostasis is that their magnitudes be uncorrelated. Correlation zero would be complete freedom, but this is a weak statement since all animal processes have some inertia so that present environment affects the animal's interior in the future. An expression is necessary that will show the statistical dependence of the two environments at every separation in time. The correlation between the present internal environment and any fixed function of the external environment for the present and every past time is denoted mathematically by the cross-correlation function. This function may help to define the distinction between systems dependent upon the pitting of internal controls against challenge, and those of the behavioral type that sacrifice a measure of freedom to achieve the desired stability. Perhaps we should not call these latter processes, which sacrifice freedom for stability, homeostatic. If so, it is essential to find a word which will not denigrate the importance of mechanisms which achieve stability at the cost of freedom, and such systems will be designated poikilostatic in the rest of this chapter.

Uses of the Word Feedback in Physiology

This brings us to the last of the three words used in the title of this chapter, feedback. This is a new word in physiology; it arose in the

engineering context to denote a type of connectivity in which the consequences (output) of an action produced by a machine are returned in
some fashion to participate in the causes (input) of the mechanical action.
A machine which has been devised to look at what it has done, contrast
it to what it should do, and act according to the distinction between the
two is a feedback system. The two essential aspects of these relationships
are, first, that a quantitative estimate of the error is made, and second,
that the responsive mechanism depends upon the error, i.e., upon the
disparity between the desired state and the existing state. The former is
called error signal generation; the latter is the general statement of the
forward loop response. The flavor of this idea has existed in physiology
for a long time. For example, Fredericq in 1885 made the following statement (quoted by Cannon, 1929), "The living being is an agency of such
sort that each disturbing influence induces by itself the calling forth of
compensatory activity to neutralize or repair the disturbance. The higher
in the scale of living beings, the more numerous, the more perfect and the
more complicated do these regulatory agencies become. They tend to free
the organism completely from the unfavorable influences and changes
occurring in the environment."

What then are the advantages in associating the term homeostasis
with the term feedback? What are the intrinsic flaws? The advantages are
few and very real. Most important, there is a very substantial identity
in the abstract formalization of the two technologies, provided we are
willing to consider fairly rigorous definitions. The gain to the science of
physiology is the hope that some of the truly powerful mathematical tools
which are evolved for engineering may shed some light upon quantitative
problems in physiology. We should observe here that the hope is not yet
realized. In some measure this is due to the fact that the tools, powerful
as they are, are inappropriate in particular, and partly to the fact that
it is all too easy to eschew the mathematics while claiming the light.

There are several serious flaws in the drawing of parallels as it is
frequently done today. Homeostasis deals with stability and freedom; the
term feedback deals with the organization of processes. Only a small fraction of systems with the feedback organization are stable. This can be
exemplified rather simply. If the portion of the output which is fed back
is manipulated so as to extinguish the error, the system is a negative feedback system. Many such systems theoretically have particular properties
of stability and form the attractive basis of the parallel. One of the
practical engineering preoccupations, however, is to discover what are
the rules which govern the occurrence of the subclass of stable negative
feedback systems. At the other extreme, if the portion of the output which
is fed back tends to augment the error, the system is one of positive feedback. Such systems taken in isolation are always unstable. In physiological
situations in which such a causal sequence is observed there is no inclination to equate feedback to homeostasis—in fact, a new term is used some-

thing like "irreversibility" or "vicious circle." Yet it should be noted that in complex systems which show stability it is sometimes desirable or even essential to have portions of the system operate in the positive feedback mode. Another and serious difficulty arises from the distinction between constructed systems and systems under observation. In a constructed system the input and output have unequivocal definition since the definitions reside in the mind of the designer. In systems under observation, if there is any appreciable complexity, the selection of input and output is not unique, but resides in the eye of the beholder. The particular consequences of this difference are that in the first place the evaluation of function will turn upon the criterion by which performance is judged, and that in the feedback frame of thought the measure of error of performance depends entirely upon the selection of input and output. Finally, in any complex system under observation there are bound to be several equations in several variables, one of which is common to more than one of the equations. It is always possible with a little ingenuity to arrange the variables so that the system is represented as one containing feedback connections. It does not take great imagination to see that, in extending the word, feedback, out of its engineering context to include all the many things that concern physiologists, we come perilously and perhaps unjustifiably close to declaring that a feedback system is any system which is describable by some set of simultaneous equations.

In spite of those difficulties, however, the utility of the parallels between the idea of homeostasis and control system theory should be emphasized. It is on this vein that we shall attempt to bring to a close some loose ends of the earlier development. By starting out with the continuity expression and insisting upon restriction of the subject to exchange of materials, we identified a fairly unequivocal expression that relates at once the environment, the animal's activity, and the composition of the inside of the animal. Moreover, the expression overall represents, in a sense, the error of performance of the animal in the exchange processes; and since materials are the object of concern, the functions described are additive in a linear sense, and the operations have the distributive property. The homeostatic mechanisms of the animal are responsive to the internal concentration, which is an error expression; therefore, in a restrictive sense, we identify homeostatic mechanisms as those mechanisms of the organism which achieve stability by feedback connections operating through the common variable of the internal composition. Whatever may be the detailed organization of each controllable output, the total system is representable as a linear, additive feedback system.

SUMMARY

If we have proposed any innovation in this chapter it is that an identity may be established between the balance of material exchange in

organisms and the balance between controlled and uncontrollable processes in the same organism. In the milieu interne the material exchanged
and the information about the transaction are contained in one and the
same entity, at least at one point. Body processes may then depend upon
translation of this information into membrane potentials, hormones, nerve
impulses, twitches, or any of a manifold number of message modalities
within the animal. But whatever the vagaries of the complex machinery
of the body, it ultimately comes back to a transaction in molecules where
the product and the coin of transaction are the same physical entity.

Since we have set aside a word, poikilostasis, for processes of obtaining intraorganism stability even more widespread than the mechanisms
of homeostasis, it is perhaps best to show how this word fits into the
scheme. We distinguished poikilostatic responses as those which exchange
freedom for stability. They are in almost every case actions that may be
termed behavioral and that involve movement and selection by the whole
organism. Two things characterize these mechanisms; in the first place,
they do not require detectors of the internal environment since they can
depend upon cutaneous sensation, or vision, or any of the many external
senses of animals. They are, therefore, not necessarily error actuated.
Secondly, responses of this sort do not necessarily alter sources and sinks,
but act by altering the class of random processes that the organism is
exposed to. There is no parallel to this in mechanical systems. Mathematically poikilostatic systems prescribe the statistical characteristics of
the input and constrain them to acceptable bounds. It is this constraint
that is identifiable as loss of freedom. Like the U. S. Supreme Court, the
poikilostatic mechanisms can refuse to consider certain situations.

By these two broad classes of approach the successful species maintain stability. We have arbitrarily divided the trees of physiology into
three woods; and although perhaps they are wrongly divided, the fences
are clear and unequivocal and can be challenged by rigorous methods.

Notes about References

A bibliography is difficult to list since thoughts of this sort are
gleaned from many sources and have become so habitual on rumination
that their origins have vanished in memory. But of course we can name a
few sources that represent that impetus from which it all began. There
is the paper, "Organization for Physiological Homeostasis," by Walter B.
Cannon (1929); the quotations here are from this source. Much of this
also appears in Cannon's book, *The Wisdom of the Body* (1939). Although
it is not the source from which the quotation about the fixity of the milieu
interne is taken, there is great pleasure in reading Claude Bernard's
An Introduction to the Study of Experimental Medicine in the English
translation by H. G. Greene (1949), for here is stated the mission of

physiology in the complex of the medical sciences. It is disturbing to me that I have yet to discover a book by Norbert Wiener that is readable or comprehensible; it reflects no doubt a great deficiency in my training. A lucid and clear presentation of the elementary essentials of the mathematical format of the chapter can be found in *Statistical Theory of Communication* by Y. W. Lee (1960). Of treatises and primers on control systems theory there is no dearth, but from long fondness and recurrent usage I shall cite *Theory of Servomechanisms,* edited by Hubert James, Nathaniel Nichols, and Ralph Phillips (1947). It is volume 25 of the Radiation Laboratory Series.

Chapter 3

A SUCCESSION OF
IMPROBABLE MODELS

WILLIAM S. YAMAMOTO

In writing both generally and abstractly about problems in regulatory biology one follows a course like the classic one between the whirlpool and the rock. One must avoid becoming so general that the abstraction is unyielding to particular demonstrations, and yet one must not be drawn down into a perpetual vortex of detail. Some detail is necessary to lighten the burden of generality of the preceding discussion about material exchanges of animals. Which details, which features of organisms are of value? To make the choice, it is necessary to do experiments of a sort that is not the usual kind performed by a physiologist: experiments in supposition. Like the ordinary type of biological experiment, most of these supposing experiments will be failures. And as against physiology, which sometimes acrimoniously has been called the "science of the disabled and dying animal," the indictment can be brought against the performance of experiments in supposition that they are oversimplified, unrealistic, or grotesque.

But just as it may be possible to reach valid generalizations by studying animals whose capability of variation has been sharply and arbitrarily restricted, so also should it be possible to examine the consequences of introducing certain kinds of ideas as elements of an idea complex, without requiring from the very first that the system include every possible realistic variation. The questions put by the investigator are of the nature of "What happens if . . .?" rather than "Does this happen in nature because . . .?" Underlying the game, however, is the belief that understanding of questions of the first kind does lend substance to explanations of the second kind. Thus one may generate a set of unending parallels between abstract conjecture and real data from living systems. We shall attempt now to illustrate how one experimentally selects concepts and examines the consequences of such choices. It is apparent that any premise imposes restrictions upon the possible modes of behavior of an ideational system under study. What, for example, can be expected from systems which have no controllable processes, but which display some of the features basic to the structure and physical organization of living organisms? Let us see if modeling even of a very simple kind gives any insight into the real biological problems.

The Simplest Model

We shall use biologically sounding words because we want them to conjure up certain images in the minds of knowledgeable biologists; and also we want to conjure up the demons of objection, since only through examination of such objections can we obtain correct abstract representations of biological objects. Our first object of study resembles a rindless Edam cheese possessing a uniformly stirred inside and a boundary between its inside and its outside, an exchange surface. The object will be called an organism and possesses a metabolism which in this instance is represented by the production of some substance J.

Let us suppose further that J undergoes no change in the organism but can exchange by passive diffusion through the skin of the organism to the exterior environment which is imagined to be boundless and well stirred. We need not ask whether such an organism exists, yet these few features do constitute the properties common to many animals.

Let us draw a picture of this organism with its parts labeled (Fig. 3-1). It has a certain size, V liters, and the rate of production of J is

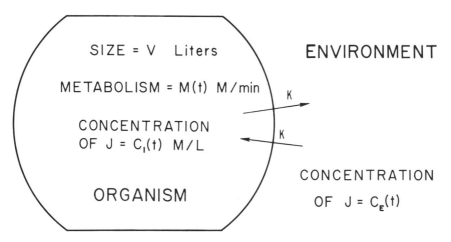

FIGURE 3-1. Diagram of a single-compartment open system, the cheese-beast, exchanging a substance J with the environment.

called M moles per minute. Exchange occurs by the process of diffusion through the boundary. For the present purpose let us suppose that the rate of efflux, $E(t)$, from a region is proportional to the concentration of J in the region.

Thus the sources of J in the organism are the sum of the metabolic rate of production $M(t)$ and the rate at which J enters through the skin by virtue of the concentration $C_E(t)$ of J in the environment. Letting K represent the proportionality constant that indicates the permeable property of the boundary, we have as ingress $I(t)$, i.e., the sum of all sources,

$$I(t) = M(t) + KC_E(t) \qquad\qquad 3.1$$

The only avenue of loss from the organism is escape through the skin; thus if $C_i(t)$ is the interior concentration of J,

$$E(t) = KC_i(t) \qquad\qquad 3.2$$

In the notation of the previous chapter, the continuity expression equation (Eq. 2.1) becomes, assuming the animal is not growing in size,

$$\frac{V\,dC_i(t)}{dt} = M(t) + KC_E(t) - KC_i(t) \qquad\qquad 3.3$$

which is a very simple sort of linear differential equation.

Since we have not supposed that the organism can in any way alter its ability to exchange according to need, that is to adjust the value of K, and since we need not suppose in this case that there is any adjustment of metabolism to the state of J inside the organism, and since the organism certainly does not control the environment in which it is found, we have on the right-hand side of this equation a sum of sources and sinks, none of which is facultative. That is to say, this is an uncontrolled organism. It does not therefore have an error-signal criterion in the sense of a match between facultative processes and random processes. There is, nevertheless, an internal concentration, and if instead of having created this organism by imagination we had come upon it in nature, we could actually proceed in our ignorance to examine the animal regarding its homeostasis of J. What would we find?

At this point it is very easy to become lost in the niceties of symbolism or the particular problems of the means of a formal solution or the justification of these means. I use the word lost because in fact the necessary methods can be intellectually difficult and yet not germane to the quest itself, much as the intricacies of design of a modern cathode ray oscilloscope are not immediately germane to examination of the action potential of a neuron as displayed on the scope. Hence, without indulging in the details of solving the equation, certain results shall be discussed directly.

First, consider the simple instance in which the metabolic production of J is a constant (M) and the environmental concentration of J is a constant C_E, and suppose that the organism enters this state of affairs abruptly with an internal composition of $C_i(0)$ at an instant in time which we shall reckon as the origin t = 0. The composition of the interior at any subsequent time is

$$C_i(t) = (\frac{M}{K} + C_E) (1 - e^{-Kt/v}) + C_i(0) e^{-Kt/v} \qquad 3.4$$

It is apparent that, as time becomes very long, the internal composition becomes

$$C_i(\infty) = \frac{M}{K} + C_E \qquad 3.5$$

That is, the internal composition is independent of time and (except for the added quantity M/K) mirrors the environmental composition. The discrepancy between $C_i(\infty)$ and C_E is proportional to the metabolism weighted by a constant which is the reciprocal of the ease of exchange across the organism's boundary. If K is small, i.e., less than 1, this means that the substance J does not exchange freely between the interior and exterior, and so the inside is more dependent upon the level of metabolism than it is upon the environment. If K is large, then the interior reflects more accurately the environment rather than the organism's metabolism; this situation we recognize is a version of the verbal description of a poikilostat as defined in the previous chapter.

Steady States of the Model: The Space of Viability

The concentration of J in the internal environment is always higher than in the external environment when J is being produced. If, contrary to the original assumption, J is consumed at a rate M, then the internal concentration of J would be less than the external environment. These two situations can be exemplified by comparing the internal oxygen concentration and the internal carbon dioxide concentration of organisms with the concentrations of the corresponding gases in the environment. The first one is lower inside than outside, while the second is higher inside than outside. The conclusion is that in steady states the internal environment of the organism parallels the external environment, and the amount by which there is a deviation between their compositions depends upon the level of metabolism of the organism. This can be shown in a pair of graphs in which are plotted concentration in the internal environment

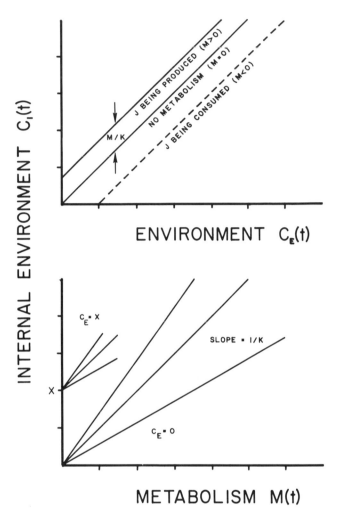

FIGURE 3-2. Graphical representation of the steady state concentration of the substance under exchange in the one-compartment open system called the cheese-beast. The upper graph relates internal composition to external environment; note that the intercept on the ordinate depends upon the steady state metabolism assumed and upon whether the substance in question is produced or consumed by the organism. The lower graph relates internal concentration to metabolic production. Two fans of lines are shown. The intercept of each fan on the ordinate represents the assumed, constant external environmental concentration of J, respectively zero and X on this graph.

against that of the external environment and also against the metabolism (Fig. 3-2). The graphs show a very complete degree of dependence between internal environment, external environment, and metabolism.

As K, the constant for permeability, becomes larger and larger, the proportionate effect of metabolic change is diminished without any substantial change in the effect of external environment. A large permeability constant, therefore, would decrease the slope of the milieu interne versus the metabolism line and would tend to stabilize the animal against the vagaries of its own metabolism without in any way protecting it from its external environment. Conversely, when K is small a large difference between inside and outside can be sustained by a small amount of metabolism. An interesting although perhaps apocryphal example of this is the case of the caribou, which is reputed to become so well insulated to withstand heat loss in the Arctic that it becomes susceptible to death through overheating during pursuit by wolves.

Obviously, Figure 3-2 illustrates only two possible correlations. To illustrate graphically all the steady states for a given K, let us plot the three variables in Cartesian coordinates (Fig. 3-3). Along the X axis is plotted metabolic production measured in some suitable dimension; along Y and Z are plotted concentration of the substance in the interior and exterior of the animal, respectively. Since neither concentration nor

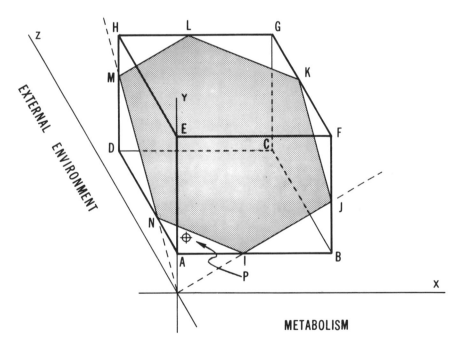

FIGURE 3-3. A graphical representation of all possible steady states in open systems within assigned limitations upon conditions of viability. The axis OY is the coordinate for internal environmental concentration of the single substance selected. The point P as shown lies upon the plane ABCD. The hexahedral volume encloses all states representing living organisms, and the shaded plane the steady states of the system.

metabolism is negative, only the positive octant need be illustrated, and in this part of the space all the possible combinations lie. There is a maximum metabolism of which any organism is capable; this is illustrated by the plane BCFG. Environments can become so hostile that organisms do not exist in them, and this may be represented by a barrier in that direction, the plane CDHG. And finally, if concentration in the internal environment exceeds a certain maximum or falls below a certain minimum, it ceases to exist, since the organism dies; these limitations are illustrated by the planes ABCD and EFGH representing the upper and lower limits for viability. It is probable that every organism may be so described in terms of comparable variables and found to be constrained to space like the one depicted. Outside this space the animal dies and ceases to exist. When many (N) metabolites are considered, each may be described in terms of three coordinates, and the space in which life persists is the intersecting segment of 3N dimensional space.

In a steady state, according to Equation 3.5, the interior concentration of J is proportional to metabolism with a slope $1/K$, as shown by the line OIJ. C_i is also directly dependent upon the external environment as depicted by line ON. The two lines intersect at O and are coplanar. The plane of these lines represents the locus of all possible steady states. It is readily seen that the plane of possible steady states intersects the space of viability, producing in the case illustrated the hexagonal plane segment shown. If an organism is found in a state off the plane, and if neither environment nor metabolism changes, subsequent observation may show that the point representing the animal's state has drifted toward and eventually lodged on the plane. In some instances, as represented in point P, however, the organism will perish under these conditions; for survival it must not pass beneath the ABCD plane, but must approach the hexagonal plane by increasing metabolism or moving to a less hostile environment. The points that are within the delineated space, but off the steady state plane, represent living but unsteady states of the organism.

Systems in these states all have the regular property of moving toward the plane of steady states. Since such behavior is regular, and since the goal state is apparently well defined, there is a natural inclination to call such phenomena regulation. This is probably not correct or useful. In the first place, it is contrary to our definition of regulation. In the second place, in physiology the term is usually applied to the mechanisms by which an organism escapes death from a point like P by altering its internal activity or by moving to a less hostile environment. It is for this activity or moving that one would like to reserve the term *physiological control*. The moot question is whether by observing trajectories of the animal's state in the whole space of viability it is possible to decide by inductive inference whether an action of this sort has occurred.

NON-STEADY STATES: THE USUAL STATES OF ANIMALS

Now although some of the foregoing ideas may sound vaguely like the statements that physiologists sometimes make about animals, we have to recognize that, even for the fictitious organism of which we are speaking, we may have been led to conclusions which are unrealistic. We have made our examination of the state of the metabolizing cheese after an indefinitely remote time during which neither environment nor metabolism changed. For some of the substances in exchange between the organism and environment, the length of time that one can maintain constant, observable conditions may be long enough to permit the final, experimentally observable state to become practically the same as the mathematically imaginable steady state. It is unlikely, however, that this will be true for many of the exchanging substances. In fact it is probable that the organism, even under the best experimental control, is never in a steady state, but moves slowly or rapidly along some trajectory in the space described in Figure 3-3. Much of experimental physiology is concerned with the problem of studying such unsteady paths, complicated further by the fact that the destination of the trajectory may be constantly changing.

The problem of physiological stability, then, has two facets. The first concerns changes in composition of the organism when circumstances of existence (such as level of metabolism or environment) change, and the second facet deals with the properties of the organism in the course of time. Although graphs depicting the steady state may show the dependence of one physiological parameter upon another, they may not reveal the temporal stability of the organism and they do not indicate how an organism may proceed from one such state to another. That is, stability or fixity of the internal environment, homeostasis, must pertain less to the time-free asymptotic behavior of organisms and more to their temporal stability.

These features are revealed by examining various trajectories in the selected space, one of which is given in Equation 3.4. For an animal during transition from one state to another, or during the occurrence of fluctuations in either metabolism or environment or both, one finds that animal size (V) and ease of exchange with the exterior (K) are particularly important. In the cheese-beast, during the transition from one environmental state A to a second environmental state B, the internal concentration of J follows an exponential time course. The speed of the change is prescribed by the exponent which is the ratio of the permeability constant and the body size, (K/V). When the animal is small, i.e., V is small, change occurs rapidly; and when the animal is large the change is slow. Similarly, animals which exchange freely with the environment follow the environment more rapidly than those which exchange slowly. With regard to metabolism these relations are also true, but since the exchange constant enters into the final expression in the denominator also, although the change occurs rapidly in animals with large K, the change also decreases in absolute magnitude. All animals having an exchange capability proportioned to body size, for example, those maintaining a constant surface-to-volume ratio for all sizes, would have the same time constant.

We see then that the internal composition lags behind the change in environment or in metabolism, and also that this composition is different from the external environment. Suppose the environment fluctuates in some regular pattern, as for example a diurnal cycle. After the transients have died down, both the environment and the internal environment will oscillate similarly. The ratio of the magnitude of the swings of the internal composition to the external composition can be determined. The numerical value of such a ratio has the same connotation as the words amplification or gain in electronic circuits and has a value for each frequency of oscillation that is chosen. If these values are calculated for our imaginary organism, the ratio is a diminishing function of the frequency and has its largest value, unity, at frequency zero; this corresponds to the steady-state condition previously discussed. A graph may be made to show the relationships of the gain to the frequency of the environmental forcing. And thus by logical similarity one can progress methodically and by easy steps into the use of gain-frequency and phase-frequency plots. The adoption of such representations by physiologists thus becomes logical and necessary rather than arbitrary.

In the example just given, a gain-frequency plot would merely show that fluctuations in composition of the inside of this organism are always smaller than fluctuations in environment, and in observing such a system without knowledge of the way in which we constructed it, one would say that the animal showed some of the features of stability. Now all this is so simple that the preceding discussion may be tedious. The ideas are those we intuitively use in attempting physically to stabilize an object. For this purpose we combine great inertia with weak coupling to the surroundings. In the case of animal composition, physical size is analogous to inertia and the exchange constant is analogous to the coupling. We can see that one of the means for achieving stability in a changing world is to become larger and larger. Perhaps this is one reason why the dinosaur represented a successful adaptation to the diurnal rigors of its time.

EXPERIMENTS ON A MODEL

At the grave risk of becoming even more tedious, let us pursue this impossibly simple cheese-organism one step further. Suppose unlike our previous approach, in which we had a completely defined set of symbols

TABLE 3-1.   SUMMARY OF OBSERVATIONS ON SEVERAL MODELS,
BASED UPON SAMPLE OF SIZE 20.

	MEAN	S.D.	COEF. OF VAR.
One-compartment			
Cheese-beast			
Inside, mm Hg	63.832	0.6111	0.00957
Metabolism, L/hr	18.872	3.530	0.1870
Outside, mm Hg	43.986	17.714	0.402
Two-compartment			
Yeast-beast			
Cells, mm Hg	67.622	0.536	0.00792
Ecf, mm Hg	64.908	0.918	0.01414
Metabolism, L/hr	17.888	3.342	0.1868
Outside, mm Hg	43.796	16.046	0.366
Controlled exchange-			
Facultative amoeboid beast†			
Cells, mm Hg	52.395	0.892	0.01702
Ecf, mm Hg	46.002	0.317	0.00689
Metabolism, L/hr	22.965	3.588	0.1562
Outside, mm Hg	39.468	16.461	0.4170
Area, m²	1.2466	0.333	0.2671

†See Chapter 4.

and could manipulate them according to precisely justified logical rules, we had come upon the organism that someone else had set to metabolizing and exchanging, and that we were to examine it as a natural historian—by sampling the inside of the animal, examining the environment, and measuring its metabolism of J. What sort of data might accrue?

How are experimental observations performed upon an improbable organism which is so simplified that no one can be deceived as to its verisimilitude? For the purpose of experiment, we can give this organism an accurate if evanescent existence in the form of a computer program. For our present purposes we need not consider just how this is done. Let the following circumstances be prescribed: That the organism shall be of the general size of a man of about 50 liters; that substance J shall be produced at a rate of about 18 L/hr, the same as a man's rate of carbon dioxide production at rest. Assume a K* of 0.45 L/min and a cyclic fluctuation in J production rate at one cycle per hour varying between the extremes of 13 L/hr and 23 L/hr. Finally, the partial pressure (i.e., the "potential") of the substance J in the environment shall be at a mean of 46 mm Hg with a cyclic fluctuation of from 23 mm Hg to 69 mm Hg. Thus we have set up the organism to have a metabolism varying over ± 28 per cent and to live in an environment which varies ± 50 per cent around its mean value. The situation thus defined is determinate for every instant in time such that we could compute and plot the complete life history of the organism.

* The dimensions of K are established by using the physical dimensions of J as shown without endeavoring to represent them as conventional measures of permeability.

Let us suppose, however, that we do not know this, but rather that, having discovered a colony of such beasts living in the field, as naturalists we have decided to observe what the metabolism and homeostasis of J are in this particular beast. To imitate this state of affairs in a real laboratory we shall suppose that we have no control over the particular instant during which we sample the parameters of an experimental animal. We achieve this illusion by taking from a large number of sequential episodes in the completely defined life of the animal a random sample. The sample is selected after we have listed the state of the animal's interior, the state of the metabolism, and the state of the environment at intervals spaced 3 minutes apart for some long interval of time. Then using a table of random numbers we pick some sequence of, say, 20 values, the thirteenth, the thirty-seventh, the fifth, and so on until 20 values have been chosen. For these we can make small sample statistical analyses of the data as though they had been selected from the hypothetical beasts living in the field. What does such a body of data look like? The first part of Table 3-1 shows such an experiment upon a given "cheese-beast." As physiologists, what might we be led to conclude about the homeostasis of J?

One could reach the conclusion from such a set of observations repeated many times that this organism had indeed a means of maintaining a very stable interior quite independent of the wide variations observable in either metabolism or environment. This set of conditions was not chosen with anything more in mind than to approximate the ratio of the rate of gas exchange to body volume which is typical of a man at rest. In point of fact, we chose these data to characterize an animal possessing no capability of adjusting its interior according to need. The outcome of the data analyses shows us that certain experiments may lead to deceptive or fraudulent conclusions about the organization and exchange capability of an organism.

Now we have seen that it is possible with the general formulation of the idea of homeostasis and continuity to set up artificial animals which embody certain facets of animal organization, and further to cause such creatures to exist in the imagination in such a manner that real physiological experiments may be performed upon them. From these we learn in part which sort of behavior emerges from a particular idea structure, and also we learn what the limitations are in the possibilities of the interpretation of data resulting from certain experimental designs. We learn that we can study the consequences of experimental design in order to learn about unknown objects of a certain kind. Once the game has been joined there are immediately a thousand things one would like to examine in this fashion, and of these thousand there are thousands of simultaneous combinations. The most difficult thing, perhaps, is to restrain our enthusiasm and to proceed in some reasonably orderly manner.

What we would like to explore are the features of animal structure which impose major characteristics upon the class of exchanging systems which we identify as animals, and in particular to learn which of these features produce the qualitative distinction to which is given the name, homeostasis. We have already discovered in passing that two major variables are the size of the beast and the nature of the coupling or exchange route between the inside of the animal and the outside, and we further know that these produce stability in a non-facultative manner.

The Idea of Compartmentation

Let us continue the search. What precisely is the interior of the organism of which we have been speaking rather glibly? So long as our animals are rindless Edam cheeses there is no problem, but the fact that animals possess interior parts is one of the oldest and also one of the most familiar facts of biology. Is this division into parts important? What is a fruitful manner of dividing the organism for our present purposes? By probe and knife it can be divided into organs; by microscope it can be divided into cells and spaces; and at a higher power or by the biochemists' centrifuge the cells, in turn, can be separated into a wondrous conglomeration of even smaller entities until one is led to the assertion that nothing is homogeneous. One may wonder whether there is any real basis for a generalization like homeostasis.

The stability concepts fostered by Claude Bernard, however, express "constancy" rather than "homogeneity." They came from his studies upon mammals, species in which the biologist is fortunate in finding a fluid matrix that is readily sampled and examined—the blood. But even here one discovers that the blood is further subdivided into a fluid matrix (plasma) and cellular portions. The stable properties referred to in the concept of homeostasis actually belong mainly to the plasma, as it exemplifies the behavior of the fluid matrix of the organism. The point to be emphasized and remembered is that the concept of stability applies only to a part of the interior of organisms. Such a part has been called a "compartment."

The idea that organisms are separated into relatively homogeneous compartments is a fruitful generalization that lends a considerable specificity to the formal description of the kinds of systems that must be homeostatic. Just what is a compartment? In its simplest conceptual state it is a space with a restrictive wall of some sort which results in the measurable segregation of material. Thus if we introduce the macromolecular blue dye known as Evans blue into the blood of a mammal it is pretty well confined to the fluid contained within the walls of the blood vessels.

Other materials are similarly confined. If instead of the dye one injects sodium or chloride ions, they too appear to be restricted, at least for a short period, to a space smaller than the whole volume of the organism. This space or volume of distribution is not exactly the same for the two ions, but is of the same order of magnitude; it is larger than the volume of distribution of the blue dye and appears to coincide with the fluid matrix within the organism but outside its cells. It includes the vascular volume and an extravascular volume, and is called the "extra-cellular space."

We see therefore that the organism which we defined as having an inside and an outside must have an inside which is further partitioned. This partitioning is made evident by the characteristic distribution of injected materials. The fact that the partitioning depends upon the nature of the materials unites the idea of compartmentation with our previously developed ideas about exchange and homeostasis. The compartments we have mentioned are measured by dilution techniques and appear to coincide sometimes with well-defined anatomical entities. There are situations, however, in which a solute may exist in different concentrations in the several fluids of the body, and then we cannot be certain through how many macroscopic compartments it is being distributed. Although neither anatomical nor operational lines for the separation of the organism into compartments are known, evidence for their existence is substantial. It is conceivable that this idea of spaces can be extended to every substance of interest, even though the volume of these spaces may not be measureable by present methods.

The importance of the idea that the body has compartments is indeed well recognized; many authors have written of such systems, whether in reference to the treatment of clinical disorders of fluid balance, or whether concerning abstract generalizations which characterize such systems. In the short compass of this chapter we cannot cover in detail even the elementary aspects of such systems as they constitute a main current in the efforts of biologists or biomathematicians (Robertson, 1962; Berman, 1962).

The Two-Compartment Model

We would like, however, to extend our primitive model one step to describe the next higher order of complexity. An interesting question to keep in mind as we proceed is this: Does increased stability follow as a natural consequence of the creation of internal partitions within the organism? In the first place let us speak specifically of a site analogous to Bernard's milieu interne, so that the organism will now at least have the property common to all metazoans and consist of a cellular inside with an extracellular fluid. The new model can be illustrated in its sub-divided form with the cell separated from the world by its own envelope of an internal environment (Fig. 3-4).

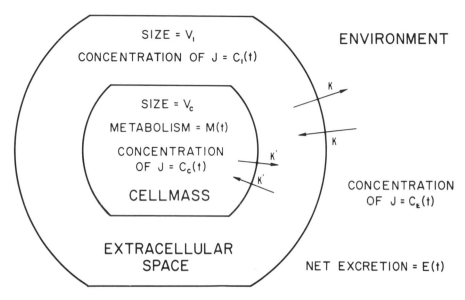

FIGURE 3-4. Diagram of an open two-compartment system exchanging a substance J with the environment. This represents the yeast-beast with fixed surface area.

Again this organism has no facultative processes. It could be a slurry of yeast in a collodion bag, since this has a metabolism and a means or manner of exchange with the exterior. The slurry contains some living cells which produce J metabolically, and these cells are suspended in a more or less inanimate liquid which is the milieu of the cells. Finally the bag delimits a distinct mass which exists in a surround. The level or rate of J production may be assumed to be an uncontrolled process, and the environmental concentration of J is also an uncontrolled variable. Further imagine that the millions of cells are in effect one improbably large cell doing the metabolism, so that the "yeast-beast" is reduced to an organism involving two distinct and well-stirred physical compartments separated by the surface boundary of the "cell." The rate of exchange between these compartments is not facultative, although it takes the form of an operation upon concentrations. Assume as before that J moves according to its concentration gradient.

The abstract description of the yeast-beast can now be set down. For each compartment a conservation equation can be expressed; thus the cellular or inner mass has a certain size V_c where the substance J exists at a concentration $C_c(t)$. The sources of that J which appears in the cell mass are, first, the metabolic production M (t), all of which is assumed to occur in the cell and not in the extracellular space; and second there is any J that will enter the cell by diffusion from the extracellular space. This diffusion is characterized by a constant K' which is representative of the area of contact and the permeability of the cell-extracellular fluid interface. The only sink, or route of escape of J from the cell, is also by diffusion into the extracellular space. The rates of diffusion are assumed proportional to the concentration within the compartment of origin; for the cell the proportionality constant is K′. The rate of change of the amount of J in the cellular mass then is the sum of all the sources and sinks, thus:

$$V_c dC_c/dt = M(t) + K'C_i(t) - K'C_c(t) \qquad 3.6$$

Similarly the extracellular space can be described. It has a volume V_i and a concentration C_i. The sources and sinks of substance J into and from the extracellular space are diffusion into and from the cell and into and from the environment which has J in the concentration C_E. The envelope between the whole organism and the environment is also the envelope of the extracellular fluid and is characterized by a diffusion constant K. The formal description of the conservation of J in this compartment is

$$V_i dC_i/dt = K' (C_c - C_i) + K(C_E - C_i) \qquad 3.7$$

There is, therefore, one equation for each compartment; each equation is formally and conceptually the same as the exchange-conservation expression for the whole organism, as developed in the previous chapter. The organism is thus described by the sets of equations representing substances in exchange, and each substance is in turn represented by a set of equations which describe the transactions in each relevant compartment. The link which holds the organism together is the fact that compartments act as sources and as sinks for each other, in relationships that include the facts of physical contiguity and the mechanisms by which transfers occur. Although thus far we have employed only passive diffusion under a concentration gradient, any other process, of course, may replace or parallel this method of transfer. For the moment, this extremely simple approach using only diffusion will suffice, since the point of interest is the pattern of the relationship and not the variety of particular mathematically representable mechanisms of transfer. The pattern, if such it may be called, from examining just two cases (the "cheese" and the "yeast"), is that with the assumptions made thus far, an organism can be described by sets of ordinary differential equations of the first order, using one equation for each compartment for each substance of interest.

Equations 3.6 and 3.7 can be readily arranged to yield inhomogeneous, linear, second-order differential equations describing the temporal behavior of either the intracellular or the extracellular composition. From the definition of the constants K and K′, as well as the constraint that neither volumes nor concentrations are negative, it can be shown that the equations always are of the overdamped type. Although the coefficients of the equations in terms of real, physiologically understandable quantities become a bit cumbersome, formal solutions of the describing equations for some prescribed conditions are not difficult to obtain. To obtain them, however, would unnecessarily festoon these pages with somewhat intricate algebraic expressions, and since it is not our purpose to expound exhaustively the oversimplified supposition of a two-compartmented organism, it therefore serves no immediate purpose to present such solutions. Suffice it to say that equations of this sort are well known in elementary mathematics, and that the development of the details of solution and their interpretation lead again by natural and gentle steps into concepts from the discipline of control systems engineering.

In lieu of a formal development, let us look at several aspects of the system which would be of interest if one were looking at it seriously as an organism whose homeostatic behavior was under investigation. Since we started with the question of whether compartmentation into a milieu interne and a cell mass is in some sense equivalent to homeostasis, let us ask of the symbols only those questions which may be meaningful in this context. The first thing of interest is the response of the milieu interne to a change in either metabolism or environment. One manner of displaying this is to determine the steady-state composition following a step-like change in the respective variables, i.e., after the transients have disappeared. In this two-compartment organism the internal environment follows the external environment, and is affected by the organism's metabolism in proportion to $1/K$. If the organism could be studied in a succession of steady states it would be observed to change in a fashion suggesting that it is uncontrolled. As in the one-compartment cheese-beast, there is an asymmetry in the manner in which the environmental concentration or metabolism affects internal composition.

Some caution is necessary. If Equations 3.6 and 3.7 are combined algebraically to yield an expression for each compartment, one can regard the unhomogeneous part as the force driving the system. Then although the coefficients of the homogeneous parts are the same for the cells and for the extracellular space, the driving forces in the two compartments are different. In terms of metabolism and environment, the forcing function for the extracellular space is

$$F_I = K'M + KK'C_E + V_c K \frac{dC_E}{dt} \qquad 3.8$$

and correspondingly for the intracellular space,

$$F_c = (K + K')M + KK'C_E + V_I \frac{dM}{dt} \qquad 3.9$$

Thus, the compartments are not affected equally or in the same manner by metabolism and environment, and even though the transient solutions are the same for both compartments, the forced solutions are different for any given set of conditions selected as the physiologically relevant states.

If metabolism M and environment C_E are kept constant for an indefinitely long time, the compositions of the two compartments are described by the expressions

$$C_I = \frac{M}{K} + C_E \qquad\qquad 3.10$$

for the extracellular space, and

$$C_c = \frac{K+K'}{KK'} M + C_E \qquad\qquad 3.11$$

for the cell mass. As one may have expected, the organism then achieves a steady state in which the net exchange of substance J across the inner barrier, marked K', between cell and fluid is just equal to the net exchange across the envelope of the whole organism, characterized by the exchange constant K. The extracellular space is higher in concentration of J than is the external world C_E, and the intracellular concentration is in turn higher than the concentration in the extracellular space. These concentration differences are in inverse proportion to the exchange constants K and K'. The above statements can be derived from Equations 3.10 and 3.11 giving

$$\frac{C_i - C_E}{C_c - C_i} = \frac{K'}{K} \qquad\qquad 3.12$$

If J should be a substance which were consumed rather than produced by the organism, the direction of net flow would be reversed but the relationship would be exactly analogous. The highest concentration occurs in the external environment, and the lowest ones, deepest in the organism. If the K and K' are equal, then the cells are as different from the internal environment as the latter is from the exterior. If K' is much larger than K, then the internal environment is much like the cell interior and quite different from the world outside. In every case, however, the concentration in the extracellular space is of a magnitude between those of the cellular insides of the organism and the external world and appears to act as a moderating cushion around the cells.

Now one of the factors that contributes to the magnitude of K values is the area of contact between the compartments. Unlike the giant "cell" that we have been using, the cell mass in reality is much subdivided. If one examines the area of surface of contact of a given volume of cells, as the number of cells increases the surface becomes quite large indeed, so that K′ is likely to be extremely large compared to K. Moreover, if one progressively divides a given volume V of material into identically shaped subdivisions, when n subdivisions have been made the area will have increased by a factor of $n^{1/3}$. Consequently, in a case in which the external integument and the cell-extracellular partition have identical permeabilities, subdivision of the cell mass with no other change will cause the extracellular fluid and intracellular fluid to have the relative concentrations given by the equation

$$\frac{C_e}{C_i} = 1 + \frac{1}{n^{1/3}}$$ 3.13

if C_E is taken to be zero. By the time 1000 subdivisions of the cell mass have been made, any concentration in the extracellular fluid will differ by only 10 per cent from that in the cellular fluid.

Another inadequacy in this particular model is thus exposed. It seems improbable that each of the many thousand subdivisions of the cell mass would be so coordinated with the whole that the single parent equation for the cell mass, Equation 3.6, would suffice. Indeed, it is probably more realistic to write one such expression for each smaller mass and its exchange with the extracellular fluid. The important fact, however, is that the pattern of the abstract construction does not change. We shall return to this later, but right now let us keep track of the possibilities in the simple system, absurd as it may seem. As in the case of the one-compartmented cheese animal, the characteristics of the two-compartment system in time are of greater interest to physiologists than are steady states. The process of transition in this model from a constant state A to a constant state B proceeds exponentially as the sum of two exponentials. The exponents are of the dimensions K/V, and relationships of size and permeability to stability are similar to those proposed in the study of the one-chambered cheese.

COMPARISON OF THE ONE- AND TWO-COMPARTMENT MODELS

In connection with the single-compartmented model the ratio of fluctuations in the inside to fluctuations on the outside was proposed as one kind of measure of stability, recognizing of course that this ratio is frequency dependent. The ratio can be formally derived from the given equations assuming that the driving force is a sinusoid. If G_1 is used to denote the value of the ratio of internal composition to environmental composition for the one-compartmented animal, and W_f represents the sinusoidal frequency at which the environmental composition oscillates, then in terms of the animal's size V and its exchange constant K_1 the following relation holds:

$$G_1 = 1 \Big/ \sqrt{1 + (W_f \frac{V}{K_1})^2} \qquad\qquad 3.14$$

The equation shows that the inside fluctuates less than the outside at all frequencies above zero and also demonstrates the effect of size and exchange constant.

For the corresponding expression describing the two-compartmented animal several preliminary definitions are necessary. Suppose the cell compartment is p times larger than the extracellular compartment (ecf) and the ease of exchange K′ is M times larger for the cell-ecf boundary than K for the ecf-exterior boundary, such that in general m and p are both greater than unity. That is,

$$V_c = pV_I \qquad\qquad 3.15$$

and

$$K' = mK \qquad\qquad 3.16$$

The expression analogous to Equation 3.14, expressing the ratio of internal fluctuations to fluctuations in the external environment, is

$$G_2 = 1 \Big/ \sqrt{1 + \frac{2\,(2\zeta^2 - 1)}{mp}\left[\frac{W_f V_c}{K}\right]^2 + \frac{1}{m^2 p^2}\left[\frac{W_f V_c}{K}\right]^4} \qquad\qquad 3.17$$

In Equation 3.17 we wish to observe only that its general form resembles the form of 3.14. The quantity ζ is a constant called the damping coefficient and is in this context defined by the expression

$$\zeta = \frac{(1 + m + mp)\,\sqrt{p}}{2\sqrt{m}} \qquad\qquad 3.18$$

where m and p are the quantities defined in Equations 3.15 and 3.16.

Now we are in a position to ask the question: If a cell mass V has a certain exchange capability K, what is the relative stability if in one case the cell mass exists free in the environment, whereas in the second it is clothed by a fluid matrix 1/p times smaller and bounded by a skin whose exchange capability is 1/m times smaller? The answer is given by comparing G_1 to G_2 for $V/K_1 = V_c/K$. The second term in the denominator of G_2 differs from the second term in the denominator of G_1 only by its coefficient, and the third term of G_2 is positive. Hence if the coefficient D has the following relation, $D = \dfrac{2(2\zeta - 1)}{mp} \geqslant 1$ for all values of m and p, G_2 must be less than G_1, and the two-compartment animal is more stable than the one having a single compartment. Since m and p are both greater than unity, from the definition of ζ the value of $D \geqslant p$, and hence $G_2 < G_1$ at all frequencies W_f. This result is a relatively trivial point as far as the derivation is concerned; however, it does answer the question. Adding an extracellular space does produce a compartment which acquires a composition different from the external environment, and which does not swing as widely about its mean position as either the environment or the level of activity as measured by the rate of production of the metabolite.

This discussion has touched upon some of the asymptotic behavior of an oversimplified non-regulating system which we called an organism with internal environment. Certain conclusions were derived from the formulation. Yet there remains the question of what can one discover about such a system, given that it exists but only in ill-defined, unsteady states. Let us again cause this imaginary creature to have a temporary existence in the environment and with the metabolic schedule which we placed upon the cheese-beast. As before, the animal comes into existence as a computer program whose output represents careful observations on the organism's insides, its metabolism, and its environment. The mass of such data is then subjected to random sampling to represent the fact that in the real situation an investigator cannot really set an origin in time. Further, this design, which is the natural history type of experiment, is not particularly concerned with temporal relations. The results of this experiment are shown as the second block of numbers in Table 3-1. The average values estimate pretty well the environmental and metabolic conditions imposed upon the system.

The yeast-beast data run quite parallel to data from the cheese-beast. The extracellular fluid of the two-compartment system and the internal composition of the one-compartment system are analogous, and we can verify the conclusion of Equation 3.12 by observing the gradient of J from the cell to the extracellular space to the exterior. Nevertheless, any effort to verify the idea that there should be an improvement in stability as Equation 3.16 implies from this data produces a Scotch verdict. One could assert that the cells are more stable and the extracellular space less stable and perhaps even defend this by some statistical test of significance. The fact of interest is that it is not difficult to think in physiological terms about systems of this sort once they can be realized in a physical form suitable for experiment. The great intellectual challenge is to discover whether conclusions arrived at deductively by analysis of the known elementary properties of a model can also be discovered by performing conceptual but physically realizable experiments upon the whole, functioning model.

The two sets of assumptions represented by the one- and two-compartment models serve to introduce one basic scheme of constraints that must apply to the general formulation of problems in studying regulation of the exchange of materials in organisms. An interior partitioned into distribution spaces is a basic pattern in the organization of animals. For each such partition a conservation equation of the same kind as we have applied to the entire organism can be written. The resulting formulation is subject only to one constraint, namely, that the systems have no mechanisms which are facultative or controllable. Any stability, in a very general physiological sense, shown by such a system must be the base line about which any real control mechanisms operate.

Although a description consisting solely of a compartmented animal in exchange is obviously inadequate, it is nonetheless desirable to set down the full generalization in symbolic form so as to give it its proper location in mathematical thought. If one considers only diffusion under the influence of a concentration gradient or material translocations which are purely functions of time, the organism is described by a set of linear, ordinary differential equations of the first order—one for each compartment. For the mythical substance J, suppose that V_i is the size of the ith compartment which is at concentration $C_i(t)$ and producing or consuming J at a rate $M_i(t)$. Further define K_{ij} as the proportionality constant determining the flux from compartment i to compartment j, and let $A_{ij}(t)$ represent time-dependent transport processes from compartment i to j. Finally, let $C_e(t)$ represent the composition which the environment has, the environment being indefinitely large. If there are n compartments, the system is, for $i = 1, 2, 3, \ldots n$

$$V_i \frac{dC_i(t)}{dt} = M_i(t) + \sum_j [A_{ji}(t) - A_{ij}(t) + K_{ji}C_j(t) - K_{ij}C_i(t)]$$

$$+ A_{ei}(t) - A_{ie}(t) + K_{ei}C_e(t) - K_{ie}C_i(t)$$

3.19

where the summation over j is $j = 1, 2 \ldots n$. Now any value of the K, M, or A may be zero for any of the n compartments. The particular combination of zeros indicates whether two compartments are adjacent and can or cannot interchange materials and also whether particular mechanisms are present or absent. One can thus represent almost any array of chambers in sequential or in parallel relationship.

The mathematical study of such systems is well advanced. To put this representation in the context of Claude Bernard's milieu interne one need only suppose that there is a well-stirred, cell-free, i.e., $(M = 0)$, commensal compartment, say the compartment F. Its equation would be

$$V_F \frac{dC_F(t)}{dt} = \sum_j (A_{jF}(t) - A_{Fj}(t)) + C_F(t) \sum_j (K_{jF} - K_{Fj})$$

$$+ A_{eF}(t) - A_{Fe}(t) + C_E(t)(K_{eF} - K_{Fe})$$

3.20

The word "commensal" is represented mathematically by the fact that every compartment j is represented by either an A term or a K term or both in Equation 3.19. Homeostasis is then identified by application of the material balance criterion and a stability criterion like the one outlined in the previous chapter to $C_F(t)$ (Eq. 2.14). One question which is physiologically interesting and which may also be mathematically interesting is whether the homeostatic criterion (Eq. 2.14) decreases as n, the number of compartments, increases.

SUMMARY

Let us summarize: The very general ideas of homeostasis and exchange are limited by their very general quality. They provide, however, a point of focus about which various sets of assumptions may be tried and examined with the idea of discovering what are the essential features that set the limits of the allowable classes of systems which behave like and include organisms. We discover that some of these features are in a sense anatomical or morphological, and that the principal and most obvious one of these is the fact of compartmentation. Each compartment in diffusion exchange with others and with the exterior can be described by a continuity expression; if no type of mechanism other than diffusion is involved the organism is describable by a large number (one for each compartment) of linear, first order, simultaneous differential equations. One should remark that such equations also describe the linear, lumped parameter, feedback networks that are the engineer's idealized descriptions for electronic and mechanical devices, and hence the great power of the mathematics developed for such systems and also some of the viewpoints of control systems engineering become pertinent to the consideration of organisms.

Notes on References

Two threads of thought need to be documented somewhat. First are those ideas centering upon systems of ordinary, linear, differential equations and their particular use in elementary control systems engineering. Many excellent books are available for the substantive consideration of these areas. Among those perhaps more useful to the student of physiology are Trimmer (1950) and Grodins (1963), which are not books on mathematics but on how mathematics are used in this context.

The second train of thought deals with compartmentation. In biological literature the impetus for the study of compartmented systems arises from the study of radioisotopes and their fate in the organism. Closed systems and experimental methods of determining an optimum configuration to represent certain experimental data are the considerations which seem to predominate in the treatment. The ideas, nonetheless, remain extremely germane in general biological modeling, as illustrated by the following: Robertson (1957 and 1962); Berman, Weiss, and Shahn (1962).

Chapter 4

THE YELLOW BRICK ROAD

WILLIAM S. YAMAMOTO

We seem to have been transported by an improbable cyclone into an absurd state of affairs, for having considered only one of the many possible general features of organisms, we have landed in a thicket of unfamiliar symbols. As in the childhood tale, the only immediate prospect of extricating ourselves is to follow a fanciful road through ever-stranger places without wandering too far from a prescribed path.

To render the idea of homeostasis into a useful and unequivocal mathematical expression, certain arbitrary restrictions were placed upon its meaning. Thus it became possible to set the concept in apposition with the error-signal criterion. The formulation narrowed the subject matter to material transactions in organisms and to physiologically familiar problems regarding material balance of organisms. Finally, in an open system like an organism having controlled and uncontrolled processes, material balance and process balance were shown to be measured by the same entities. At this point the situation was rigorous but too general to serve for particular inquiry; so casting about for features of animals that would lend a focus adequate for pursuit of particular questions, we selected the general feature of compartmentation. This is a feature already familiar to physiologists and is readily formalized with regard to material balance. What one can learn from such systems is the nature of the substratum upon which any physiological control system must operate.

Consideration of systems engaging in material exchanges and possessing this one feature, compartmentation, opened to us the possibility of a new game which is a paradigm of physiology. Two absurdly unreal organisms, a cheese-beast and a yeast-beast, were imagined; to characterize them we produced abstract patterns like those patterns of composition change that occur in some real organisms. In the absence of other assumptions, the compartmented yeast-beast in the steady state had an internal environment that mirrored the external world. The models could be generalized to show that compartmented systems are representable by simultaneous, ordinary differential equations of the first order. They gave rise to the question of whether in such systems it can be proved that the milieu interne becomes more stable as the number of compartments increases. As the next step in the argument the make-believe animals were given an evanescent and equally make-believe existence in the form of a computer program. During this existence it was possible to perform experiments of a physiological design, such that the schemes of the investigator were tested against the prescience of the creator. This state of affairs could be realized because the investigator and the creator were the same person.

A mathematician may raise some very difficult and perhaps unanswerable questions about the whole game. Are the assumptions independent, consistent, or unique? The physiologists' bias is that even though another set of abstractions might be devised from which all subsequent deductions would flow in identical fashion, there is no virtue in exchanging the abstract analog of a physiologically identifiable entity for one which is not so identifiable. Another question could be whether, given a set of assumptions of sufficient complexity, it is possible by any finite set of experiments upon the system to divine an equivalent set of assumptions. Finally, given an experiment performed in ignorance of the assumptions, can one decide whether the data are deceptive or non-informative as to the possible range of choices of abstract systems which may be represented?

But physiologists do perform experiments upon real organisms and do attempt to interpret them in terms of whole organisms. Particularly in the study of control and regulation do we attempt to assign some configuration of relationships (a model) to the animal in the laboratory. In the study of the improbable models in unsteady state, the simple-minded design of merely making many measurements of the organism in the natural state generated both relevant and deceptive data. The occurrence of such adverse mathematical verdicts need not imply that the whole venture is in vain. The purpose of the preceding chapter was to introduce the technique of examining the whole process of examining finite systems by the experimental method. Having come this far, we must proceed further, however inauspicious the start, in the hope perhaps that at the end will be an Oz from whence one can return home to the intelligent study of real animals.

Introduction of Facultative Processes

Let us first examine the change in mathematical forms produced by modeling an organism that can alter its capability for exchange according to its interior state. A compartmented system in which the exchange constants are dependent upon internal composition behaves differently from a similar collection of passive compartments. Many kinds of mechanisms may be described. However, only particular instances can be modeled, and their selection is based upon laboratory experimentation. The diversity of mechanisms observable in real animals is great, and many kinds of conceivable models turn out to be reasonable representations of actual physiological mechanisms.

Suppose the two-compartment system which was called the yeast-beast is now perpetuated and endowed with the power of amoeboid extension in such a manner that it can increase or decrease its surface by sprawling or by curling up into a ball. This means that the exchange constant K, governing the rate of movement of substance J between the organism and the environment, is a function of the internal composition $C_i(t)$. Of the many possible such functions let us elect for illustration a proportional one: i.e., the exchange constant K will be proportional to the internal composition $C_i(t)$. A constant A is added to imply that the organism shall not be permitted to shrink beyond the minimal surface which its volume would permit. Symbolically this is expressed as

$$K = K_1 C_i(t) + A \qquad\qquad 4.1$$

The material balance expressions for the two compartments, Equations 3.6 and 3.7, should be repeated here for completeness. Recall that for the cellular mass V_c the rate of change of the amount of J was

$$V_c \frac{dC_c(t)}{dt} = M(t) + K'C_i(t) - K'C_c(t) \qquad\qquad 4.2$$

And for the extracellular volume V_i the rate of change of the amount of J was also the sum of sources and sinks:

$$V_i dC_i(t) / dt = K'(C_c(t) - C_i(t)) + K(C_E(t) - C_i(t)) \qquad\qquad 4.3$$

Substitution of K from Equation 4.1 in Equation 4.3 produces a term which is of the second degree in $C_i(t)$; this equation no longer fits into the generalization that compartmented systems are representable by first-order, linear differential equations. One has instead a relatively simple, non-linear differential equation. Secondly, one now can divide the terms of the equations into facultative and non-facultative processes and thereby place this scheme of symbols into the larger framework of homeostasis (see Chap. 2).

The formal solution of Equations 4.1, 4.2, and 4.3 is somewhat more difficult than that of the linear two-compartment system. It is of interest to compare the steady-state solutions of the amoeboid system with the steady-state solution of the linear two-compartment yeast-beast. Such a comparison is shown in Figure 4-1 for models representing organisms of the same size. In each case the external environment or metabolic production of J is varied similarly. The location of the lines on the graph is based upon the arbitrary selection of the constants K_1 and A. When steady states of metabolism are plotted against either the extracellular or the

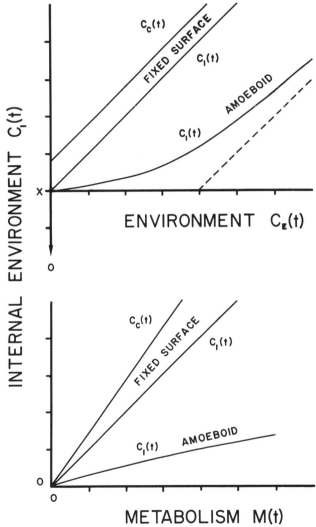

FIGURE 4-1. Graphical comparison of the fixed-surface and the amoeboid two-compartment models with respect to the steady-state concentration of J in the internal environment. The upper half shows the change in internal environment as the concentration in the external environment increases. Note that the origin of the coordinates is displaced in the positive direction of the ordinates. This reflects the fact that at zero external environment the internal environment has a positive concentration. The lower graph relates the internal concentration of J to changes in metabolism.

intracellular composition, the linear model displays a diverging set of
straight lines starting at the origin. The extracellular compartment maintains a composition which rises in direct proportion to the level of metabolism. Since exchange between the cells and extracellular space is
passive, the gradient between these compartments is always proportional
to the gradient from the inside of the organism to the exterior. Consequently, the intracellular composition follows a straight line which di-
verges from that of the extracellular. When the facultative ability to
change surface area is added, the internal composition follows the flat-
tened parabolic line, and over the range of metabolic effort the change
in internal composition is now only a small fraction of the change observed in the non-controlled model.

When the external environment changes, the internal compositions of
the two models also follow different patterns. For a given level of metabolic production of J, the model with a fixed surface area responds to
an increase in external environmental concentration of J with an exactly
equal increase in internal environment concentration. This is shown in
Figure 4-1 by a straight line slope of 1. At zero external environmental
concentration the internal composition has a concentration at which the
efflux of J matches the metabolic production. These matters were detailed
fully in the preceding chapter. The system which can vary its surface
area in response to the internal concentration of J responds an increase
in environmental composition by increasing its internal composition along
a line which is concave upward. At high external concentrations of J the
internal concentration approaches asymptotically a line which parallels
the response of the fixed system. Both the origin of the curved line and
the asymptote represent lower concentrations in the inside of the responsive model than in the corresponding unresponsive model. The position
of the asymptote depends upon the values selected for the constants in
the equations. The appearance of curvature in the line is an essential
result of the new assumption. Moreover, the concavity of the curve is such
that if one could make observations on such organisms in the real world,
one would assert that the amoeboid animal was more stable than the one
with a fixed surface area.

As a final exercise in comparison, the amoeboid model can be caused
to have a temporary existence in a computer. In this case the equivalent
of a nine-hour existence was examined. Random samples were chosen.
The data in condensed statistical form appear as the third block of numbers in Table 3-1 of the preceding chapter. The internal gradient of concentration in the organism seems larger, but the overall gradient of the
concentration of J from the cells to the outside appears smaller. In terms
of the coefficient of variation, the addition of a facultative process does
not produce an increase in observable stability.

With any increase in the complexity of expression for facultative
processes, the increased complexity in obtaining a solution becomes for-

midable. The generation of such models remains relatively simple, however, and the avenue of computer simulation remains the most practical possibility. The prospect seems harsh. As we develop a scheme of compartmented organisms which have material balance, any change which permits the inclusion of a facultative process requires the introduction of non-linearity into the system. One may perhaps be able to avoid this difficulty by inventing new forms of variables which perpetuate the policy of the system without preserving physiologically recognizable and separable variables. This is called transformation of variables. Sometimes by judicious choice a simple system of non-linear equations like the amoeboid beast can be re-represented as a linear system in some unfamiliar terms. But such a maneuver is only successful in fairly simple systems, and we have now to consider another large source of non-linear difficulties.

INTRODUCTION OF CHEMICAL PROCESSES

In all the discussion that dealt with the exchange, distribution, and handling of J, it has been tacitly imagined that J passed through the organism as an independent entity and had no relationship to any other substance in the body. There is probably no substance in the animal body that has this independence. In fact, the other extreme may be fairly asserted—that every substance in the animal body depends upon the state of every other substance in the body—and perhaps a fairly strong case can be made for that position. Neither extreme, however, is very fruitful or realistic. It is easy to demonstrate that some chemical substances are more closely and directly coupled to each other within the organism than others are. For example, the relationship between liver glycogen and plasma glucose is much closer and more explicit than that between plasma glucose and the plasma bicarbonate ion. Many such examples suggest themselves, and one may have the immediate reaction that all such statements are trivial. What is not trivial is the endeavor to define in some common and definable terms all the gradations from the obviously coupled chemicals to those only remotely related.

In the sets of assumptions previously examined, exchanges were formulated in terms of sources and sinks. These terms imply the appearance and disappearance of the material from the location or compartment under consideration. When a substance enters or leaves the space, obviously a source or a sink is implied. However, a substance may be synthesized or converted chemically into other substances; for the material-balance equations these processes are also effective as sources and as sinks. In particular the uncontrolled source, the metabolic production of J, refers to the synthesis of J from some precursors. And if J participates in some further chemical process which changes its nature, then without leaving the compartment physically the substance J has entered a sink. At this point, therefore, the particular knowledge of biochemical events becomes meshed with the macroscopic problems of exchange and control.

Suppose we then start to write down the chemical reactions that describe the anabolism and catabolism occurring within each compartment with respect to J. The reactions appear to fall into certain categories. There are first a number of reactions in which J appears explicitly as a reactant or product. Then there is a group of reactions which govern the supply or reactivity of substances appearing in the first set. To define this second set of reactions will in turn require more new reactants, and thus a third group of reactions becomes necessary. From several considerations one can predict that the rate of increase of the number of reactions in successive groups will be very large. Since all the chemical reactions in the animal constitute a finite set, eventually all the chemical relations in a compartment would be present in the list. When all the compartments of an organism are so considered, all the reactions of the organism relevant to the exchange of J would be included. For any substance J, therefore, the description of the organism extends in one direction over all the compartments and in a second direction over all the reactions. While such a system is readily conceivable, it is quite unmanageable. The distinction between the physiologist and biochemist is to a large measure the fineness of grain with which the latter examines a representation of this sort. There seems to be no way at present to bridge the discontinuity in conceptualization that occurs when one proceeds from a few to many reactions. Perhaps the difficulty lies in the nature of our thought processes, since the pattern of concept levels is a very general one in science.

A convenient and practical method of approach is to work with only the innermost circles about a particular J. If some caution is exercised in the selection, this method is successful because the overall effect of many of the chemical events in the periphery of a web is negligibly small; that is, they are only weakly coupled to the system under study. Sometimes it is found that two substances, which have common chemical reactions and should obviously be closely related for physiological reasons, have a sufficient degree of autonomy that the animal's state regarding one substance need be considered only casually when studying the second substance. For example, an endocrinologist studying the intricacies of glucose homeostasis does not find it essential at every step to observe simultaneously the consumption of oxygen and its distribution. Conversely, studies of man's behavior at hypoxic altitudes can be conducted more or less independently of his glucose metabolism.

We are led to conclude from the success of our practices that there are substance-centered problems of regulation in the organism which are coupled only weakly to others. Yet in certain instances a substantial degree of dependence is observed between systems of material exchange. This represents a coupling which is weak enough to permit the isolated study of each member, but which is strong enough to require the simultaneous control of observations on other members of the set. Contemporary interest in physiological problems of this sort is considerable and is expressed in numerous publications in which the word *interaction* plays a prominent role. The sufficient size of the biochemical web of reactions which constitutes a physiological system suitable for the laboratory study of animals is much smaller than the possible size of the whole description. It is because weak couplings exist between such subsystems that the systematic approach to physiology is both practical and interesting.

But why is all this included in a description of the constraints under which the general definition of homeostasis must operate? We learned in the case of compartmentation that the physical nature of a compartment specifies the kind of mathematical expression that must be used. Compartments generate linear, first-order, ordinary differential equations. Any attempt to represent control mechanisms dependent upon composition introduces non-linearity. Now we shall see that chemical synthesis and the degradation of material within the compartments impose similar changes upon the choice of equation systems.

For homogeneous chemical reactions the mathematical machinery is well developed. The necessary variables, apart from the chemical identity of the reactants and the compartment volume, are the rate constants of the reactions and the concentrations of the reactants. In those reactions in which J may be present explicitly, its concentration is a variable common to the equations describing the rate of reactions of all its neighbors. Except for the simplest kind of reaction, ordinary differential equations of the first order but polynomial in the dependent variables are required. The particular form of the polynomial depends upon the stoichiometric relationships. The number of non-linearities is large, and it is no longer possible by any simple transformation of variables to deal with the system as a linear system. Consideration of even a few of the biochemical steps in material exchange, therefore, compels one to abandon any serious attempt to represent organisms in terms of linear differential equations. One should not deny, however, that a certain valuable insight may still be obtained from the study of linear compartment systems. Conceptually, as one examines particular examples of physiological systems, any distinction between the metabolic pool of the biochemist and the physiological idea of content of a compartment simply vanishes.

Perhaps some comment is needed in regard to heterogeneous reactions which are so common in living systems. Representation of such processes is considerably more difficult and uncertain than in the case of homogeneous reactions. Both the idea of a compartment and the idea that a continuous function can represent transformations are subject to challenge—the former because it seems to require at times the identification of an intramolecular site as a "compartment" (Goodwin, 1963). The concept of continuity is under challenge when the number of reactant molecules falls to such a small value that reaction must be regarded as a discrete process. As in many of the matters that remain to be discussed we can only identify these difficulties without offering a solution.

THE POSSIBILITIES FOR AN ALGORITHM
OF MATERIAL-FLOW GRAPHS

Serious as the problem of non-linearity is, there remain two other major areas of difficulty which must be considered. They concern the means of representation and condensation of complex control networks and the proper representation of systems able to utilize events in the past for the purpose of making present adjustments. We shall now attempt to point out some of the features of these problems. Suppose we represent on a diagram the related chemical reactions and also the processes of physical dislocation or transport. Select a simple symbol for the substance J in the compartment and represent it as a labeled node or dot on pieces of paper, using a different piece of paper for each compartment in which J is present. Then let us represent by means of an arrow the phrase "the atoms go to form a physical part of," much as arrows are used in writing chemical reactions. Those sheets which represent compartments that exchange J are joined by such arrows. Within each page there will be one or more substances not-J which will be connected to J. In the case of reversible reactions the arrows will be in pairs between the neighbor node and the node J; in the case of irreversible reactions, or reactions in which the reverse reaction is many orders of magnitude slower than the forward reaction, only one arrow will appear between adjacent nodes. If the arrow is directed toward J it will represent a source of J; if the arrow is directed away from J it will represent a sink.

Now between pages or compartments one can draw arrows joining corresponding symbols only, and that only when there exists the possibility of physically moving the molecule from one compartment to another. After all such connections are made one would have a complicated diagram indeed for even the simplest number of reactants. There should appear to be two groups of substances. One is a group for which a source or a sink or both terminate outside the organism; these are the materials in exchange with the environment. The second type has all the termini of sources and sinks within compartments of the body; these are substances in the internal economy. A material of the second type will at some point connect to the exterior to nodes of the first type. The source or sink passing outside the animal terminates upon an uncontrolled node, viz., the environment.

Let us define a "principal path" as follows: It begins at any node not-J, follows the directed arrows in succession through as many J nodes as possible, and terminates on a not-J node. It includes, however, a second class of allowable termini, namely, J nodes in the external environment. The collection of principal paths which includes every J node at least once is a "principal set." For any given node the number of sources and sinks (arrows) may exceed the number necessary for a principal set. The number of these excess sources and sinks plus the termini of the principal paths define the independent variables of the principal set. It is clear that the excess may not be unique, but in every case represents the coupling which imbeds the J system into the whole organism and the organism into the environment.

It is further plain that the principal set represents the paths of material flow through the organism. Now if one examines any principal path, one discovers that over portions of it, and through adjacent nodes or through nodes several times removed from it, one may find paths which parallel the principal path. Some such paths are oriented in the reverse direction to the principal path and indicate processes which diminish the magnitude of a preceding node at the expense of a subsequent node. Also, paths will exist which tend to augment subsequent nodes at the expense of preceding ones. Principal paths furthermore may be joined together in a fashion that includes one or more irreversible steps. If such a string of paths originates or terminates in the exterior world, one may conjecture that it represents a major path for the study of material balance. There is a considerable conceptual similarity between paths so joined and the forward loop of a feedback control system. All the other paths which parallel this string can then be represented as feedback or feed-forward loops. Moreover, if one sums the molecular exchanges over the principal strings for a substance J, then one arrives at an expression for the error-signal criterion developed earlier.

There remain such problems of detail to be worked out as the most desirable manner of representing material flux as arrows and concentrations as nodes (since in the latter case the algebraic sum of fluxes yields the time derivative of the concentration). The fact is clear that the flow of material through the organism may be represented by a multi-layered graph consisting of directed arrows and named nodes. The diagram is very similar to the signal-flow graph of feedback theory. The graph which we have evolved, however, is concerned with the location, transformations, and flow of materials in organisms and could properly be called a material-flow graph.

Let us pursue this matter a step further. A signal-flow graph portrays certain kinds of physical systems in more detail than the familiar block diagram. It is intended to be a visual representation of signal flow through a network. It evades, however, certain problems of the block diagram. For example, the block diagram requires that the blocks represented must be non-interacting and that the transfer functions of tandem blocks be the product of the transfer functions of the individual blocks.

Clearly a certain ambiguity is possible, when one considers the large, complex systems of physiology, in the choice and manner of study of what is included in any single block. The feedback paths are all represented in a signal-flow graph; in fact, the signal-flow graph presents exactly the same information as the corresponding set of simultaneous equations. The signal-flow graph is useful particularly in analyzing complicated systems, in establishing the basic relationships of feedback theory, and finally in investigating the role of particular parameters in the overall system. Finally, one should note that signal-flow graphs can be manipulated in a rigorously defined manner to yield a basis for block diagrams. For linear systems the theory of signal-flow graphs is rather complete and furnishes substantial advantages in visualization and hence in the solution of problems of complex systems.

Now the material-flow graph which we have just constructed shows one manner of connectivity which holds the animal together as an entity, and it also provides a means for deciding to some extent what constitutes the subsystems separable for purposes of study. To tie this view of the organism to the general body of feedback theory there seems to be here a logical point of connection. It seems plausible to derive signal-flow graphs appropriate to the organism from the material-flow graph. This connection also seems the starting point in discovering the method by which macroscopic concepts may be derived from a miscroscopically detailed description. Although the problems of relating material flow to control theory and the problem of isolating macroscopic groupings from a microscopic representation are distinct problems, the material flow diagram is a suitable common starting point. In particular, many physiologically useful ideas require the concept of a signal. For example, the physiologist employs the idea of a nerve impulse pertinent to some chemical change in the body rather than the set of chemical reactions which themselves constitute the impulse. He also employs the idea of hormones which represents the bodily state of balance of some other chemical substances. The general words *stimulus* and *response* are indeed well-entrenched, macroscopic ideas.

Both condensation and orientation in the approach using control-system engineering may depend upon the success one has in effecting the derivation of a signal-flow graph from a material-flow graph. The derivation is beset by difficulties which are of such severity that we cannot offer a solution to them; but let us examine a few of them. In the usual signal-flow graph the nodes are variables and the arrows are linear operations, such as multiplication by a constant, integration or differentiation with respect to time, and so on. Interpretation of the graph proceeds by performing upon each node the operation indicated by the arrow and summing the result into the subsequent node as indicated by the direction of the arrow. The material-flow graph, in contrast, has only fluxes and concentrations.

For systems such as the imaginary beasts invented earlier in these chapters, the representation in terms of signals is simple; one need only use arrows to indicate diffusion constants and integration over time as the operations needed to make a complete system. This kind of system can be treated correctly and exactly by the rules for signal-flow graphs and becomes a signal graph when each material is asserted to be a signal. However, as soon as chemical reactions of an order higher than zero are included it promptly becomes necessary to represent products and powers of variables. These operations are outside the province of present linear, signal-flow-graph theory and so require further invention. Each node and flux must be translated into a graphical symbol representing relationship rather than material flow. One sees that two kinds of nodes or two kinds of arrows may suffice if the multiple representation of variables is permitted. Ambiguity is difficult to avoid, and an already complicated diagram becomes even more complicated. One can see that what is required at the very least is a simple, graphical representation of all algebraic operations. It is not clear whether such a system can be fruitfully evolved, or whether when it is done it will have the useful property of being macroscopically reducible. It is certain that the resulting graph will be quite different from any signal-flow graph now in use, and while it may map the organism faithfully it may also approach the organism in complexity. A successful map is not just one which is faithful in detail but one which successfully omits the 99 per cent of detail which is irrelevant or confusing. The problem of representation is, then, the first major unsolved problem in this line of thought, which endeavors to establish an unequivocal transformation between material flow and signal flow.

While the graph can be imagined and perhaps even constructed, to attempt to do so may be as impractical as to set out to compute the pressure of a gas in a container from the individual momenta and positions of the molecules. On the other hand, the diagram might show in detail the link between biochemical and physiological understanding. But to be worthwhile there must be rules for the unambiguous manipulation of the graph into fewer parts.

Even in the absence of a correct signal-flow graph some reasonable conjectures may be made about its use. In the material-flow graph one may propose to proceed by a process of condensation, selecting for instance individual J nodes or strings of J nodes, treating their appended arrows as arbitrary functions. This is precisely the method we employed in studying compartmented systems with and without control. In general there are two kinds of relationships in a material-flow graph. The first deals with the compartments and with the reactions which directly involve J and only immediate precursors and successors. These relationships must appear in any serious control-system representation that purports to represent physiological understanding of bodily function. Secondly, a material-flow graph has large numbers of reactions and compartments which do not involve J but which may be connected in definite patterns. Some of the sequences of the second group, nevertheless, do constitute representations of J, or can assume configurations relevant to the state of J at some site in the graph. These seem to be appropriate candidates for macroscopic representation. Such patterns resemble physiological mechanisms, and one may speak of them in terminology which is physiological rather than biochemical.

Although we have now pointed out a manner in which the knowledge one has about animal organization might be diagramed, it seems unlikely that one would ever do so. A complete graph would require an extremely fine mesh of compartment and reaction. The effort seems neither practical nor promising of understanding. Rather, condensation and organization into macroscopic concept must proceed without the actual construction of the graph except in terms of its rules of existence, the rules of ordinary, algebraic differential equations.

Problems of Time Lags in Systems

We may now turn to the second of the added complexities beyond the appearance of non-linearities. If one ascribes to organisms either the capability of acting upon past events in time or sufficient size so that communication between parts takes a finite time, then the whole edifice of ordinary, algebraic differential equations becomes imperiled. The simplest example is circulation of the blood. Because of the limitations of diffusion processes, large organisms can maintain an existence only through some system of convection. There are several successful means of achieving this end, of which the most familiar and probably the most effective is that which involves a vasculature, a pump, and a fluid tissue, the blood. All the organs and tissues then become riparian sites. And just as primitive communities on the banks of a stream draw commerce from the stream and dispose of effluent waste into the stream, so the tissue cells draw nourishment from, and excrete waste into the blood stream. For material exchange each tissue is autonomous except for the commerce that proceeds via the blood stream, and it is this stream that binds together the whole organism and gives it continuity and coherence. It should be plain that no two riparian sites see the same stream, and that each modifies the blood as it flows by. The differences in material composition between different parts of the body are obvious and can be represented graphically by drawing the grain of the compartments progressively finer until a satisfactory representation can be made.

However, the remaining difficulty is not correctible by this means. Suppose you and I lived on the shores of a common stream. We could communicate in a fashion by your putting messages in bottles and sending them downstream to me. For some purposes this would be a satisfactory system. Suppose that at the downstream site I had a dam and sluice valve by means of which you expected me to govern the rate of flow and hence the level of the stream at the upstream site. In this situation, the stream's rate of flow would depend upon information received via the stream. It is plain that such a system has inherent in it a delay which is the time between the launching of the bottle and the time of receipt of the message; I would always be controlling the rate of flow of the stream on the basis of past information. And even though we should attempt to refine the bottle technique to the ultimate and transmit an continuous concentration of some sort of dye, there would be a finite discontinuity between the time of your observation and the time of my response.

Returning to organisms: If a chemoreceptor exists downstream from an excretory site or an anabolic site, the chemoreceptor must function on the past actions of the upstream organ. Endocrine and target organ alike bear such relationships. Circulation then introduces the necessity of dealing with continuous processes which depend upon other continuous processes separated from them by finite time intervals. This is true, of course, of any communication system in which there is a conduction time. For example, when one inadvertently grasps a hot object, the possibility of response does not exist until the conduction time and minimum processing time of the central nervous system have elapsed. Control-system theory, which takes its strongest impetus from electrical engineering in relatively unextended systems, could proceed quite adequately at first without paying great attention to problems created by time lags; but in the organism in which the fastest mode of communication has the slow speed of the nerve impulse, and many other processes are in communication by the even slower circulation, and finally when actions may occur on the basis of a remote past, as by means of a remembered event, finite time lags become numerous and of significant magnitude. Thus, as one strives to create a symbolic image of the processes of exchange between compartments and between chemical forms, one finds that even the very extensive realm of ordinary differential equations is not sufficient.

Of the many types of equations which have been used to deal with systems involving not only the present state but also the past, the simplest which will represent fixed circulatory lags is the difference differential equation. To illustrate, suppose that the respiratory minute volume $\dot{V}(t)$ is governed by the composition of arterial blood arriving at the brain. A finite time lag representing the circulation time from lung to brain exists; let us call it w. The formula for the idea that the present ventilation is proportional to the past blood composition is the simple, linear difference differential equation,

$$\dot{V}(t) = k\ C_j(t-w) \qquad\qquad 4.4$$

where k is a constant of proportionality and C_j (t—w) is the concentration of j, w time units in the past. This simple equation expresses behavior quite different from the comparable differential equation,

$$\dot{V}(t) = k\ C_j(t) \qquad\qquad 4.5$$

One can imagine the logical extension of this idea to all the forms of equations which we have previously used, in which the spans or retardations, w, are several in number and magnitude. Further, if one assumes that the vascular volume is relatively fixed and the blood flow variable over several fold, in any actual representation the retardation is itself variable. This leads to even more complex types of functional equations. The upshot of the preceding paragraphs is that we can perceive why the more familiar and widely used tools of elementary control-system theory are inadequate for an even modestly realistic representation of biological systems.

In addition to the difficulty introduced by the use of non-linear differential equations, time lag operations introduce other profound difficulties while increasing the types of systems which can be described. Some of the difficulties are of the following sort: In the first place it is usually impossible to express solutions of even the more simple types of difference differential equations in the form of the simple functions in explicit form. If one attempts to compose the solution out of a sum of exponential solutions one finds that generally there is an infinity of such solutions. The boundary or initial condition that it is necessary to specify in the process of solution consists of the system's behavior over a continuous interval of time equal to the span. Although one may be able to solve particular equations, the applied uses of such mathematics are uncommon and unfamiliar to most of us. Therefore, the physiologist who wishes to see whether his chain of biological reasoning is coherent, or self-consistent, or produces a system which resembles the organism realistically, can do so only by undertaking a large-scale simulation on a computing machine. But even here the methods for correctly instructing a machine to render the behavior prescribed by the equations are only now coming under fairly intensive investigation.

Summary

Like the travelers on the yellow brick road, we have passed by many scenes which seem fortuitously selected and only superficially understood. Therefore we should summarize the long excursion to show that it was not without design. We wished to show the kind of detail that needed to be included if one set out to construct an abstract but accurate description of a physiological control system, a description which focused fruitfully upon the concept of exchange balance as the measure of performance. In the first place a map was necessary for determining the location of the substances in the organism; in the simplest form the maps were those of systems of linear differential equations which defined compartments. From them one could describe the kinds of behavior which are a consequence of such organization independent of any active processes of exchange.

The next logical extension was the inclusion of the chemical reactions which produce or destroy substances, and also the inclusion of macroscopic representation of those processes which do not contain particular substances but which reflect their status. These are ideas very close to compartmentation on the one hand, and to signal-flow graphing on the other. The latter introduced a distinction between material flux and control information pathways. A useful definition for the term feedback in naturally occurring biological systems was suggested. To describe these ideas, systems of ordinary, non-linear differential equations which are polynomial in the concentrations of chemical substances were introduced.

Finally and very inadequately, we considered systems capable of making present behavior depend upon past history. In the simulation of such systems it is not only necessary to state the rules for the system— namely, the describing equations—but also a plausible and appropriate history. The extension of models to such capabilities required the introduction of both difference differential equations and functional equations.

In making these brief statements we have included a large chunk of modern mathematical analysis. A serious and detailed pursuit of specific problems in this direction requires considerable mathematical erudition. But the surprising and perhaps tantalizing thing is that many of the physiologically conceived notions already have names and representative counterparts in modern mathematics. It even seems possible that a devoted and single-minded biomathematical approach to some of the modern physiological problems would be entirely successful. One may wonder whether a mixture of equation types in a system of sufficient magnitude could be found that satisfied both the domains mathematically prescribed for the equations and the physiologically sound descriptions.

Such a system of representation might turn out to be more valuable than are the constantly growing libraries needed to accumulate the primary ideas and their numerous permutations that constitute the research literature of physiology and biochemistry. We still ask of the novitiate in physiology the ability to reason, "If a and b and c, then d" of a reasonably complex chain, adding also that "Neither a nor b nor c is certain." The surprising capacity of the human mind is that an experienced physiologist can remember quite a long chain of uncertain conditional assumptions and predict the consequence of their concurrent action.

A physician can do the same, and in circumstances in which he has the added burden that his judgment will affect the welfare of his fellow man.

The approach which our long discussion has proposed is somewhat unfamiliar, and like the yellow brick road, may lead to a disappointing result. But at present it is less complicated than an endless compendium of words would be. It starts at a justifiable, rigorous, macroscopic level of formalism which is simpler than the written language. It cannot abolish the uncertainties of the data, but neither can the written word. It has one important article of faith: that the number of primitive ideas in the biology of an organism is finite and probably small. Hence, all the richness and wonder of modern biology must consist of ever-compounded permutations of history, process, and constraint. The great challenge of contemporary physiological investigation must lie, not in the observation of further phenomena, but in the discovery in the growing mass of verifiable experimental detail of a set of basic laws from which we can regenerate the accumulated observations of laboratory science. Then alone can it be said that we have some wisdom about the body.

Notes on References

Many elementary sources exist for consideration of the simpler mathematical problems posed in this chapter, any of which should be adequate to extend the particulars of the treatment. In the context of control systems, however, one should examine a treatise of intermediate level complexity to discover the pertinence of signal-flow graphs, parametric feedback, and discontinuous operation. Two satisfactory sources for this purpose are Truxal (1955) and Tou (1959).

The importance of time lags in the abstract modeling of physiological systems is brought out by writers as early as Lotka (1924, reprinted 1956). More recently, attention has been directed to the importance of such phenomena by, for example, Bellman, Jaquez and Kalaba (1960). A difficult but rather complete survey of the difference differential equation is to be found in Bellman and Cooke (1963). An interesting attempt to employ functional equations as the general basis for non-linear control systems with pertinent overtones for biological readers is embodied in two papers by Katzenelson and Gould (1962, 1964). A consideration of control problems within cells with particular attention to heterogeneous reactions is to be found in Goodwin (1963).

This selection of references is heavily weighted in the direction of helping to find further mathematical information, since probably to many readers of primarily physiological training the degree of unreality of the physiological models will be all too clear, while the mathematical fabric from which the models are cut may be less familiar.

Chapter 5

NEURONS AND
TEMPERATURE REGULATION

H. T. HAMMEL

John B. Pierce Foundation
New Haven, Connecticut

Without the central nervous system, the regulation of body temperature is no more than an equilibrium between heat loss and heat production in which heat production is a basal metabolic state. Under this condition the body temperatures passively adjust so that the core-to-skin-surface temperature difference times the body tissue conductance, and the skin-surface to ambient temperature difference times the cooling coefficient become equal to heat production. A thermal stress imposed upon such a system passively affects the body temperatures and indirectly, through the activation energy of chemical reactions, alters the heat production so that a new equilibrium occurs at a new level of body temperatures and heat production.

In normal homeothermic mammals, the central nervous system intervenes by way of its thermal sensing elements, afferent pathways, integrative components, and effector pathways to activate responses which modify the rates of heat production and heat loss so that the core temperature does not vary beyond narrow limits when thermal stress is imposed upon the system. Many of the relationships between thermal regulatory responses and body temperatures for many species are well known and have been reviewed from time to time throughout the history of physiology; recent reviews are those of Bazett, (1949), Hensel (1952), von Euler (1961), Hardy (1961), and Andersson, Gale, and Sundsten (1963). Without exception, each reviewer has recognized the sensing, integrative, and effector roles of the CNS, and using the information available has constructed a highly simplified "black-box" model, identifying inputs, outputs, and feedback loops, as well as important anatomical sites represented by the "black-box" components.

Almost without exception, no one has dared to detail the properties and interconnections between neurons required to achieve the known characteristics of the black box and for a very good reason, namely, that little is known about the neurons in those parts of the CNS known to be essential to the accomplishment of temperature regulation. Consequently, it is not difficult to construct models unhampered by facts about how neurons act and interact upon each other to yield regulation. Such models can give the impression that they accomplish the job at hand and account for the known relationships between thermoregulatory responses and body temperatures and other inputs. After we have reviewed the phenomenon of temperature regulation as we understand it, we too will suggest a working model composed of neurons which we think could accomplish the job.

THE CLOSED-LOOP SYSTEM

We shall attempt an analysis of temperature regulation on the assumption that the system behaves like a closed-loop control system involving negative feedback as an essential feature. Thus the basic black-box representation of temperature regulation can be diagramed as in Fig. 5-1.

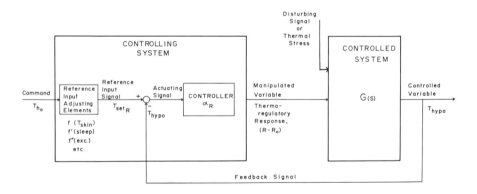

FIGURE 5-1. Block diagram for regulation of hypothalamic temperature.

We shall assume that the controlling system originates within the hypothalamus and in the preoptic region between the optic chiasm and the anterior commissure, and extends by way of the CNS to thermal effector organs within the body. The output of the controlled system is assumed to be the temperature of the body as it is represented by the temperature of the hypothalamus, T_{hypo}. The control system serves to keep the output signal T_{hypo} equal to or close to the reference input signal, T_{set_R}, at all times in the face of varying environmental disturbances, changing properties of the controlled system, and other inputs. Since the transfer function $G(s)$ for the controlled system is defined by the laws of heat transfer from within the body to its surface and from the surface to the environment, and does not directly involve the CNS, no further attention will be given to it. The controlling system, on the other hand, depends upon the CNS in an essential way, i.e., without the CNS there could be no controlling system.

There is abundant evidence in any current review of temperature regulation (Andersson et al., 1963; von Euler, 1961; and Hardy, 1961) that the preoptic region of unanesthetized mammals responds to local heating by inducing panting and vasodilation, and to local cooling by evoking shivering and vasoconstriction. Electrical stimulation and ablations in and around the hypothalamus leave no doubt that this region is involved in and, in fact, is essential for the regulation of body temperature.

Figure 5-2 is, in a sense, a summary of the evidence that the rostral hypothalamus is sensitive and responsive to small displacements of its

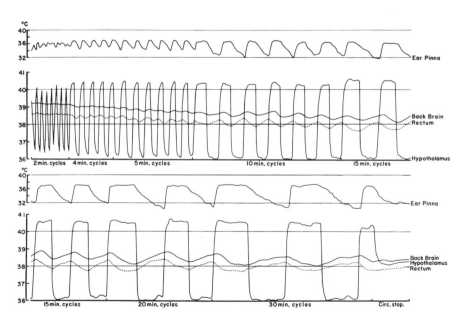

FIGURE 5-2. Cyclic heating and cooling of hypothalamus with water perfusing thermodes alternately at 41.0°C and 35.0°C. Ambient temperature was 25°C. The 30 minute cycle equals 15 minutes heating and 15 minutes cooling. (Hammel, H. T., Strømme, S. B., and Cornew, R. W., *Life Sci.* 12:933, 1963.)

own temperature. The widely swinging temperature (from 36.0° to 40.5°C) in this record is that of a thermocouple in the middle of the anterior hypothalamus while the hypothalamus is alternately heated and cooled at several frequencies. The low-amplitude traces in the middle of the widely swinging trace are those of the cerebellar ("back brain" in Fig. 5-2) and rectal temperatures swinging in consonance with the hypothalamic temperature. Above these temperature traces is the trace of the ear pinna temperature of the dog in an ambient temperature of 25°C. When the hypothalamus was alternately cooled and heated, the blood vessels of the ear pinna were alternately constricted and dilated, resulting in an alternately falling and rising ear pinna temperature. Likewise, the rectal and cerebellar temperatures were alternately rising and falling at the same rate and by about the same magnitude while the hypothalamic temperature was displaced by equal amounts below or above its unperturbed level. We conclude from this that the anterior hypothalamus is responsive to both heating and cooling and with equal sensitivity to moderate heating and cooling.

The Law of the Controlling System

An important next step in elucidating the nature of the system is to obtain the so-called "law of the controlling system," that is, the relationship between the inputs and the outputs of the controlling system. We shall assume that the law of the controlling system is an equation* of the form

$$R - R_0 = -\alpha_R(T_{hypo} - T_{set_R}) \qquad 5.1$$

where $R-R_0$ is the thermoregulatory response (metabolism, vasomotor activity, sweating, panting, and so forth), R_0 is the basal level when $T_{hypo} = T_{set_R}$, T_{hypo}, the actual hypothalamic temperature, is the feedback signal, T_{set_R}, the functional† set-point temperature for response R, which is the reference input signal, $(T_{hypo} - T_{set_R})$ is the actuating signal, and α_R is the proportionality constant for the response $(R-R_0)$.

Although the evidence is scanty, it is adequate for proposal of this law (Eq. 5.1) as a working hypothesis. It says that a given thermoregulatory response is proportional to an actuating signal which is the difference between the actual hypothalamic temperature and some functional set-point temperature for that response. It should be recognized

---

* *Ed. Note:* Observe that R and R_0 are quantitatively undefined except in the very general sense of having the dimension of heat flow, **cal/hr.**

† *Ed. Note:* The word functional is intended to convey the meaning "effective in determining the level of function at the selected time."

at once that the properties of the system, α_R and T_{set_R}, may be different for each type of thermoregulatory response R.

CHANGE OF THE SET POINT VERSUS CHANGE OF HYPOTHALAMIC TEMPERATURE

This hypothesis suggests that if we could design an experiment whereby we could maintain the functional set-point temperature unchanged while displacing the hypothalamic temperature over a range of values, the response, say shivering, would occur as follows: $(M—M_o)$, the increase in metabolic rate above the basal level, would be zero for hypothalamic temperatures above T_{set_M}, while for T_{hypo} below T_{set_M} it would increase in proportion to the difference between the two temperatures. The results obtained from one attempt to perform this ideal experiment are shown in Figure 5-3.

In our hypothetical statement of the law of the system, we intended to imply that the reference input signal T_{set_R} is not an invariant term, but may be adjusted by a variety of factors including skin temperatures, core temperatures, exercise, sleep, pyrogens, and possibly humoral agents which excite, inhibit, or depress hypothalamic neurons. Therefore, in our experiment conducted in a neutral environment of 23°C, we attempted to maintain the dog in a resting state by training, to prevent excitement by isolating the animal in the environmental chamber, to prevent sleep by occasional conversation through a speaking tube, to maintain constant environmental temperature, and to initiate a test of metabolism at each hypothalamic temperature only after the rectal temperature had returned

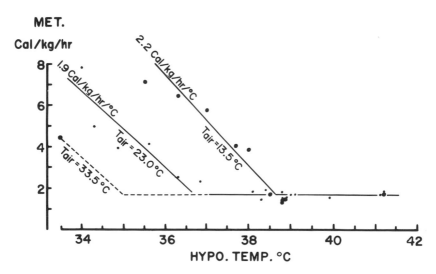

FIGURE 5-3.   Heat production as a function of hypothalamic temperatures for a quiet, resting, wakeful dog at three air temperatures, 13.5°C, 23°C, and 33.5°C. (Hammel, H. T., Strømme, S. B., and Cornew, R. W., *Life Sci.* 12:933, 1963.)

to 38.1°C. To obtain the results shown in Figure 5-3 at 23°C and 13°C, the hypothalamic temperature was artificially displaced by varying amounts above and below its normal level by perfusing six thermodes surrounding the preoptic region with water at suitable temperatures. The period of artificial hypothalamic temperature displacement was limited to about 10 minutes and the corresponding metabolism was taken as the average over a 2-minute interval of the period of highest oxygen consumption during each period of thermal stimulation.

The need to relate the hypothalamic temperature to the peak metabolism during a short interval was obvious, since any artificial shivering or panting caused by the thermal stimulation changed the core temperature and in turn reduced the magnitude of the response. For example, with prolonged hypothalamic cooling, initial shivering was vigorous, resulting in a rising rectal temperature and eventually in an elevated steady rectal temperature with the shivering decreasing to zero. In a neutral environment, only vasoconstriction was sustained with prolonged hypothalamic cooling. Similarly, panting could not be sustained by prolonged hypothalamic heating. Referring back to Figure 5-2, we see that the rising or falling rectal temperature levels off during the 30-minute hypothalamic heating-cooling cycle.

From these studies it does seem possible to represent the shivering response of a resting, wakeful dog in a neutral environment by a pro-

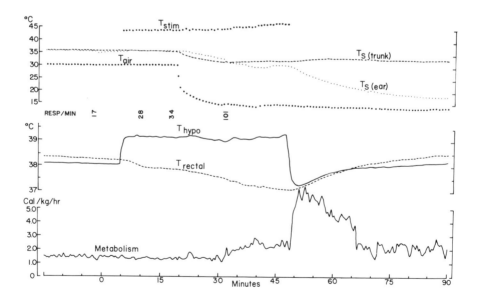

FIGURE 5-4. Body temperatures and heat production of resting, fasting dog exposed to neutral and cold environments during manipulation of its hypothalamic temperature. (Hammel, H. T., Jackson, D. C., Stolwijk, J. A. J., Hardy, J. D., and Strømme, S. B., *J. Appl. Physiol.* 18:1146, 1963.)

portional equation of the form we have hypothesized in which the properties of the system are $\alpha_M \simeq 2$ kcal (kg hr °C)$^{-1}$ and $T_{set_M} \simeq 36.8°C$ in a 23° environment. The properties of the system for the same quiet, resting, wakeful dog in a cold (13.5°) environment, as determined by the same method, are $\alpha_M \simeq 2$ kcal (kg hr °C)$^{-1}$ and $T_{set_M} \simeq 38.8°C$ in a 13.5°C environment.

By measurement of the water loss from the respiratory tract in the same experiments from which Figure 5-3 was drawn, the functional set-point temperatures for panting in the neutral and cold environments were found to be $T_{set_p} \simeq 38.8°C$ in 23°C environment and $T_{set_p} \simeq 41°C$ in 13.3°C environment.

We have suggested that the law of the control system for temperature regulation is a zero-order, proportional equation, that is to say the response of the controlling system is a function of T_{hypo} but not a function of the time derivative, $\dot{T}_{hypo}$. Although the evidence is meager, this assumption is based on an experiment such as that illustrated in Figure 5-4. For this experiment the hypothalamic temperature of a dog was held at an elevated temperature (39.3°C) for a half hour while the ambient temperature was maintained at 15°C, with the result that the core temperature steadily decreased. When the artificial elevation of the hypothalamic temperature was discontinued, it fell very rapidly by about 2°C to a level just above the core temperature. There followed a large increase in metabolism, but the increased metabolism appeared to be due to the low hypothalamic temperature and not to the very rapid rate of decrease in hypothalamic temperature. If the shivering response were a function of $-\dfrac{dT_{hypo}}{dt}$, we would have expected to see a large burst of shivering during the interval when the temperature was falling. There was none. An additional value can be drawn from this experiment. By comparing the metabolism and the hypothalamic temperature just before the drop in T_{hypo} with the metabolism and hypothalamic temperature just after the drop, and assuming that there was no change in the functional set-point temperature in the short interval between, the proportionality constant for shivering was computed to be $\alpha_M = 2.0$ kcal (kg hr °C)$^{-1}$, or the same as that obtained in Figure 5-3.

We have reviewed evidence which suggests that the regulation of body temperature may be described as if the hypothalamus were responsive to changes in its own temperature and as if the set-point temperature for each response were adjusted by the environmental temperature, i.e., by its effect upon the skin temperature. Next we ask, what changes in hypothalamic temperature do actually occur when a resting animal is placed in a hot, neutral, or cold environment and do these changes contribute to the thermoregulatory response? The answer may be found in Figure 5-5 which illustrates that the hypothalamic tempera-

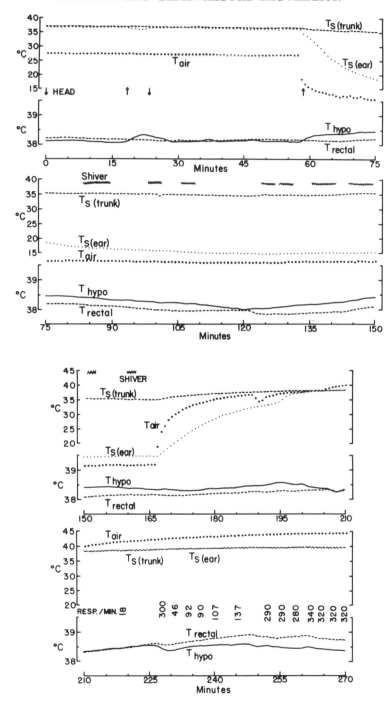

FIGURE 5-5.   A continuous record of hypothalamic temperature, T_{hypo}, and other temperatures of a resting, fasting dog exposed to neutral, cold, and hot environments. Shivering and respiration rates are noted. (Hammel, H. T., Jackson, D. C., Stolwijk, J. A. J., Hardy, J. D., and Strømme, S. B., *J. Appl. Physiol.* 18:1146, 1963.)

ture changes very little, if at all, when the environmental temperature is changed from neutral (27°C) to cold (12°C) to hot (44°C).

At the end of 60 minutes the air temperature was dropped to 12°C. The immediate increase in hypothalamic temperature seen here was due to the awakening of the animal as the air temperature fell. At 85 minutes the dog was shivering vigorously and the hypothalamic temperature was 38.4°C. Intermittent shivering continued for the next 80 minutes while the hypothalamic temperature varied between 38.0 and 38.4°C. After 165 minutes the air temperature was increased to 44°C. After an hour in the hot environment the dog began to pant, and for 40 minutes panted vigorously even though its hypothalamic temperature was between 38.4 and 38.5°C; that is, no higher than it was when the dog was shivering vigorously. We have often observed the same dog shivering vigorously when its hypothalamus was at a temperature as much as 0.3° higher than when it was panting vigorously.

Since the hypothalamic temperature may not change in a useful way in response to external thermal stress, we may suppose that either the hypothalamus is needlessly sensitive to changes in its own temperature and that the entire regulatory response is initiated by and regulated by a signal derived from the skin temperature, or that the actuating signal for driving the regulatory response may be achieved by offsetting the set point according to the thermal needs of the animal. By the latter thesis, which we prefer, when the skin temperature falls in a cold environment, the steady-state and phasic firing rates of cold receptors in the skin increase and elevate the set point so that the hypothalamic temperature, without changing, is below the set point and drives heat conservation mechanisms or increases heat production. Conversely, when the skin temperature rises in a hot environment, the steady-state firing rate from the cold receptors diminishes to zero, and the steady-state and phasic firing rates of the warm receptors increase and may, thereby, lower the set-point temperature below the hypothalamic temperature so as to drive the mechanisms that promote heat loss.

Even though the hypothalamic temperature may change very little in response to the change from a cold to a hot environment, or may even increase in a cold environment or decrease in a warm environment so that the hypothalamus appears to be insensitive or needlessly sensitive to its own temperature, it is, nevertheless, essential that it be responsive to changes in its own temperature and probably equally sensitive to warming and to cooling.

Next we shall attempt to show that other factors such as exercise, sleep, and fever have an effect upon temperature regulation which may also be interpreted as if there is an adjustment of the set-point temperature. In the explanation of the experiments that follow, the emphasis will be on showing that the results may be plausibly interpreted *as if there had been an adjustment of the set-point temperature.* No one can believe that these results exclude all other interpretations.

ADJUSTMENT OF THE SET POINT IN NORMAL ANIMALS

Measurements were made of respiratory heat loss, hypothalamic temperature, oxygen consumption, and rectal and skin temperatures on dogs resting and running on the level at 4 mph at several environmental temperatures. Figure 5-6 is a composite of results on three dogs showing the relationship between respiratory heat loss and hypothalamic temperature at rest and in exercise. Several conclusions may be drawn: (1) In resting dogs C and D, there was only insensible respiratory heat loss at hypothalamic temperatures up to about 38.6 or 38.7°C. Above this temperature there was active panting. In resting dog B there was also active panting at 38.7°C and higher, but there was somewhat more than the insensible amount of respiratory heat loss down to 38.3°C. (2) In all exercising dogs, there was a several-fold increase in respiratory heat loss over the same range of hypothalamic temperatures as in the resting dogs.

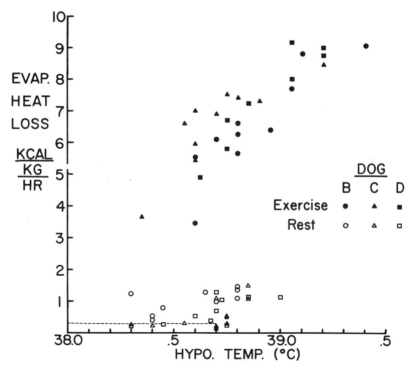

FIGURE 5-6.   Evaporative heat loss as a function of hypothalamic temperature at rest and in exercise. (Jackson, D. C., and Hammel, H. T., USAF AMRL-TDR-63-93, 1963.)

(3) The respiratory heat loss increased with the increasing hypothalamic temperature in the exercising dog. (4) The exercise points lie on a line shifted to the left of the resting points. These results are consistent with the view that the functional set-point temperature has decreased by about 0.8°C for the level of exercise maintained in these experiments.

There is also no indication of an increase in the proportionality constant relating respiratory heat loss to hypothalamic temperature. From calorimetric data obtained in our laboratory some years ago on normal, resting dogs, the proportionality constant for respiratory heat loss was found to be a 3.8 kcal (kg hr °C)$^{-1}$ increase in rectal temperature. When a least square line is drawn to represent the data during exercise of dogs B, C, and D, a proportionality constant, or slope, of 3.4 kcal (kg hr °C)$^{-1}$ increase in rectal temperature is obtained. Thus the major change in the rate of evaporative heat loss during exercise may be interpreted as if the set-point temperature were decreased at the onset of exercise.

Another situation in which there is suggestive evidence that a shift in the set-point occurs is in the transition from the wakeful state to sleep. The hypothalamic temperature of a rhesus monkey exposed for 24 hours to a hot environment (35°C), a neutral environment (30°C), and a cold environment (20°C), is seen in Figure 5-7. On each of these three days the light in the climatic chamber was turned off at 6:00 P.M. and turned on again at 9:00 A.M. After 6:15 A.M., daylight from the laboratory could also enter through a small uncovered window in the chamber. The

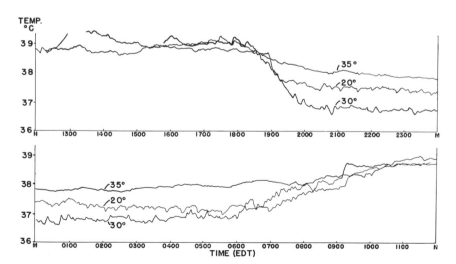

FIGURE 5-7. Hypothalamic temperatures of a rhesus monkey restrained in a primate chair in hot (35°C), neutral (30°C), and cold (20°C) environments (50 per cent relative humidity) for 24 hour periods with normal day-night lighting. (Hammel, H. T., Jackson, D. C., Stolwijk, J. A. J., Hardy, J. D., and Strømme, S. B., *J. Appl. Physiol.* 18:1146, 1963.)

monkey had been living in a primate chair for two months after the thermodes were implanted, and was trained to feed itself at will from a food pellet dispenser. During the hours of light the hypothalamic temperature was almost constant at $39.1 \pm 0.2°C$ for all environmental temperatures. In each instance, soon after the light was turned out, the hypothalamic temperature fell to another level by an amount depending upon the ambient temperature. In the neutral $30°C$ environment, the hypothalamic temperature fell within two hours to $36.8 \pm 0.2°C$ and remained so throughout the night. In the hot environment, the hypothalamic temperature fell to only $38.0 \pm 0.1°C$ and remained so with only small fluctuations throughout the night. In the cold environment ($20°C$) which produced vigorous shivering day and night, the temperature fell to an intermediate level of $37.5 \pm 0.2°C$ and stayed so throughout the night. In each instance, the hypothalamic temperature returned to the daytime level more slowly than it fell the evening before. The onset of the rising temperature occurred at about 6:00 A.M., and the daytime temperature was achieved by about 10:00 A.M.

The changes in the hypothalamic temperature that occur when an animal goes to sleep and the effect of these changes upon the thermoregulatory responses strongly suggest that a dramatic change has occurred in the regulation of body temperature. The only question is whether the change can better be described as a decrease in the set-point temperature, or as a decrease in the responsiveness of the controller, α_R. Suppose that

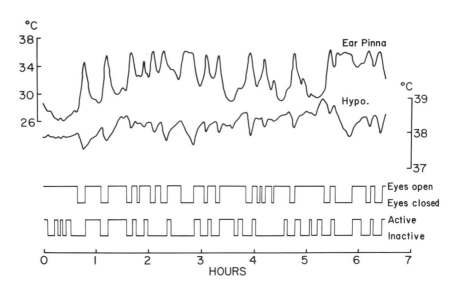

FIGURE 5-8. Hypothalamic and ear pinna temperatures of a rhesus monkey in a primate chair in cool environment (22-$24°C$). Open or closed eyes and activity are noted. (Hammel, H. T., Jackson, D. C., Stolwijk, J. A. J., Hardy, J. D., and Strømme, S. B., *J. Appl. Physiol.* 18:1146, 1963.)

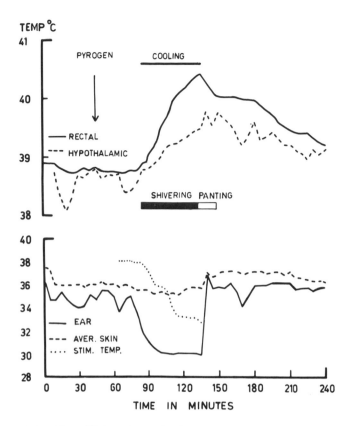

FIGURE 5-9. The additive effect of injected pyrogen and hypothalamic cooling. A "hyper-fever" was produced. (Andersen, H. T., Hammel, H. T., and Hardy, J. D., *Acta Physiol. Scand.* 53:247, 1961.)

the onset of sleep does not lower the set point, but that the gain of the regulatory mechanism is reduced. Then, in a hot (35°C) environment, a reduced gain at night would predict that the hypothalamic and core temperatures would passively increase to a level above the day temperatures. The fact is, the hypothalamic temperature is 1°C lower at night than in the day in the 35°C environment. Therefore, it appears that the observations of the hypothalamic temperatures in the monkey may be accounted for as a set-point shift at constant gain rather than by an assumption that the set point is unchanged with the onset of sleep and that only the gain of the thermoregulatory mechanism is reduced.

This conclusion receives further support in Figure 5-8. Here the hypothalamic temperature and the temperature of the ear pinna of a rhesus monkey exposed to a cool environment (22 to 24°C) during the

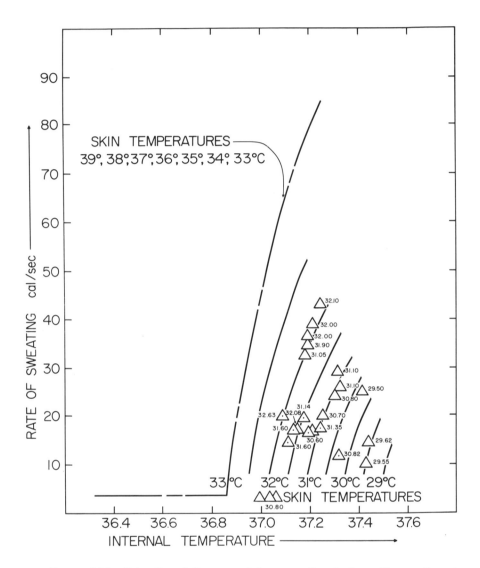

FIGURE 5-10. Intensity of thermoregulatory sweating during cold reception at the skin. Sweating rates were plotted against internal cranial temperatures. Measurements obtained at similar skin temperatures were connected with "best lines." At any given cranial internal temperature, sweating rates are seen to be diminished by approximately 40 cal/sec for every degree of decrease in level of skin temperature. The line for 33°C skin temperature is from Figure 5-12. Figure 5-10 contains no resting observations, since these paradoxical conditions cannot be produced in steady states at rest. Work rates were mechanically equivalent to 6 cal/sec (△) or 11 cal/sec (△) respectively. Increase in work rate enlarges the range of observations to the right (low skin temperature with high internal temperature). Experiments were carried out between April 4 and June 5, 1961, with one subject, D. D., nude, age 26, weight 88.6 kg, height 176 cm. (Benzinger, T. H., Kitzinger, C., and Pratt, A. W., In *Temperature—Its Measurement and Control in Science and Industry.* Reinholt, New York, 1963.)

day were recorded while it was also noted whether its eyes were open or closed and whether or not it was active. This monkey was also confined to a primate chair for 5 weeks after the re-entrant tubes were implanted. Its hypothalamic temperature was about 1.5°C lower at night than during the day in a neutral environment. When isolated in the climatic chamber but under continuous observation through a half-silvered mirror, the monkey's hypothalamic temperature fell several tenths of a degree each time it closed its eyes, and conversely, its hypothalamic temperature increased each time its eyes opened during seven hours of observation. Each time its eyes closed, vasodilation occurred so that the ear pinna temperature increased to 36°C. Likewise, each time it opened its eyes, vasoconstriction occurred, causing the pinna temperature to fall to 30°C or lower. So with falling hypothalamic temperature, the animal's heat loss is increased, and with rising hypothalamic temperature the heat loss is decreased. These results may also be interpreted as if the temperature regulator were modified by a decrease in the set-point temperature at the onset of sleep and an increase upon awakening.

Adjustment of the Set Point in Fever

The transition from a normal to a fevered state provides another example illustrating how thermal regulation may be modified by a shift in the set-point temperature. The description of fever as a condition involving an elevation in the set-point temperature enjoys wide acceptance although the concept is still rejected by a few. After waiting the usual latent period following the administration of an exogenous pyrogen, hypothalamic cooling was instituted; its effect is shown in Figure 5-9. A "hyper-fever" was produced during the chill phase, marked by vasoconstriction and shivering. While still in the fever phase, the hypothalamic cooling was terminated, with the result that panting and vasodilatation occurred and brought the "hyper-fever" down to the usual fever level. Again, the suggestion that a pyrogen acts to elevate the set-point temperature is plausible.

The quantitative data from which we have derived a law of the controlling system for temperature regulation have been obtained from experimental animals in which hypothalamic temperature may be readily obtained and easily displaced by artificial means. Man is not to be neglected in developing a quantitative understanding of temperature regulation. The data from Benzinger's laboratory, as summarized in Figures 5-10 and 5-11, may be interpreted by the same relationships between input and output which we have already described for experimental animals. We interpret the results in Figure 5-10 to mean that in a neutral environment when the skin temperature is 33°C, heat loss by sweating increases proportionally as the hypothalamic temperature rises above some functional set-point temperature for sweating $T_{set_{sw}}$ where the proportionality constant $\alpha_{sw} \simeq$ 12 kcal (kg hr °C)$^{-1}$ and $T_{set_{sw}} \simeq$

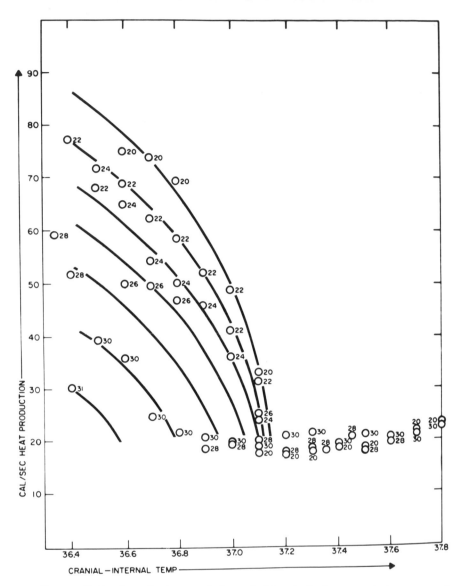

FIGURE 5-11. Quantitative resolution of "chemical" heat regulation (cranial internal temperature plot). This figure gives the metabolic responses to almost any physiological combination of skin and cranial internal temperatures. The dependence of metabolic rate upon cranial internal and skin temperatures appears quantitatively from this graph. The thermostatic function of the internal thermoreceptive system is clearly visible: when cranial internal temperature falls below the set point, 37.1°C, the metabolic response to cold begins to rise with falling cranial internal and falling skin temperatures. (Benzinger, T. H., Kitzinger, C., and Pratt, A. W., In *Temperature—Its Measurement and Control in Science and Industry*. Reinholt, New York, 1963.)

36.85°C $+ \triangle$ where $\triangle$ is some small difference between hypothalamic and tympanic membrane temperatures. Figure 5-10 also suggests that the functional set-point temperature increases as the skin and ambient temperatures decrease.

Similarly, Figure 5-11 may be interpreted to mean that in a neutral or slightly cool environment when the skin temperature is 31°C, heat production by shivering increases proportionally as the hypothalamic temperature drops below some functional set-point temperature for shivering, $T_{set_{sh}}$, where the proportionality constant is $\alpha_{sh} \simeq 4$ kcal (kg hr °C)$^{-1}$ and $T_{set_{sh}} \simeq 36.6°$ C $+ \triangle$. Again the functional set-point temperature is seen to increase as the skin and ambient temperatures decrease. It also appears that the proportionality constant α_{sh} increases to as high as 10 kcal (kg hr °C)$^{-1}$ for extremely low average skin temperatures. If we assume this to be true, then the apparent non-linear relationship between shivering and ($T_{set_{sh}}$ — T_{hypo}) for very low skin temperature may be explained by suggesting that, as the metabolism increases to four times basal, there is a leveling off, since it is difficult to produce heat faster than four times the basal rate by shivering.

There appear to be some discrepancies between the results obtained from man and those obtained from dogs. For instance, Benzinger has interpreted his results obtained for a resting or exercising man, Figure 5-12, to mean that the set-point temperature for sweating is the same in exercise and at rest. He notes that if the lower average skin temperature during exercise (1.5°C) were to have the same effect upon the sweat rate as a similar drop in skin temperature in a resting man as shown in Figure 5-10, this would have lowered the exercising man's sweat rate by 60 kcal/sec at the same brain temperature. Perhaps these results may be interpreted to mean that the lower skin temperature during exercise does in fact raise the functional set-point temperature for sweating, as in Figure 5-10, and at the same time the exercise per se lowers the set-point temperature, so that there is no apparent change in the set-point temperature in Figure 5-12. Recent data do suggest that the functional set-point temperature decreases during exercise in man (Minard, 1963).

Other apparent differences between the results on man and on dogs are the larger proportionality constants in man than in dogs (twice as large or more) and the smaller effects of ambient temperature and exercise upon the functional set-point temperatures in man. It cannot now be stated whether these differences are more apparent than real, because there are important differences in the methods by which the brain temperatures were manipulated in order to relate the response to the brain temperature, and differences in the way the skin temperatures were affected by the environment. Moreover, there is some uncertainty in relating hypothalamic temperature to tympanic membrane temperature. If the differences are real, then we might suppose the greater thermal capacity, and consequently greater thermal inertia, of man enables him

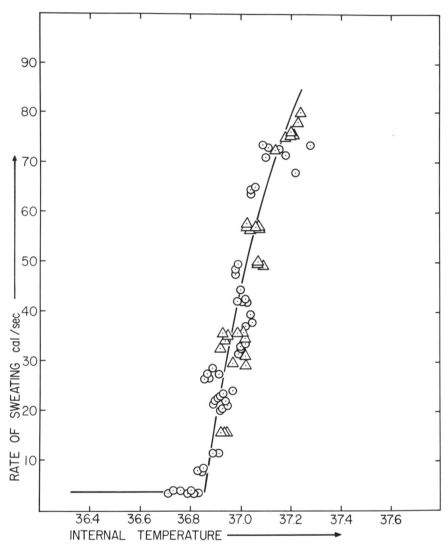

FIGURE 5-12. Intensity of thermoregulatory sweating during warm reception at the skin. Sweating rates were plotted against internal cranial temperatures in steady states of rest (⊙) or exercise (△6 cal/sec and △11 cal/sec). Sweating began at a sharply defined internal cranial temperature, the "set point" of the human thermostat, a result reproduced over a period of two months. The intensity of thermoregulatory sweating was inseparably related to the level of internal cranial temperature. There was no visible effect of the drastic differences in skin temperature (average, 2°C) between the resting and the working observations. This difference (average, 2°C) would have lowered the triangles by an estimated 80 cal/sec in comparison with the cycles measured at the same internal cranial temperature if the inhibiting effect of cold observed in Figure 5-10 had extended with similar intensity into the range of warm reception. Experiments were carried out between April 4 and June 5 with one subject, D. D., nude, age 26, weight 88.6 kg, height 176 cm. (Benzinger, H. T., Kitzinger, C., and Pratt, A. W., In *Temperature—Its Measurement and Control in Science and Industry.* Reinholt, New York, 1963.)

to enjoy a more sensitive control mechanism without the risk of an unstable system.

NEURONS IN THE CONTROL SYSTEM

Having speculated this far in our effort to arrive at a formal statement or a "law of the controlling system" for temperature regulation, we feel there will be only an imperceptible transition in our story if we attempt to offer at least one plausible scheme showing how neurons may be connected to achieve the relationships stated in the formal equations.

Our model will be based on the following assumptions:

1. There are neurons in the rostral hypothalamus having spontaneous firing rates which are strongly temperature dependent, i. e., $Q_{10} >> 1$, over the range of normal deep body temperatures. These are designated as hi-Q_{10} primary sensory neurons.

2. Axons of these sensory neurons synapse with neurons within the hypothalamus which ultimately activate thermoregulatory responses. These latter are designated as first stage or primary motor neurons.

3. The primary motor neurons may or may not have spontaneous firing rates depending upon the choice of models to be preferred. Their firing rates are assumed to have little or no temperature dependence except as influenced by the sensory neurons.

4. Synaptic terminations on cell bodies of both primary sensory and primary motor neurons may either facilitate or inhibit the neurons.

5. Although studies employing experimentally induced lesions and electrical and thermal stimulation may indicate that primary sensory and motor neurons are found in high concentrations in certain hypothalamic sites, these results cannot be interpreted to mean that neurons of a given type are located only in a small, circumscribed region.

These are not unreasonable assumptions to make regarding neuronal activity and are generally accepted as working assumptions on the limited evidence available (Hardy, 1961; von Euler, 1961). An additional assumption will be made at this time; its justification, or rather its desirability, will become apparent later.

6. We shall assume that another set of primary sensory neurons designated as lo-Q_{10} sensory neurons is located in the rostral hypothalamus in the same region as the hi-Q_{10} sensory neurons. The lo-Q_{10} sensory neurons are assumed to have spontaneous firing rates over the range of deep body temperature. Further suppose that the cells are either not strongly temperature dependent, i.e., $Q_{10} \simeq 1$ or, as suggested by Bazett (1949), increase their firing rate with decreasing temperature, i.e., $Q_{10} << 1$. Like the hi-Q_{10} sensory neurons, the lo-Q_{10} sensory neurons are assumed to synapse with and facilitate or inhibit the action of the primary motor neurons which activate regulatory responses.

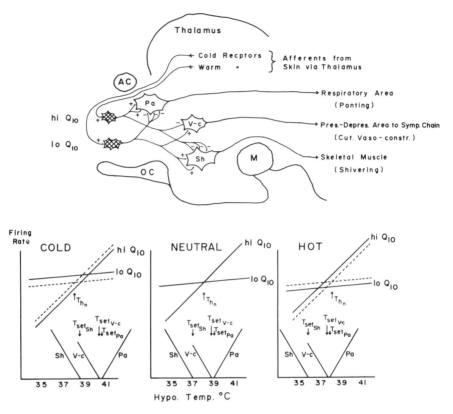

FIGURE 5-13. A physiological model for establishing a set-point temperature and illustrating possibilities for adjusting the set point. AC, anterior commissure; OC, optic chiasm; M, mammillary body; Pa, primary neuron for panting; Sh, primary neuron for shivering; v-c, primary neuron for vasoconstriction; crosshatched cell bodies, low-Q_{10} and high-Q_{10} primary sensory neurons.

In Figure 5-13, we are suggesting one way in which the hi-Q_{10} and lo-Q_{10} sensory neurons are connected with three classes of primary motor neurons which respectively activate panting, vasoconstriction, and shivering. In this figure we have included two more assumptions, one essential and the other trivial. It is essential to assume that the hi-Q_{10} sensory neurons facilitate primary motor neurons increasing heat loss, e.g., panting, and at the same time inhibit primary motor neurons which lead to vasoconstriction and shivering. Conversely, the lo-Q_{10} sensory neurons must inhibit panting and at the same time facilitate vasoconstriction and shivering. The other assumption is that the primary motor neurons for panting and shivering in Figure 5-13 have no spontaneous firing rates. In order to obtain different set-point temperatures for panting and shivering, we have shown more inhibition than facilitation from the sensory neurons synapsing with the neurons for panting and shivering.

The same condition could also be achieved by assuming these motor neurons to have low but not zero thresholds so that more facilitation than inhibition would be required to activate panting and shivering.

Reference to the activity curves of each of the sensory and motor neurons below the diagram of Figure 5-13 will indicate how the controlling system is presumed to function. First, examine the set of activity curves for the neutral environment. For the temperature at which the firing-rate curves of the hi-Q_{10} and lo-Q_{10} sensory neurons are equal, i.e., intersect, the facilitation and inhibition from these sensory neurons upon the vasoconstriction (v.c.) motor neurons nullify each other so that the v.c. motor neuron is active at its own spontaneous firing rate. The temperature at which the firing-rate curves of the hi-Q_{10} and lo-Q_{10} neurons intersect in a neutral environment for a resting, wakeful, normal animal is designated to be T_{h_n}.* If the hypothalamic temperature drops below T_{h_n} in the neutral environment, then the v.c. motor neuron is more facilitated than inhibited and vasoconstriction is increased. If the hypothalamic temperature rises above T_{h_n}, then vasoconstriction decreases. The temperature at which vasoconstriction becomes zero is designated as the functional set-point temperature for vasoconstriction in a neutral environment.

As shown in the activity curves for the neutral environment, when the hypothalamic temperature equals T_{h_n}, both the panting and shivering motor neurons are more inhibited than facilitated, so there is no panting or shivering. As the hypothalamic temperature increases, the facilitation of panting increases faster than the inhibition. For temperatures above $T_{set_{pa}}$, facilitation exceeds inhibition and panting results in proportion to $(T_{hypo} - T_{set_{pa}})$. Similarly, as the hypothalamic temperature drops below T_{h_n}, inhibition of the motor neuron mediating shivering decreases more rapidly than does facilitation. For hypothalamic temperatures below $T_{set_{sh}}$, facilitation exceeds inhibition, and shivering increases in proportion to $(T_{set_{sh}} - T_{hypo})$.

So far, we have considered a neuronal model of temperature regulation located within the hypothalamus and have considered how it may function without input from outside itself, i.e., no input from peripheral receptors in the skin, from extrahypothalamic core receptors, from the reticular activating system, or from any other source. The model postulates that the primary sensory and primary neurons alone can activate thermoregulatory responses and, to do so, require only an appropriate displacement of the hypothalamic temperature from the functional set-point temperature for each response. The model, thereby, conforms with our experimental results obtained from displacing the hypothalamic

---

* T_{h_n} may differ from the intrinsic hypothalamic set-point temperature T_{h_0} of Figure 5-1, since in the wakeful dog in a neutral environment there may be and very likely are afferent inputs into the hypothalamus from the thermal receptors in the skin and from the reticular activating system.

temperature and finding that the response is proportional to the difference between the actual hypothalamic temperature and the functional set-point temperature for that response.

Our results then went on to show that placing the animal in hot or cold environments did not actually lead to a useful displacement of the hypothalamic temperature although there were appropriate regulatory responses. We did not choose to interpret this to mean that the hypothalamus is needlessly sensitive to changes in its own temperature, but rather that an afferent input into the hypothalamus from thermal receptors in the skin somehow shifts the functional set-point temperature for each regulatory response. Our results obtained by displacing the hypothalamic temperature with the dog in neutral, cold, and hot environments (Fig. 5-3), support this interpretation. They also suggest that the environmental temperature shifts only the functional set-point temperatures and not the proportionality constants.

The effects of afferent input into the hypothalamus consistent with our experimental results may be readily and simply achieved by running afferent fibers to the primary sensory neurons where they either facilitate or inhibit the spontaneous and temperature-dependent activities of these sensory neurons. In Figure 5-13, afferents from the cold receptors in the skin are shown to facilitate the lo-Q_{10} sensory neurons and afferents from the warm receptors in the skin are shown to facilitate the hi-Q_{10} sensory neurons. In the cold environment, the activities of the primary sensory and motor neurons are presumed to be as shown in the lower left graph of Figure 5-13. The increased firing rate from the cold receptors in the skin is shown to facilitate the activity of the lo-Q_{10} sensory neurons, and the decreased firing rate from the warm receptors in the skin is shown to reduce the activity of the hi-Q_{10} sensory neurons with the combined effect of raising all functional set-point temperatures. Thus, although there may be no change in the actual hypothalamic temperature—and, in fact, it may increase a little—it is below the functional set-point temperatures for vasoconstriction and shivering and will drive these responses in proportion to the differences.

In like manner, the activities of the primary sensory and motor neurons in the hot environment are presumed to be as shown in the lower right graph of Figure 5-13. Reduced firing rates from cold receptors in the skin are shown to reduce the activity of the lo-Q_{10} neurons, and increased firing rates from warm receptors in the skin are shown to increase facilitation of the hi-Q_{10} neurons, with the combined effect of lowering all functional set-point temperatures. The hypothalamic temperature, without changing, still drives panting and reduces vasoconstriction.

Simply by suggesting that all afferent connections to the temperature regulatory mechanism within the hypothalamus act by facilitating either the hi-Q_{10} or lo-Q_{10} sensory neurons, it is possible to account for the apparent shifts in the functional set-point temperatures that occur in the transition from wakefulness to sleep or in exercise, and without any apparent adjustment of the proportionality constants. For example, we may visualize that connections from the reticular activating system terminate on the lo-Q_{10} sensory neurons and facilitate these during the waking hours. At the onset of sleep, the facilitation may rapidly diminish with a resultant immediate drop in all functional set-point temperatures —without changing any of the α_R's.

We recognize that terminating all inputs to the hypothalamus upon the sensory neurons is not the only way to affect temperature regulation. It is possible that some or all of the peripheral inputs feed into the motor neurons directly and facilitate or inhibit these neurons. But to do so would require that the peripheral inputs would also have to exercise antagonistic control over the classes of primary motor neurons. For example, in order to shift all functional set-point temperatures in the cold environment, as occurs experimentally, it is necessary to suppose that the cold receptors in the skin not only facilitate the primary neurons for shivering but at the same time inhibit the primary neurons mediating panting. Similarly, the warm receptors in the skin must not only facilitate the primary neurons for panting but also inhibit the neurons for shivering. If the peripheral inputs do go directly to the primary motor neurons rather than to the primary sensory neurons, then the hypothalamus may be thought of as needlessly sensitive to changes in its own temperature because the hypothalamic temperature does not change in response to external thermal stress.

Since we know of no experimental evidence as to whether the afferent inputs go directly to the primary motor neurons* or go directly to the primary sensory neurons, we have favored the latter view because it is a simpler arrangement and because it does not render the hypothalamus so needlessly sensitive to changes in its own temperature.

It is worth noting that whatever afferent connections are made with the hypothalamus, or for that matter whatever agent acts upon the primary sensory or motor neurons within the hypothalamus, appears to affect regulation in the same way—namely, by adjusting the functional set-point temperatures rather than by changing the proportionality constants. Should a body of evidence accumulate which demonstrates that the proportionality constants do change with thermal stress, for example, $\alpha_{shivering}$ increasing in cold exposure or $\alpha_{panting}$ increasing in heat exposure, then it will be necessary to suggest that within a class of primary motor neurons there is a wide range of thresholds or a range of levels of spontaneous activity so that there is a range of functional set-point temperatures for each response. Thus, as the hypothalamic temperature

---

* *Ed. Note:* Primary sensory and motor neurons in this context refer to the initial (hi-lo-Q_{10}) and terminal cells of the model rather than cells of the spinal cord.

deviates toward its extreme limits, there would be a recruitment of themo-regulatory response and therefore increasing α_R. At present, the evidence is too meager to speculate further on this possibility.

Benzinger has proposed that there are two central sites involved in temperature regulation and that they "differ basically in their main characteristics." The central site for the regulation of vasodilation and sweating is placed in the anterior hypothalamus; it acts as a terminal sensory receptor organ for temperature and acts independently of heat stimulation of the skin. On the other hand, the central site for regulating metabolism is placed in the posterior hypothalamus and acts like a synaptic relay station for afferent impulses from cold (not warm) receptors in the skin. This synaptic relay station, he supposes, is not affected by its own temperature, but it may be influenced by the anterior hypothalamus, which is responsive to its own temperature; that is to say, a warm anterior hypothalamus may diminish shivering by depressing the activity of the posterior hypothalamus which is relaying the impulses from the cold receptors in the skin to the muscles.

In an opposite way, cooling the anterior center is said to release the normal depressing effect of the anterior upon the posterior hypothalamus. Thus, the function of the cold receptors in the skin is to "elicit (not to gradate and regulate) the metabolic response to cold," and the function of the central thermoreceptive system is to "either depress or release the metabolic response to cold receptor impulses from the skin precisely to such an extent as is required to maintain or restore homeostasis" (Benzinger, Kitzinger, and Pratt, 1963). Andersson holds a similar view with regard to the role of the thermoreceptive center in controlling shivering (Andersson et al., 1963).

If the posterior hypothalamus serves as a relay station for impulses from the cold receptors in the skin which are presumed to elicit the metabolic response to cold, and if the only role of the thermoreceptive center in the anterior hypothalamus is either to depress or release the metabolic response to cold receptor impulses from the skin, then it is difficult to understand how it is possible to elicit shivering in an animal when its skin temperature is normal, as in a neutral environment, or when its skin temperature is high, as in a hot environment. It is, in fact, easy to elicit shivering in a neutral environment by dropping the hypothalamic temperature of the dog below about 37°C, and it is even possible to more than double the metabolism in a hot environment by dropping the hypothalamic temperature to about 34°C (Fig. 5-2). Therefore, we are inclined to the view already expressed that there are two sets of primary sensory neurons with widely different firing-rate Q_{10}'s, one set with $Q_{10} >> 1$ and the other with $Q_{10} \simeq 1$ or possibly $<< 1$, and that the regulatory response is a result of the difference between these two antagonistic sets acting upon the primary motor neurons.

SUMMARY

For want of information, we have had to ignore many relevant and interesting questions. We cannot elaborate upon the mechanism by which a primary sensory neuron achieves a $Q_{10} >> 1$ except to suggest that such a unit may be in fact a cascade of neurons each with a more normal Q_{10} of 2 or 3. We do not attempt to say how a given sensory neuron may facilitate on the one hand and inhibit on the other. It may be that one half the class of hi-Q_{10} sensory neurons facilitates and the other half inhibits their respective target motor neurons. We are unable to say whether or not the thermal drive from the skin needs to be divided into regions, with each region having a different effectiveness upon temperature regulation. Common experience does suggest that raised or lowered temperatures of the skin of the face, hands, and feet yield sensations different from those derived from similar temperature changes of trunk surfaces.

We have assumed five basic classes of neurons in our model, two primary sensory and three primary motor neurons. The direct evidence for these is meager, indeed, and the indirect evidence from electrical stimulations, ablations, and thermal stimulations was only partially reviewed here. Nakayama, Hammel, Hardy, and Eisenman (1963) have evidence that at least in the anesthetized animal there are neurons in the rostral hypothalamus whose firing rates increase with increasing temperature, and neurons whose spontaneous firing rates do not change with changes in temperature (Nakayama et al., 1963). But we can only suppose that at least some of these are involved in regulating temperature. The supposition may be wrong. Furthermore, it cannot be stated whether those neurons with Q_{10} as high as 10 are primary sensory neurons or perhaps primary motor neurons for panting. However, microelectrode studies in this rostral region of the hypothalamus in the unanesthetized preparation should reveal a difference between a hi-Q_{10} sensory and a "panting motor neuron." The activity of the former should increase continuously over the range of a few degrees below body temperature to a few degrees above; whereas "panting motor neurons" should be silent at temperatures below body temperature and increase only at higher temperatures. Another useful microelectrode study would be an attempt to record activity from "shivering motor neurons" in the posterior hypothalamus while cooling in the rostral hypothalamus in the unanesthetized animal.

A recent provocative study by Keller and McClaskey (1964) has led them to conclude, "The neural integration of heat dissipation is wholly dependent upon the anatomical integrity of the subthalamic and cephalic midbrain level of the brain stem. *Except for a possible permissive function, heat dissipation is quite independent of the hypothalamus.*" Also they conclude, "The neural integration of resistance-to-hypothermia is wholly dependent upon the anatomical integrity of the hypothalamic grey *and is completely independent of tissues lying cephalad to the hypothalamus."* These views are in stark contrast to those held by the author throughout this entire chapter.

Much thought and effort may be required to rationalize the interpretations of those who seek insight into the nature of body temperature regulation by ablating or sectioning segments of the central nervous regulator and examining the fragments of normal physiological function that may be left, and those who choose to molest the hypothalamus by perturbing its temperature. A few suggestions are made which, hopefully, will be steps toward mutual understanding:

1. Keller and McClaskey (1964) found in the dog ". . . that diencephalonectomized or high-midbrain preparations retain the ability physiologically to resist overheating in a sufficiently adequate manner to prevent a critical rise in body temperature." They believe this to be evidence "that not only the anterior but the entire hypothalamus is not essential for effective dissipation of heat" and that "the only permanent deviation from the normal which may be exhibited by these preparations is a 'raised heat-dissipation threshold.'" These results seem rather to demonstrate that there are internuncial neurons which reside in the midbrain and which connect the regulatory neurons in the preoptic region with the respiratory neurons in the medulla and that these have thermal properties of their own which include a spontaneous firing rate with a not unlikely Q_{10} of 2 or 3. If this were the case, then these midbrain neurons, when isolated from the rostral controller, might very well activate some panting if their temperature were elevated sufficiently. On the other hand, assuming reasonable values for the tissue conductance, cooling coefficient, and insensible evaporative heat loss for the dog, it may be shown that, had Keller and McClaskey anesthetized their high-midbrain preparations, they would have found the rectal temperature of these dogs in a 38°C environment to rise no more than a few tenths higher than that of the same preparation unanesthetized. We do not consider the residual capability of these preparations as consistent with a "raised heat-dissipation threshold," for it lacks the essential characteristics of regulation. Such a preparation makes no response at all to a greatly reduced body temperature.

2. Keller and McClaskey also found that dogs with high hypothalamic sections and even dogs with low hypothalamic sections from which "anterior hypothalamic tissue lying dorsal and cephalic to the midlevel of the chiasm was aspirated" were able to pant and shiver with threshold shifts of no more than 2 or 3°C. These results may demonstrate that although the primary neurons (including sensory and motor neurons) of the controlling system are found in the highest concentration in the preoptic region, they are also found, albeit more sparsely, in more caudal parts of the brain. If such is the case, we would expect a variety

of bizarre results ranging from little to severe deficit depending upon asymmetry in the ablation and on the presence of even small remnants of primary tissue.

It is only too plain at this stage that neuronal models for temperature regulation are, to a high degree, guesswork and that they may be justified on the grounds that making them is pleasurable. They might also lead to useful investigations.

Research in the author's laboratory is supported in part by Contract AF-33-(657)-11103 with the 6570th Aerospace Medical Research Laboratories, Wright-Patterson Air Force Base, Ohio.

Chapter 6

THE "SET-POINT" CONCEPT IN PHYSIOLOGICAL TEMPERATURE REGULATION

J. D. HARDY

John B. Pierce Foundation and Yale University School of Medicine

In Chapter 5 the problem of the body temperature of animals has been discussed in detail, and a hypothesis of regulation by adjustable or varying "set points" has been presented. It seems appropriate, therefore, to extend this discussion to explore the meaning of the term "set point" as it is used by systems engineers and to see how the concept can be applied to temperature regulation in physiology. In this situation it is clear that, although there will be problems associated with the rigid adaptation of an engineering concept to a physiological system, yet engineers have had long experience and much discussion concerning complex regulating systems, and the physiologist can gain increased understanding of the functioning of biological regulators by looking carefully at these concepts. In regard to set-point regulators, the American Standards Association (1963), in their brochure on terminology for systems engineers, defines the set point as "a fixed or constant input . . . established by means external to and independent of the automatic control system . . . which sets the ideal value of the controlled variable." Since regulation is brought about in a system and by a system, it is necessary to begin with a discussion of what makes up a "system." A system has been defined as a "collection of components arranged and interconnected in a definite way" for which there can be identified an output which is related to a known input (Grodins, 1963; LaJoy, 1958). The number and complexity of the components is not theoretically limited and thus may

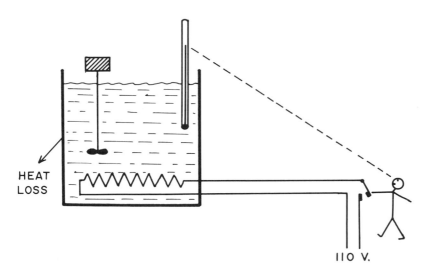

HEAT
LOSS

11O V.

FIGURE 6-1.   A closed-loop system including a human operator.

include electrical, mechanical, chemical, biological, and other elements, taken singly or in any combination. A not unusual regulating system is one in which a man will change the heat input to a bath to maintain either a constant or a planned variable temperature (Fig. 6-1). Inasmuch as the system includes a human operator, it is sometimes spoken of as semi-automatic rather than completely automatic. However, the system as a whole can be identified by the fact that the actions of the operator (i.e., the input) can be predictably related to the bath temperature (i.e., the output).

The components of a system may consist of simple or complex sub-systems, each with its own input-output relationship, and certain components may be common to more than one subsystem. Thus, in Figure 6-1, the performance of the human operator (a very complex subsystem!) in following instructions will have a certain input-output value which may vary with his alertness, state of health, motivation, and so on. The study of this input-ouput relationship is of great interest to psychologists and is sometimes termed the "human transfer function." The switching system is, of course, common to the electrical and biological elements of the system.

The sorting out of the subsystems and their transfer functions is one of the difficult and exciting problems of biological research. Often a beginning has to be made by lumping many of these sometimes poorly understood subsystems together while studies are carried out simultaneously on the entire system and the subsystems and their interrelationships. This "black box" approach, while confessing ignorance of subsystems, permits useful research in biology at various levels. However, the black box always contains unanswered questions, and the questions must be answered before a complete characterization of the system is possible. That is, the input-output functions of *all* subsystems must be known.

Keeping in mind that physiologists start their studies by observing a stimulus-response (or input-output) relationship without being able to understand the actions of all the various contributing subsystems, we can venture to introduce into our discussion of systems the block diagram (black box) as shown in Figure 6-2. Physiologically, the "system" can mean that a group of structures or other components can be identified *functionally* by an output response related more or less predictably to what is being done to the organism—i.e., the input stimulus. Whenever possible, it is the usual practice to measure the input in terms which are quite arbitrary (i.e., volts, concentration, calories, and so on) and to observe the related output in as quantitative a manner as possible (i.e., contraction magnitude, excretion rate, sweat rate, and so on). The studies are usually done over a period of time which may be long enough for the input and output to be relatively unchanging (i.e., a steady state), or so short that the output is changing rapidly after the input has been shifted suddenly to a new selected value (a transient or *phasic* response), or during periods in which the stimulus is cycled in a controlled manner (frequency analysis).

These responses have been long used by physiologists and are of great importance to an understanding of the system. As shown at the bottom of Figure 6-2, the physiologist's problem is that of determining the "law" of his system based on his information about its input and output. Since the law of the system in this sense is the result of the combined actions of components and subsystems, it is necessary to break the organism

PROBLEM	AVAILABLE INFORMATION		REQUIRED INFORMATION
Scientific predictions	Input	Law	Output
Diagnosis	Output	Law	Input (cause)
Research	Input	Output	Law Components; Properties

FIGURE 6-2.   The closed-loop system and its problems.

down into these elements and study each in detail. The magnitude of the biological research problem is at once apparent, extending, as it must, from the atomic and molecular systems through cellular systems to multicellular systems of increasing complexity and including the interactions of such systems upon each other. Where one begins in this problem is often a matter of preference in respect to the predictability of the input-output relationships.

The statistical nature of the results of psychological and sociological studies of the intact organism and groups of individuals is to be compared with the statistical but much more predictable results of the biophysical chemist. It is perhaps possible from Figure 6-2 to understand how some of the preclinical disciplines combine their efforts to study the system, "man." The anatomist provides essential information as to the form and interconnections of the components and their development; the biochemist provides detailed information about the subsystems at the molecular level; the physiologist provides information regarding cellular and multicellular subsystems and also the performance of the intact system, the whole man. The problem of component and subsystem failure, which is the ultimate cause of death, is the province of the pathologist in the areas in which abnormal changes in component structure and in system function are involved. The physiologist and pharmacologist work essentially on the same systems problem, i.e., the providing of input-output information on cellular and multicellular complexes.

A major problem in the study of biological systems is the isolation of the system to be studied; that is, can all other systems and components be held constant while the input-output relationship of a single subsystem is studied? This is a standard condition for the engineer who tests each component and subsystem separately before assembly into a complete system of known performance. The engineer is essentially dealing with the first of the problems indicated in Figure 6-2. He knows the inputs and he designs his system to have certain components, which are tested individually, so as to have properties that will fulfill his design specifications as to output. The physiologist on the other hand cannot study his system or its components in isolation, that is, in freedom from interaction with other systems.

There are isolation difficulties even in the study of molecular systems; in complex multicellular systems complete system isolation is not possible. Not only may cells which furnish inputs serve several systems at the same time, but cells which affect the output may also serve more than one physiological function. For example, in the control of body temperature, the sympathetic nervous system controls circulation through the skin blood vessels and thus serves as a major factor in the control of the flow of heat from the muscles and the internal organs of the body to the skin and into the environment. These same nerves and blood vessels are also under the control of the system which regulates blood pressure, and thus

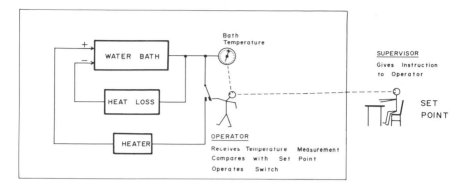

FIGURE 6-3. A diagrammatic representation of a control system with a set-point regulator.

to study circulatory regulation it is necessary to minimize the effects of the temperature regulator or to measure the temperature effects. In theory, the best way to evaluate the interaction effects of one system on the other is to measure all the inputs and outputs to all elements of both. The practical difficulties of accomplishing this multivariable recording are being gradually overcome as instrumentation becomes more highly developed and computer analyses become adapted to the problems of physiology. As an aid to the understanding of the plethora of data and analytical relations derived from them, the concept of the "system" is invaluable because it provides a framework into which the data and analyses can be fitted in a way which is meaningful to the human mind.

Having stated our problem as to systems and subsystems, we can return to our consideration of the physiological regulator of temperature and inquire as to the possible existence of a subsystem "external to and independent of the . . . control system . . . which sets the ideal value" of body temperature. To illustrate the set-point function, Figure 6-3 shows an extension of our semiautomatic controller of temperature in which a person outside the "control loop" gives instructions to the switch operator in the loop. In comparing Figure 6-1 with Figure 6-3, we can identify as "external to and independent of the control system" the supervisor who provides a continuing input into the system to identify the desired bath temperature. The set-point temperature may be constant in time or variable, and perhaps the most significant aspect of the supervisor is that he is not affected by the bath temperature. Of course, one may say that Figure 6-1 is essentially equivalent to Figure 6-3, inasmuch as the operator of Figure 6-1 may have received his instructions to maintain a constant or a programmed temperature variation at an earlier time, and thus the "supervisor" became a part of the control loop. However, in such a case it could be argued that the memory of the operator is then serving as the supervisor and is independent of the control loop itself.

In automatic control systems, the set point might be built in mechanically, electrically, or otherwise, but would have the characteristics which are contained in our original definition. For our physiological temperature regulator, the physiologist's task is to characterize the temperature-insensitive neurons of the thermoregulating system that furnish the continuing input signal to the controller itself so that this signal may be compared with other signals from temperature receptors and adjustments made to minimize the deviation of the actual body temperature from its set point. It should be borne in mind that the set-point temperature need not be constant but could change with any physiological conditions which alter the signals from the set-point elements. In approaching this question the physiologist will want information as to the overall operation of the thermoregulator and will need to explore the activities of its subsystems in his attempt to identify the set points for temperature regulation.

In man one finds a highly developed capability for the regulation of body temperature; this is illustrated in Figure 6-4 which shows the ranges of normal temperature regulation and the limits of survival in respect to body temperature. From the figure it is seen that the body temperature has a normal thermoregulatory range of 36°C to 40°C, and thus, assuming there is a set point, one must conclude at once that the set point in man is variable, or that there is a limited capacity of the regulator to adjust for deviations from a fixed set point, or perhaps that the system

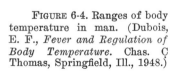

FIGURE 6-4. Ranges of body temperature in man. (Dubois, E. F., *Fever and Regulation of Body Temperature.* Chas. C Thomas, Springfield, Ill., 1948.)

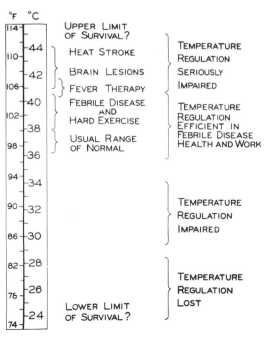

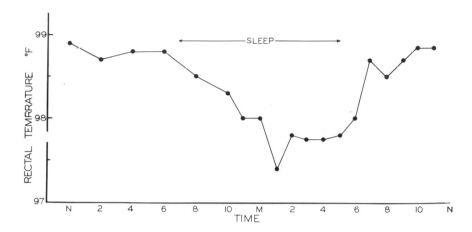

FIGURE 6-5.   Internal body temperature of a man on bed rest—an average over 7 days.

is non-linear, or some combination of these possibilities. Even under rather constant conditions, variations in body temperature are observed throughout the day in what has been called a diurnal rhythm. Much speculation has centered about whether this change (shown in Fig. 6-5) is due to a shifting of the set point or due to the action of a controller.

It would thus seem appropriate at this time to look briefly at the action of the basic controller functions to enable one to evaluate their possible actions on temperature regulation. From the point of view of "control" there are two major categories, the "open loop" and the "closed loop." The former is characterized by the fact that the input to the system is not dependent upon the output of the system. Any arrangement of components, such as levers, pulleys, amplifiers, and so on, in which there is no "feedback" is an open-loop system. Biologically, an open-loop system may be considered as a strict stimulus-response effect without feedback, such as the movement of an arm or leg under motor stimulation after all afferent nerves have been out. However, even in the simplest intact biological system it is difficult to exclude some feedback effects so that the final response does not in some way affect the input to the system.

A closed-loop control system is one in which the input to one or more of the subsystems is affected by its own output. In fact, control such as "servo" control depends upon feedback to achieve the rapid and accurate control of a large power output with the expenditure of a very small power input. Such systems are well known in physiology in the control of the motion of the extremities by nervous impulses; signals are fed back from muscle and tendon propioceptors into the central nervous system and in some way compared with the voluntary demands so that accurate and smooth movement of the large body levers is possible with small expenditure of energy by the controller. It is clear that a closed loop is essential for an automatic control system although the presence of a feedback does not necessarily imply that the system is an automatic controller. In fact, some systems in dynamic equilibrium may have strong feedbacks and yet not be controlled in the usual sense. (It is appropriate to mention here that engineers do not speak of *regulation* but prefer the term *control,* whereas physiologists use both control and regulation; their meanings as they are used here are given on page 5.)

Automatic Controllers

In studying the problems of automatic regulation, it is convenient to consider separately the controlling elements and the various elements of the passive or controlled system. A block diagram of an automatic regulator which brings together the elements of the total system in a form which is convenient for study is shown in Figure 6-6. A major part of the problem in analyzing the physiological regulator lies in the characterization of the passive system. Fortunately, much of the research in the past has been directed toward obtaining data on the passive system —i.e., the measurement of the metabolic rate, the body-surface area, and the heat-transfer coefficient of the skin, to list but a few. These data represent the values of the basic properties of the system and have been the results of studies extending almost 200 years from the initial experiments of Lavoisier in the latter part of the 18th century. The studies on the controlling system were begun independently in Philadelphia by Ott and in Paris by Richet (about 1885) who demonstrated the im-

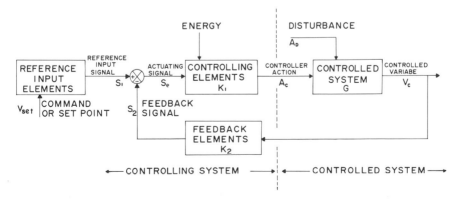

FIGURE 6-6. A block diagram of a regulatory system.

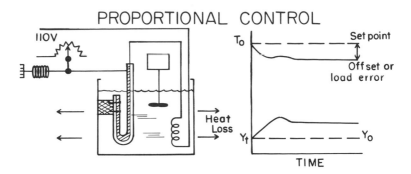

FIGURE 6-7.   Temperature regulation with proportional control.

portance of the hypothalamus in temperature regulation. It has been only recently that studies have been attempted which combine the data on the controlling elements with those on the passive system in a systems-analysis approach. We will confine our attention in this chapter to a discussion of the controlling elements and their possible relationship to fixed-point signals.

In Figure 6-6 the controlling elements are indicated by the box which receives the input $(S_e = S_1 - S_2)$ which is the error signal representing the deviation of the feedback signal from the command or set-point signal. The controller action is represented as A_C (i.e., vasomotor action, sweating, or shivering) which directly controls the body temperature, and the open-loop coupling function of the controlling elements is thus A_c/S_e in terms of units of controller function per unit change in the error signal. There are three basic types of control action which may be found singly or arranged in combination to maintain the body temperature within a prescribed range. It will be useful to describe these three actions briefly before asking specifically about the set-point signals.

CONTINUOUS PROPORTIONAL CONTROL

A very common type of regulation is one provided by proportional control; that is, one in which there is a continuous and, in the simplest case, linear relation between the deviation of the temperature from its set point and the magnitude of the effector action. This may be expressed simply as

$$\epsilon = T_t - T_o = -\alpha (A_t - A_o) \qquad 6.1$$

in which ϵ = load error, T_t = temperature at any instant, T_o = the temperature of the bath at some standard heat input and heat loss, A_t = controller output, A_o = controller output for the standard condition (i.e.,

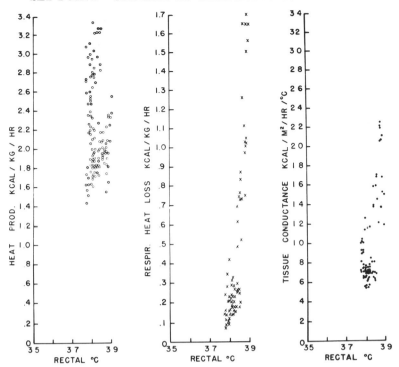

FIGURE 6-8. Responses of the three primary effectors of temperature regulation in the normal dog as a function of rectal temperature.

reference value), and α = proportionality factor.* The mode of control is gradual or modulating and is sometimes called "throttling." In Figure 6-7 there is represented a water bath with a bimetallic detector actuating a rheostat or other electrical device to control the amount of electrical energy supplied to the heater. In a small range of temperature the change of input to the heater will be proportional to the displacement of the temperature from the set point. A characteristic of this system is "offset" or load error; that is, under load the control temperature must change. To the right in the figure is a schematic plot of the response of the system to a sudden and sustained increase in heat loss. The response is shown with some oscillations due to thermal lags. The system shown in Figure 6-7 is not a set-point regulator in the sense of the ASA definition, inasmuch as there is no input independent of the system which sets the value of T_0. T_0 is determined solely by the balance of heat loss and heat production at some standard air temperature, and thus, T_0 could be called a balance point.†

---

* *Ed. Note:* Since A_0 and T_0 are constants, equation 6.1 reduces to proportionality between the output and the bath temperature.

† *Ed. Note:* Conversely the word "standard" implies that someone or something external to the system "decided to set" this balance point as the standard. Not every air temperature is a "standard."

This form of regulation has greatly appealed to physiologists as being the nearest approach to the type of regulation characteristic of physiological temperature regulation. Among the data supporting such a concept are those showing that body temperature during exercise in man increases in proportion to the rate of work. Thus, the cooling system of the body seems to be driven in proportion to the difference between the body temperature in exercise and the normal resting body temperature. In Figure 6-8 are shown the results of a series of calorimeter experiments on a normal dog. In this figure, the three primary effector actions of the physiological temperature regulator, namely, change in metabolic rate, evaporative heat loss, and vasomotor state, are plotted as functions of rectal temperature. The animals were studied at a series of environmental temperatures (between 8° and 35°C) which caused minor displacements of the body temperature.

We note that in the warm experiments there is a reasonably linear relationship between the change in internal body temperature and the evaporative heat loss from the respiratory tract and the tissue conductance as changed by the peripheral blood flow. These two responses clearly support the concept of proportional control. However, to the left in Figure 6-8, we note that the heat production due to shivering appears to *increase* slightly with *increased* rectal temperature, the exact opposite of what should be expected. The conclusion that must be drawn from the response to cold is that the controller action is not proportional control, or that the rectal temperature does not represent the "regulated" body temperature and that the thermal stimulation for proportional regulation must originate in other parts of the body, perhaps in the skin, the preoptic region, or elsewhere. It is a matter of common observation that the internal body temperature in man rises during mild exposure to cold, and that at the same time the physiological temperature controller is being vigorously stimulated to decrease heat loss by vasoconstriction and to increase heat production by shivering. Thus, although the physiological temperature regulator probably does have a type of proportional control, it is probable that many body temperatures participate to provide input signals.

INTEGRAL CONTROL

As can be seen from the above discussions, proportional control maintains the temperature near the set point but rarely at it. In some instances, it is desirable to have the controller bring the temperature back to the set point during exposure to thermal load and thus provide a more precise regulation. For this purpose, "floating control" or "automatic reset control" or "integral control" has been devised. This controller is one in which there is a fixed relationship between the deviation

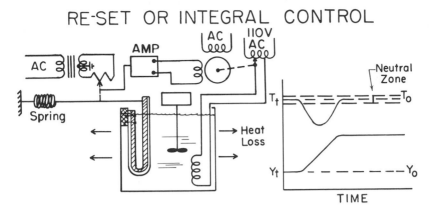

FIGURE 6-9. A diagram of temperature regulation with integral control.

of the temperature from its set point and the rate of application of the effector restoring action. The control action can be expressed in the case of a linear relationship as:

$$\frac{dA}{dt} = -\beta \, (T_t - T_o) \qquad\qquad 6.2$$

In this type of controller there is no fixed relation between the load error and the amount of effector action at any time. By integrating the above equation one obtains:

$$A_t - A_o = -\beta \int_{t_o}^{t} (T_\epsilon - T_o) \, dt \qquad\qquad 6.3$$

in which $\beta =$ the proportionality constant and $T_t - T_o$ is the deviation as a function of time. The value of A_t thus depends upon both magnitude of the load error and the time the temperature has been displaced from the set point.

Using the water bath regulation as an illustration, one can represent the action of the integral controller graphically as shown in Figure 6-9. The controller indicated consists of a reversible motor driving an auto-transformer which supplies the power to the heater. To the right in the figure is a representation of the controller response to a change in heat loss, based on the assumption that the temperature deviation responds in the form of a cosine function of the time. The temperature of the bath is raised until the set-point temperature is reached and the motor stops driving. If there is no overshoot, the temperature is held in the neutral zone. If there is an overshoot, the transformer will reverse the direction of the motor, which will start driving the power transformer to decrease

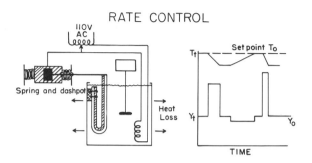

FIGURE 6-10. A temperature controller indicating rate control.

the input. The system will settle down if it has stability. This type of controller is a set-point controller determined by the point of reversal of the motor.

The evidence of reset control action in physiological temperature regulation could be in the action of the body in shifting from shivering and the inhibition of sweating (panting) during a period of heat debt, to vasodilatation and sweating during periods of heat storage. On the other hand, a deviation in internal body temperature such as that seen in fever or in exercise does not seem to drive the sweating and vasomotor systems to increasingly high levels of activity simply because a slightly elevated temperature has obtained for some minutes or even hours. In fact, the converse seems to be true; a rather constant rate of sweating and vasomotor action appears to be called for during exercise at a fixed rate. Also, the physiological controller does not drive the body temperature back to its exact neutral level (i.e., eliminate load error or droop) during long exposures to external thermal loads. These considerations seem sufficient to rule out a large degree of control action of the integral type in physiological temperature regulation, although a significant action of this type appears likely. In general, endocrine systems seem to operate approximately as integral controllers, since the rate of hormone production is usually proportional to the hormone deficit, that is, to the error of the system.

RATE CONTROL

The type of controller action which "quickens" the response of a regulating system is the "rate" mode of control; that is, the effector action

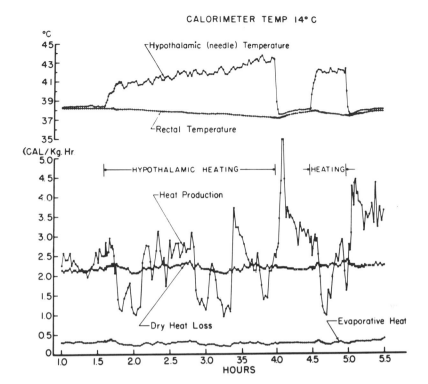

FIGURE 6-11.   A calorimetric study of an unanesthetized dog in a cold environment during heating of the preoptic region.

is proportional to the rate of change of temperature. This can be expressed mathematically as follows:

$$A_t - A_o = -\gamma \frac{dT}{dt} \qquad\qquad 6.4$$

in which $A_t - A_o$ = the effector response; γ = the proportionality factor; $\frac{dT}{dt}$ = the rate of change of the regulated temperature. This control is the converse of the reset control in that the effector action is proportional to the rate of change of the temperature rather than to the integral of the temperature as a function of time. It is seen from the equation for the rate controller that this type of action will not regulate temperature at a fixed level because the effector is responsive only to rate and, thus, this type of effector is not a set-point controller but operates from any value of T to resist any change in the *variable*. This mode of control is usually combined with proportional and reset control to provide a system more quickly responsive to disturbances. As rate control is sensitive to the rate of change of temperature, it responds to the magnitude of an imposed thermal load, and for this reason rate control is sometimes known as "anticipatory" control and, interestingly, *slows* the movement of the controlled temperature, thus providing additional stability to the control.

The system is represented schematically in Figure 6-10. In this figure we see that, if the bath temperature is rapidly lowered, the bimetallic detector actuating the piston in the oil-filled cylinder will move the cylinder an amount proportional to the rate of movement of the regulator arm. This movement will alter the energy input into the bath as long as the cylinder remains displaced. With the cessation of movement of the bimetallic strip, and therefore of the plunger, the springs will restore the cylinder to its neutral position regardless of the bath temperature level. This control action has some of the characteristics of the alert human operator, who seeing a large displacement in a tracking task, initially overcorrects the controls in order to hasten the return to the position of equilibrium. Such a control mode may introduce instability into the regulating system, and the advantages of anticipatory action can best be seen when they are combined with slower acting reset and proportional controls.

There is evidence that rate control is of importance physiologically in the regulation of body temperature in both man and animals. For example, such action can be seen in Figure 6-11. In this experiment the dog was exposed in a cold calorimeter while, at the same time, the hypothalamic temperature was increased by radio-frequency heating. After continued heating for two and a half hours, the heating was suddenly terminated, and the hypothalamic temperature fell several degrees in a few minutes. Associated with this drop in hypothalamic temperature, there was a large increase in the metabolic rate which had decreased even though the hypothalamic temperature had increased only slightly. The large increase in heat production associated with the fast rate of change of the hypothalamic temperature is the type of response that would be expected from a rate controller. Since most sensory receptors show the phenomenon of adaptation, rapidly changing stimuli are more effective than steady or slowly changing ones; this provides a sort of built-in response of the rate-control variety. It appears that vasomotor activity and sweating in man are in large degree under rate control from skin thermal receptors (Gagge, 1964).

NON-LINEAR CONTROLLERS

In the foregoing discussion of controller actions, it has been assumed that the coupling function of the controller was constant, i.e., was not dependent on the magnitude of the input. Thus, with proportional control, we assumed in Equation 6.1 that $\alpha (A_t - A_o) = T_t - T_o$ and that α was a constant. However, we can visualize a situation in which the relationship of A to T could be variable and could change with the value of T. For example, experiment may show that

$$-K f (T_t) (A - A_o) = T_t - T_o \qquad 6.5$$

We now have T on both sides of the controller equation, and the system will not respond as indicated for simple proportional control. Such a system is called non-linear, and complicates the problem to the extent that a rigorous mathematical treatment of the system is not generally possible. The effects of this system can be seen easily for the situation in which $f(T) = (T_t)^{-1}$ and the coupling coefficient is $-K(T_t)^{-1}$. In this case the action of the controller would be very weak for small values of T_t and would become progressively stiffer with the value of T_t.* Inasmuch as the "gain" of the system, $-K(T_t)^{-1}$, is automatically changed with the value of T_t, it is sometimes called "automatic gain control." It has often been suggested that the physiological controllers for heat production, vasomotor action, and sweating are non-linear controllers. In respect to automatic gain control, it should be mentioned that, by the appropriate choice of the properties of a system, a good regulation can be achieved without recourse to controlling elements. The voltage regulation which is obtained from saturated core transformers affords an example of such regulation.

SUMMARY

In summary, physiological temperature regulation contains elements of all the basic types of control actions. Thus, one might write the control equation for physiological temperature regulation as:

$$A_t = A_o - \alpha (T_t - T_o) - \gamma \frac{dT}{dt} - \beta \int_{t_o}^{t} (T_t - T_o) \, dt \qquad 6.6$$

The solution of the equation can be obtained in some cases, providing the values of α, β, and γ are known. It should be pointed out that this equation is not the completed closed-loop systems equation for the physiological temperature regulation, but rather, combines the controller actions in simple form. The completed closed-loop equation would, of course, contain the expressions for the passive system as well as those for the controller actions.

---

* *Ed. Note:* By algebraic transposition of 6.5 after substituting $f(t) = (T_t)^{-1}$, one observes that A is very large for large T_t and becomes small for very small T_t.

In considering the controller equations, we see that the set-point temperature enters twice and has the significance of being that temperature at which the regulator switches from a cooling action to a heating action. One would say, therefore, that the system performs "as if" it were a set-point system, but it could also be a combination of two balanced, dynamic systems with proportional control only.

It is necessary to proceed further and determine the location and action of the elements which are temperature insensitive but which may furnish the set-point information to the regulator. For this purpose certain evidence is presented from a study of the activity of the neurons of the anterior hypothalamus, an area of the brain which is known to contain a major element of the physiological thermostat. In these experiments, cats and dogs under urethane anesthesia were prepared by implanting stainless steel tubes on both sides of the hypothalamus and by exposing the brain directly over the anterior hypothalamus so that the region between the anterior commissure and the optic chiasm could be explored with micro-electrodes. The stainless steel tubes permitted the circulation of warm and cool water so that the temperature of the hypothalamic region could be controlled at will (Nakayama, Hammel, Hardy, and Eisenman, 1963).

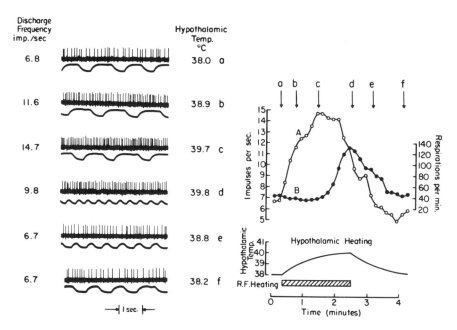

FIGURE 6-12. Changes in the frequency of discharge of a neuron in the preoptic region during local heating. Curve A represents electrical activity, curve B represents respiration.

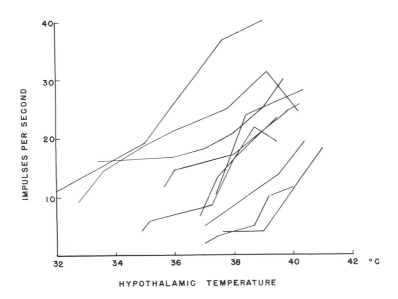

FIGURE 6-13.   Impulse frequency versus hypothalamic temperature for thermally responsive units.

In this experiment the input was, of course, a change in hypothalamic temperature, and the output was the rate of discharge of single neurons of the hypothalamus and the preoptic region. It was observed that these neurons were in continuous activity with discharge rates between 2 and 30 impulses per second, and that when the hypothalamic tissue temperature was raised, the discharge rate of some of the neurons was increased. An experiment of this type is shown in Figure 6-13 in which the thermode temperature was increased from 38°C to 39.7°C. The rate of firing of the neuron was increased from 7 to 15 impulses per second, and after a delay of about two minutes, the respiration rate increased as might be expected in response to heating. About 20 per cent of the cells studied in the preoptic region were found to be sensitive to heating, and the response of several of these cells is shown in Figure 6-12. A few cells were located which responded to cooling by increasing their firing rates. The great majority of the cells showed no change in firing rate in the temperature range from 32 to 42°C, as shown in Figure 6-14. It might be that some of these neurons serve as the set-point signal generators for temperature regulation, inasmuch as they are not sensitive to temperature and could furnish a signal to the regulator for comparison with temperature signals from the hypothalamus and from other neurons.

The functions of these neurons, temperature-sensitive or not, in respect to temperature regulation is as yet unproved. However, the

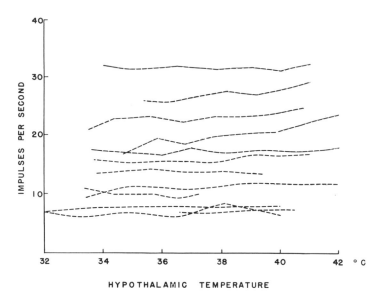

FIGURE 6-14. Impulse frequency versus hypothalamic temperature for thermally unresponsive units of the preoptic region.

evidence is suggestive in that neurons whose activity may be interpreted as thermosensitive and "set-point" functions have been found localized in the area of the brain which has an important thermoregulatory function. The task of studying these neurons further and of evaluating their possible role in temperature regulation is a currently interesting and exciting research problem in neurophysiology. Whether or not, within the normal physiological range of body temperature, the set point alters because of sleep, hormonal influences, pyrogens, or other bodily changes as discussed in a previous chapter, can now be examined in terms of the neuronal firing pattern. Present evidence would indicate that shifts in the set point for temperature regulation do occur in response to nonthermal influences. How these shifts are brought about and whether the neurons in the preoptic region participate in the shifts directly or indirectly are interesting questions. Although the traditional concept of temperature regulation in physiology is that the thermal system operates as a balanced dynamic system, it would seem that a considerable body of experimental evidence supports the concept that the physiological thermoregulator may be a set-point regulator. At the present time the question cannot be definitely settled.

Research in the author's laboratory is supported by Grant NB-04655 from the National Institutes of Health.

Chapter 7

NEURAL CONTROL OF THE PITUITARY-ADRENOCORTICAL SYSTEM

GERARD P. SMITH

In the past decade the nodal point for study of control of the pituitary-adrenocortical system has shifted from the anterior pituitary gland to the medial tuberal region of the hypothalamus. This area of the hypothalamus contains neurons which secrete a corticotrophin-releasing factor (CRF). The hypothalamus is accessible to humoral and neural stimuli of diverse origin, and its dense meshwork of fibers and small cells serves to organize the numerous inputs into a neural signal of definite intensity and sign which then acts upon the CRF neurons. In this view, the pool of CRF neurons forms the final common pathway for the pituitary-adrenocortical system.

In this essay I shall discuss the characteristics of the final common path, the structural and functional organization of the hypothalamic control units, the major neural influences on the control units, and the evidence for a variable set point and negative feedback in the system.

HYPOTHALAMIC FINAL COMMON PATHWAY

For any discussion of hypothalamic function, a detailed, anatomical analysis of afferent and efferent connections within the hypothalamus is not yet available. The problem is technical and requires careful application of silver staining and microelectrode techniques. Most of the hypothalamus is a dense, reticulated network of cells and fine fibers which pursue complicated courses of varying length. Szentágothai and his group (1964) have begun to clarify the anatomy of this region. The portion

117

of the hypothalamus which forms the final common pathway for neural control of anterior pituitary hormones is composed of small cells which surround the third ventricle in the medial ventral region of the hypothalamus. In the rat these cells occupy the areas allotted to the arcuate nucleus and to the ventral part of the anterior periventricular nucleus. Other nuclei, particularly the ventromedial nuclei, may also contribute to the pathway.

From these cells a system of fine fibers arises and runs towards the pituitary stalk and median eminence region; this is called the tuberoinfundibular fiber system. Some of its fibers terminate directly on the capillary loops of the median eminence. The capillary loops in turn lead into the hypophysial portal veins which distribute blood to the capillaries of the pars distalis (Fig. 7-1). However, the majority of the fibers of the tuberoinfundibular system terminate in the outer rim of the median

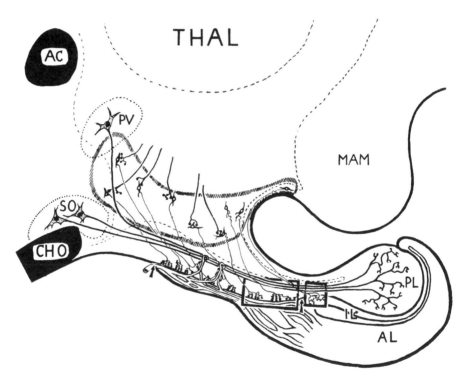

FIGURE 7-1. Schematic drawing of the tuberoinfundibular fiber system. Cell bodies of the system and the "hypophysiotrophic" region coincide within the area outlined with stippled lines. AC indicates the anterior commissure; AL the anterior lobe; CHO, the optic chiasm; IL, the intermediate lobe; MAM, the mammillary body; PL, the posterior lobe; PV, the paraventricular nucleus; and SO, the supraoptic nucleus. The broken lines indicate the outline of the third ventricle (Szentágothai, J., In Bargmann, W., and Schadé, J. P., [Eds.]: *Lectures on the Diencephalon, Progress in Brain Research.* Elsevier Publishing Co., Amsterdam, 1964.)

eminence which is known as the zona palisadica and which is entirely covered by the pars tuberalis. A dense vascular network ("mantelplexus") is found on the border between the zona palisadica and the pars tuberalis. The mantelplexus provides, for the fiber terminals in the zona palisadica, the vascular connection to the pars distalis which the capillary loops and portal hypophysial veins provide for the fiber terminals in the central region of the median eminence.

Corticotrophin-Releasing Factor (CRF)

It is generally agreed that hypothalamic control over anterior-pituitary ACTH release is exerted by way of a peptide liberated into the hypophysial portal system and the mantelplexus. The peptide is not vasopressin, although vasopressin has CRF activity (McCann and Brobeck, 1954). The chemical structure and properties of the peptide are not known, but are under active investigation. (See Guillemin, 1964, for a review of current research on CRF.)

Although the origin of this peptide is also unknown, the possibility that it is formed by the parvicellular system of neurons by a process of neurosecretion is logically attractive. Viewed this way, the basic organization of hypothalamic control of the anterior lobe and posterior lobe of the pituitary gland would be very similar. The major evidence against the suggestion is that the usual histochemical signs of neurosecretion (which are clearly demonstrated in the magnocellular system) are not present in the parvicellular system. However, the experiments of Halasz, et al. (1962) do support the possibility of neurosecretion in the parvicellular system. They implanted rat anterior pituitary tissue homografts into the hypothalamus, the thalamus, the corpus callosum, the lateral ventricle, and under the renal capsule. Two to three weeks later the engrafted rats underwent hypophysectomy. At the end of ten more weeks the hypophysectomized, engrafted rats were sacrificed; the histologic status of the grafts and the target organs was determined; and the completeness of the hypophysectomy was verified. The anterior pituitary grafts in the ventral medial hypothalamus surrounding the third ventricle retained their normal microscopic appearance and staining properties. This portion of the hypothalamus coincides with the parvicellular neuronal system (see Fig. 7-1). In this "hypophysiotrophic" region, pituitary grafts were maintained without contacting the capillary loops or mantelplexus of the median eminence region. None of the other grafts, including those in other portions of the hypothalamus, retained normal histologic characteristics. Halasz et al. concluded that the material from the hypothalamus which is essential for the maintenance of anterior pituitary structure is not simply a synaptic mediator discharged by nerve terminals into the hypophysial portal circulation, but is a true neuro-

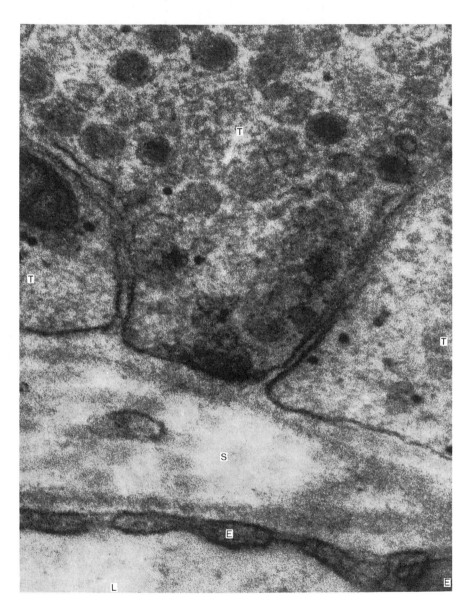

FIGURE 7-2.   Electron microphotograph of the median eminence of the rat. Note the lumen of the capillary (L), the thin fenestrated layer of capillary endothelium (E), the perivascular space (S), and the nerve terminals (T) containing synaptic vesicles as small vacuolated ring-shaped structures. The nerve terminal also contains more solid-looking opaque masses of neurosecretory material, larger than the synaptic vesicles. Magnification X 141,000. (Harris, G. W., *Triangle, The Sandoz Journal of Medical Science* 6:242, 1964.)

secretory substance produced by, and available from, small neurons in the hypothalamus.

Although the normal structure of the grafts was preserved, there remains the question of their function. Only four of 21 rats with intra-hypothalamic grafts maintained adrenal gland weight at a level significantly higher than host animals with non-hypothalamic grafts. Moreover, in these four rats the adrenal gland weights were only approximately 40 per cent of normal. Gonadal weights tended to be maintained better, but none approached normal, and the same applied to the thyroid glands. Consequently, it is clear that the rats bearing histologically normal intra-hypothalamic grafts were in a subnormal endocrine state. This does not necessarily mean, however, that the grafts were functioning at a low level, since the subnormal state may have been due to the small size of the graft and subsequent loss of parenchyma during the "take" period. It may also be that the graft's trophic hormones were producing very strong negative feedback on the parvicellular final common pathway because of their unusual position.

In at least one species, the duck, there is evidence for neurosecretion by the tuberoinfundibular system. In this species, Assenmacher and Benoit (1958) described a finely granulated, Gomori-positive, neurosecretory material accumulated in the surface zone of the median eminence.

The concept of neurosecretion by the parvicellular system is supported by the electron microphotograph of Bradbury and Harris (Harris, 1964), which shows opaque masses, larger than the synaptic vesicles, in nerve terminals ending on the capillary regions of the median eminence in the rat (see Fig. 7-2). Presumably these masses are neurosecretory material.

THE CONCEPT OF TONIC AND PHASIC NEURAL SYSTEMS

Before discussing the various neural influences on the hypothalamic final common path, it is useful to consider the kinds of responses the system can make. Table 7-1 depicts the plasma cortisol response of an unanesthetized monkey to a new situation in which he must learn to avoid electric shocks by pressing a lever at least once every 20 seconds (Sidman avoidance procedure). The data are from three different training hours on successive days. The plasma cortisol response to the hour of training was similar on all three days. This similarity of response is striking when one considers that during this period the animal was mastering the behavioral task; this is obvious from the great decrease in the number of shocks it received during the third session as compared with the first session. Furthermore, the plasma cortisol response to the first session was similar to the other two sessions despite the marked increase in the initial cortisol values on the second and third days. Finally, it is clear that the cortisol response was independent of the physical discomfort produced by the shocks. Thus, it must have been related in some way to the training situation as such.

The change in the initial cortisol values between the first session and the second and third sessions is significant. Because the experiment was designed so that the monkey was not disturbed for at least 16 hours prior to obtaining each of the initial blood samples, the higher initial cortisol values in the second and third sessions were not the result of any recent, abrupt stimulation. Rather, I believe that these higher initial values were due to a higher level of background stimulation of the adrenocortical system which resulted from the introduction of the training situation into the otherwise quiet environment of the monkey.

From these data it is clear that the adrenocortical system exhibits responses over two quite different time scales. Rapid responses occur in a period of a few minutes up to one or two hours and can be related temporally to a definite stimulus. The slower responses occur over a period of hours, even days, and usually cannot be related to a single factor in the environment. From Table 7-1, it is clear that the rapid,

TABLE 7-1. PLASMA CORTISOL RESPONSE OF ONE UNANESTHETIZED RHESUS MONKEY TO THREE AVOIDANCE TRAINING SESSIONS OF ONE HOUR EACH ON THREE SUCCESSIVE DAYS*

| SESSION | PLASMA LEVEL, μG/100 ML | | | NUMBER OF SHOCKS |
	INITIAL	FINAL	RISE	
1	18	48	30	334
2	35	64	29	184
3	38	72	34	8

* *Note:* The number of shocks administered was inversely related to the correct performance of the behavioral task. It is important to note that the plasma cortisol response was not correlated with the number of shocks. (Smith et al., 1964.)

stimulus-bound response is independent of the initial value, and that the neural control of the two types of responses may be analyzed separately. Sherrington (1947), in facing an analogous situation in his study of the neural control of movement, proposed a useful distinction between the phasic and tonic neural systems in these words: "Two separable systems of motor innervation appear thus controlling two sets of musculature: one system exhibits those transient phases of heightened reaction which constitute reflex movements; the other maintains that steady tonic response which supplies the muscular tension necessary to attitude." I find this to be a useful distinction in ordering the abundant data concerning the neural control of the adrenocortical system, and shall use it throughout the remainder of this essay.

Tonic Neural System

If the pituitary gland is isolated from hypothalamic influence by pituitary stalk section and the insertion of a barrier between the pituitary gland and the tuber cinereum, both the pituitary gland and the adrenals maintain secretory activity, but at a reduced level, and the adrenal gland weight decreases. This decrease in adrenal weight is seen also when pituitary glands are transplanted to any place in the body away from the tuber cinereum (such as into the anterior chamber of the eye or under the renal capsule).

Harris and Jacobsohn (1952) were the first to show that when isolated pituitary glands were replaced beneath the tuber cinereum, the weight of the adrenal glands was normal. This return of normal adrenocortico-trophic function by the pituitary occurred only when the hypophysial portal vessels regenerated and thus re-established continuity between the pituitary and the hypothalamus. If pituitary glands were transplanted beneath the temporal lobe region, they did not support adrenocortico-trophic function. Thus, it was specifically medial, tuberal hypothalamic influences which maintained the pituitary ACTH secretion, and not a non-specific product of brain activity.

Nikitovich-Winer and Everett (1958) later showed that pituitary grafts originally transplanted beneath the renal capsule and then re-transplanted beneath the tuber cinereum were able to maintain normal adrenal weights and to mediate the normal adrenal hypertrophy response to unilateral adrenalectomy. These studies constitute crucial evidence for the essential role of the medial tuberal hypothalamus in pituitary-adreno-cortical function, an influence "tonic" to the pituitary-adrenocortical system.

An important question is whether the neuronal system providing this tonic stimulation is intrinsic to the hypothalamus or whether it involves other neural loci as well. Ganong (1963), attempting to answer this question in regard to dogs, isolated the hypothalamus from its con-nections with the rest of the brain, but preserved its neurohumoral link with the anterior pituitary gland. Under these conditions the adreno-cortical secretion rate was maximal. This result suggests that intrinsic hypothalamic influences are strongly excitatory to the pituitary and that other tonic neural inputs are mainly inhibitory. If so, then the tonic neural system does include extrahypothalamic structures playing upon the hypothalamic final common path. Reports of increases in the level of plasma corticoids under "resting conditions" following decortication, lesions of the hippocampus, and midbrain section support this hypothesis.

Phasic Neural System

The list of agents, environmental conditions, and subjective states which activate the adrenocortical system is as varied as it is extensive. A common feature of these diverse stimuli is that all are relatively abrupt disturbances* of the internal or external environment which impose anticipated or actual demands for somatic or visceral responses. A phasic increase in plasma corticoids may be seen in situations where the animal's response is one of approach, e.g., self-stimulation in the monkey, (McHugh et al., 1963), as well as in situations characterized by avoidance (Table 7-1).

The median eminence region is the final common path of the phasic neural system for stimulation which can arrive over pathways from structures located above and below the hypothalamus. The amygdala, orbital frontal cortex, septum, and lateral preoptic region have strong excitatory influences in the phasic system. Hippocampal input is inhibitory, whereas input from the reticular formation can be either excitatory or inhibitory. The inhibition observed from the hippocampus and reticular formation might be due to the blockade of ascending excitatory influences, possibly within the reticular formation. Or there may exist a pool of hypothalamic neurons which are inhibitory to the CRF neurons and which are stimulated by the hippocampus and reticular formation. I believe that such a pool of inhibitory neurons exists, and that it is concerned with negative feedback by plasma corticoids as outlined below.

AUTOMATIC CONTROL OF PLASMA CORTICOIDS

Having analyzed the elements of the neural control of the adrenocortical system, let us now consider the possibility that the neural control systems are organized so as to provide regulation of the pituitary-adrenocortical system. The discussion will be restricted to the control of the inflow of corticoids, and will not include the self-regulation that the liver contributes to the system by removing plasma corticoids in a manner approximating first-order kinetics. Self-regulation of this latter type is not of itself an adequate explanation for the behavior of the system.

WHAT IS REGULATED?

Faced with the bewildering assortment of actions which corticoids mediate and permit, Yates and Uhrquhart (1962) suggested that the

---

* Disturbance seems a better word than stress in this context, because stress commonly connotes something noxious.

control system regulates not one of these actions, but in fact, the plasma level of corticoids. As they point out, this kind of regulation is common in industry. Their suggestion is reasonable and is supported to some extent by the negative feedback effect which exogenous corticoids can produce. The suggestion also has the force of inviting experimentation because of the existence of reliable methods to measure plasma corticoids.

SET POINT

Yates et al. (1961) were the first to suggest that neural control of the adrenocortical system involves a set point; furthermore, they claimed that the set point varies. Strong evidence for this view comes from experiments with adrenalectomized rats (Sydnor and Sayers, 1954). In these animals the plasma ACTH concentrations are high, and plasma corticosterone is absent. Yet stimuli which usually increase ACTH release in intact rats produce in these adrenalectomized rats an increase in the high plasma ACTH concentration already present. There is no possibility that changes in the plasma corticosterone initiate these changes in ACTH, since the animals have no adrenal glands. The excitatory signal must come through the hypothalamic final common path, and it must include some neural mechanism which selects a new and higher level of activity for the adrenocortical system. Abrupt changes in the set point activate units of the phasic system, whereas gradual changes in the set point utilize units of the tonic system. It should be understood that the same neurons probably function as units of both systems, and that some other characteristic, such as position in the sequence of activation, is likely to be the critical factor for differentiating which system is brought into play.

NEGATIVE FEEDBACK

It is well known that injections of corticoids can block the adrenocortical response to a wide variety of stimuli. Sayers and Sayers (1948) interpreted their original data to be evidence of negative feedback by plasma corticoids. They proposed that the increased utilization of corticoids by peripheral tissues decreases the plasma levels of corticoids, and that this releases the pituitary gland to stimulate adrenocortical secretion until the plasma level is restored. Thus, the system would be activated by a fall in plasma corticoids and would be inhibited at the pituitary gland by a rise in plasma corticoids.

When methods of measuring plasma corticoids became available, increased peripheral utilization was not demonstrated, and activation of the pituitary-adrenal system was shown to occur concomitant with rising plasma corticoids. The importance of negative feedback by plasma corticoids was rejected on the basis of these findings. Yet the data of Sayers and Sayers remained, and more data accumulated which were difficult to explain without the concept of negative feedback.

Yates et al. (1961) attempted to reinstate negative feedback as an essential aspect of the control mechanism by suggesting that the pituitary-adrenal system is activated by the difference (error) between the set point for plasma corticoids and the existing plasma level of corticoids. An increase in the set point produces a difference (error) between the set point and the plasma level of corticoids. This error activates the system. The system remains active until the error is reduced by an increased level of plasma corticoids close to the new set point. (It seems unlikely that the system is so sensitive that plasma corticoids rise or fall exactly to the new set point.) This hypothesis restores negative feedback to the control of the system, and it provides an explanation for negative feedback operation in the many situations in which the system is activated while the plasma level of corticoids is increasing. At least two characteristics of the control system are implied in this view of negative feedback:

1. The plasma corticoids must have access to neural elements which inhibit CRF secretion directly or transynaptically.

2. There must be a neural mechanism which senses some function of the plasma corticoids and compares this with the set point so that a signal can be sent to the neurons of the final common path to activate the system appropriately.

The first implication has been verified by the discovery that large doses of cortisol prevent the increase in CRF activity of median eminence extracts in rats under ether anesthesia at 1.25 minutes after sham adrenalectomy (Vernikos-Danellis, 1965).

The mechanism by which corticoids inhibit CRF is not known. It is possible that corticoids depress neural units of the control system or that they stimulate a neuronal pool whose activity inhibits the final common path. The known tendency of corticoids to increase neural excitability is compatible with the latter suggestion and with the usual principles of neuronal organization. A microelectrode study of the response of units in this region of the hypothalamus to injections of corticoids would be very helpful in deciding this important question.

Although there is convincing evidence that corticoids can inhibit CRF, the critical question is whether changes in plasma corticoids within the physiological range inhibit phasic CRF release appropriately—that is, so that the plasma corticoids approximate the momentary set-point value of the system. The existence of this kind of behavior in the system requires a neural mechanism which can detect some function of a change in the level of plasma corticoids and compare this with the set point. At the present time there is considerable debate over the existence of this kind of behavior in the system, and the remainder of the essay will review the dispute.

EVIDENCE FOR A NEURAL MECHANISM WHICH SENSES A FUNCTION OF
PLASMA CORTICOIDS AND COMPARES IT WITH THE SET-POINT VALUE

Yates et al. (1961) performed experiments designed to test for the
existence of such a neural mechanism. They phrased their hypothesis in
the following manner: "According to this hypothesis, reset of the controller could account for the elevations in plasma corticosteroid concentration observed following noxious stimuli, as well as for the diurnal
periodicity of plasma corticosteroid concentration. If this view is correct,
then a dose of corticosteroid sufficient to produce an increment in plasma
corticosteroid concentration equal to the increment produced by a noxious
stimulus should prevent entirely the secretion of endogenous corticosteroids, if it is given immediately before the noxious stimulus. The
exogenous steroid would raise the plasma corticosteroid concentration to

TABLE 7-2. PLASMA CORTICOSTERONE CONCENTRATION IN ANESTHETIZED
RATS FIFTEEN MINUTES AFTER INTRODUCTION OF EXPERIMENTAL
VARIABLES*

GROUP	NO.	PLASMA CORTICOSTERONE CONCENTRATION μG/100 ML
Corticosterone, iv		
5 μg	15	24.5 ± 2.02
7.5 μg	21	29.0 ± 1.37
12.0 μg	30	32.8 ± 1.15
15.0 μg	11	34.5 ± 1.49
30.0 μg	13	49.5 ± 1.94
35.0 μg	8	54.2 ± 1.73
ACTH, iv		
0.125 mU	8	23.0 ± 2.70
0.250 mU	10	34.1 ± 1.56
Histamine, iv		
100 μg	26	34.9 ± 1.12
Laparotomy	25	40.4 ± 1.37
Corticosterone + ACTH		
5 μg + 0.125 mU	27	33.4 ± 1.75
12 μg + 0.250 mU	15	51.3 ± 2.00
Corticosterone + histamine		
12 μg + 100 μg	12	35.9 ± 2.20
Corticosterone + laparotomy		
5 μg + laparotomy	7	40.3 ± 1.32
12 μg + laparotomy	24	38.6 ± 1.28
35 μg + laparotomy	8	56.7 ± 2.29

* *Note:* The central column refers to the number (No.) of rats in each group.
(Yates et al., 1961.)

the new level at which the noxious stimulus resets the controller: no discrepancy between the existing concentration and the new high set point following the noxious stimulus would then be detected by the controller. In the absence of a discrepancy between the existing value of the steroid concentration and the set point value, the negative feedback controller could not initiate an increase in endogenous corticosteroid release.

"In contrast, if a noxious stimulus provokes a release of adrenocorticotropin (ACTH) in a manner independent of feedback control of plasma corticosteroid concentration, then the increments produced by the exogenous corticosteroid and by the noxious stimulus should summate partially or completely."

To test the hypothesis, these workers injected varying amounts of corticosterone intravenously 15 to 30 seconds before laparotomy or before

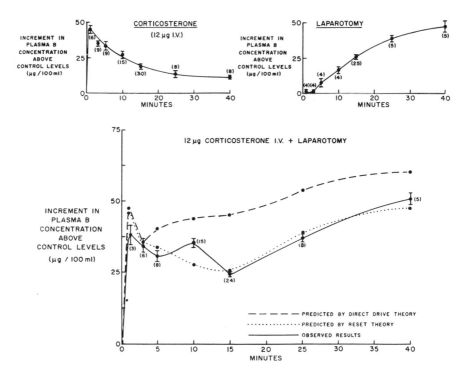

FIGURE 7-3. The two smaller graphs in the upper portion of the figure depict the plasma corticosterone values at varying time intervals after an intravenous injection of 12 μg of corticosterone and after laparotomy. The large graph shows the plasma corticosterone values at varying time intervals after an intravenous of corticosterone plus laparotomy. Observed results are compared to what the direct drive theory predicts (summation) and to what the "reset" hypothesis predicts. Except for a small deviation at 10 minutes, the observed results follow very closely the values predicted by the "reset" hypothesis. (Yates, F. E., Leeman, S. E., Glenister, D. W., and Dallman, M. F., Endocrinology 69:67, 1961.)

intravenous injections of ACTH or histamine. Their results (see Table 7-2) may be summarized by the following statements:

1. Injections of corticosterone summated with injections of ACTH.

2. Injections of corticosterone did not summate with injections of histamine or with laparotomy.

3. The level of plasma corticosterone following laparotomy was not changed by injections of corticosterone which, if given alone, would have produced plasma levels equal to or less than the level produced by laparotomy.

4. When a large dose of corticosterone was injected prior to laparotomy, the resulting increase in plasma corticosterone equalled that seen after corticosterone alone.

These results were predicted by the hypothesis. Further support for the hypothesis came from experiments in which the increase produced by 12 μg of corticosterone and by laparotomy, separately and together, was studied from 3 to 40 minutes following the introduction of the experimental variable. Figure 7-3, taken from their work, shows that the curve formed from the observed increments following corticosterone injection and laparotomy follows closely the curve predicted by the reset hypothesis. The theoretical curve predicted by the direct-drive theory in Figure 7-3 is derived from the summation of the corticosterone values determined when 12 μg of corticosterone and laparotomy acted separately, as shown in the two smaller graphs at the top of Figure 7-3. The direct-drive theory would predict complete summation, because it postulates that

TABLE 7-3.   EFFECT OF INTRAVENOUS PRETREATMENT WITH CORTISOL ON THE PLASMA CORTISOL RESPONSE TO AMYGDALOID STIMULATION IN UNANESTHETIZED *Macaca mulatta**

Monkey no.	AMYGDALOID STIMULATION (1 HR)			1000 μG HYDROCORTISONE IV PLUS AMYGDALOID STIMULATION (1 HR)			
	Initial (μg%)	Final (μg%)	Rise (μg%)	Initial (μg%)	15 min after hydrocortisone	Final (μg%)	Rise (μg%)
180	34	70	36	30	76	68	38
337	36	58	22	54	88	72	18
				27	48	40	13
99	23	48	25	39	74	67	28
109	20	41	21	25	60	35	10

* *Note:* One hour of intermittent electrical stimulation of the amygdala produced an increase in plasma cortisol. (The stimulation was applied for five seconds each five minutes for one hour.) An intravenous injection of 1 mg of hydrocortisone 15 minutes before the hour of amygdaloid stimulation did not change the increment of plasma cortisol which resulted from the amygdaloid stimulation alone. Thus, the exogenous cortisol did not summate with the endogenous cortisol produced by the amygdaloid stimulation. This is interpreted as evidence of negative feedback by plasma cortisol. (McHugh and Smith, 1964.)

TABLE 7-4.   EFFECT OF INTRAVENOUS PRETREATMENT WITH CORTISOL
ON THE PLASMA CORTISOL RESPONSE TO HYPOTHALAMIC STIMULATION IN
UNANESTHETIZED *Macaca mulatta**

	HYPOTHALAMIC STIMULATION (1 HR)			1000 μG HYDROCORTISONE IV PLUS HYPOTHALAMIC STIMULATION (1 HR)			
Monkey no.	Initial (μg%)	Final (μg%)	Rise (μg%)	Initial (μg%)	15 min after hydrocortisone	Final (μg%)	Rise (μg%)
809	45	70	25	22	53	76	54
109	26	46	20	22	44	70	48
99	25	42	17	38	53	84	46
429	22	44	22	40	60	84	44

* *Note:* Intravenous injection of 1 mg of hydrocortisone 15 minutes before one hour of intermittent electrical stimulation of the hypothalamus resulted in a higher increment of plasma cortisol than that produced by hypothalamic stimulation alone. Thus, the exogenous cortisol added to the endogenous cortisol produced by the hypothalamic stimulation. There was no evidence of negative feedback in these experiments. (McHugh and Smith, 1964.)

laparotomy provides a direct drive of adrenocorticotrophin release, independent of the plasma corticosteroid concentration.

McHugh and Smith (1964) explicitly tested the reset hypothesis in experiments using unanesthetized rhesus monkeys maintained continuously in primate chairs. The monkeys had bipolar electrodes chronically implanted in the basolateral amygdala and in the medial tuberal region of the hypothalamus. Intermittent, electrical stimulation of the amygdala to the point of after-discharge produced a significant increase in plasma cortisol (Table 7-3). If an intravenous injection of 1 mg of cortisol (Solu-Cortef, The Upjohn Company) was given, and this was followed 15 minutes later by the same intermittent, electrical stimulation, the increase in plasma cortisol level at the end of the stimulation period was the same as that following amygdaloid stimulation alone. Thus, there was no summation between exogenous cortisol and the endogenous cortisol produced by stimulation. This result is what the reset hypothesis predicts.

By contrast, in experiments of identical design in which the hypothalamic electrode was stimulated, there was a significant summation (Table 7-4). We inferred that the electrode in the medial tuberal region of the hypothalamus stimulated the neurons of the final common path directly and that this was not affected by changes in the plasma cortisol. Furthermore, the comparator mechanism and the site of corticoid feedback must lie, at least in a functional sense, between the amygdaloid and hypothalamic electrodes.

Our interpretation of these data was supported in another series of experiments in which the amygdala or the hypothalamus was stimulated for several hours. Here a prolonged hypothalamic stimulation gave a steady increase in plasma cortisol. However, after two or three hours of amygdaloid stimulation, the plasma cortisol stopped rising and even fell slightly. Yet if an intravenous injection of ACTH, or if medial tuberal stimulation was introduced when the plasma cortisol had reached its plateau, a further increase was easily demonstrated. This is unequivocal evidence that the plateau of plasma cortisol during prolonged amygdaloid stimulation is not explained by the exhaustion of CRF, ACTH, or cortisol, or by the decreased responsiveness of the pituitary or adrenal glands. Electroencephalographic monitoring of the amygdaloid after-discharge throughout the experiment revealed no change in the threshold or form; this means that, by this crude estimate, the electrical stimulation continued to produce the same neural event in the amygdala. By exclusion,

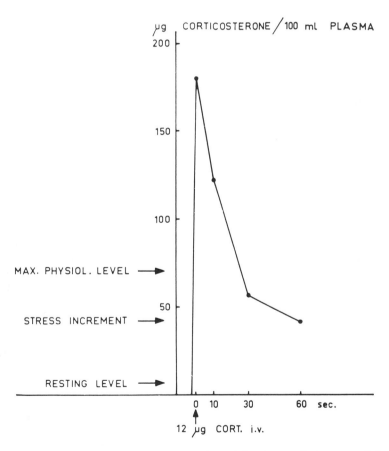

FIGURE 7-4. Plasma corticosterone values following an intravenous injection of 12μg of corticosterone into a rat. The approximate levels attained during stress by endogenous corticosterone production are indicated, and it is obvious that the injection produced values of plasma corticosterone above the physiological range for at least 10 minutes. (Smelik, P. G., *Acta Endocr.* 44:36, 1963a.)

then, we interpreted these experiments as further evidence of negative feedback by the plasma cortisol upon the phasic input from the amygdaloid stimulation.

Under the conditions of the experiments which provided the data for Table 7-3, several inferences can be made concerning the control system. The first is that the initial level seems to be the result of function of the tonic neural system, while the response of the phasic mechanism following amygdaloid stimulation is independent of the level of plasma cortisol which prevailed at the beginning of the stimulation. This independence is support for the functional dissociation between the tonic and phasic systems as postulated.

The close agreement between the change of plasma cortisol levels in response to amygdaloid stimulation in the same monkey on different days supports the hypothesized resetting of the controller's set point. Finally, the data from the experiments preceded by intravenous cortisol provide strong evidence of a definite set point, because the difference between the cortisol level after the injection and at the end of the hour of stimulation varied widely, but the difference between the cortisol level before injection and the final level showed very little variation. In these experiments and those of Yates et al. (1961), the neurohormonal system was found to be unexpectedly sensitive.

Smelik (1963a), however, demonstrated a flaw in the experimental design of Yates et al. by showing that a single intravenous injection of 12 μg of corticosterone into rats produced plasma corticosterone concentrations as high as 180 μg/100 ml of plasma immediately after the injection (see Fig. 7-4); this level exceeds the maximal physiological concentration of about 70 μg/100 ml of plasma in the rat. Consequently, Yates was testing the influence of stressful stimuli at a time when pharmacological levels of plasma corticosterone were present. As a result, some of his data cannot be used to support his hypothesis that changes of plasma corticoids in the physiological range perform an effective negative feedback function.

The criticism, however, applies only to those experiments in which Yates and his associates found no endogenous corticosterone response, and is not relevant to those experiments in which exogenous corticosterone only partially suppressed the endogenous corticosterone response (see Table 7-2). Similarly, Smelik's criticism does not explain the variable fall in plasma cortisol to the set-point value at the end of the hour of amygdaloid stimulation in the pretreatment experiments with unanesthetized monkeys (Table 7-3). McHugh and Smith considered this variable fall as an indication that a variable negative feedback was operating within the system to approximate the set-point value.

In a second series of experiments, McHugh and Smith (unpublished) evaluated the criticism of Smelik by using constant intravenous infusions of cortisol. After approximately one hour, a plateau level of cortisol was reached near the usual set-point value for one hour of amygdaloid stimulation. At that point the amygdala was stimulated intermittently for one hour. There was a marked rise in cortisol, very similar in magnitude to that seen in the monkeys without cortisol. Thus, with a constant intravenous infusion of plasma cortisol there was no evidence that a negative feedback operates on the phasic input produced by amygdaloid stimulation. This result is similar to Smelik's observation that constant subcutaneous infusions of corticosterone in anesthetized rats did not prevent the corticosterone response to intraperitoneal histamine (Smelik, 1963a).

In another series of experiments, Smelik (1963b) compared the time course of plasma corticosterone levels following a large subcutaneous injection of corticosterone with the time course of the inhibition of the endogenous corticosterone response to handling and the transferral of rats to a strange room (Fig. 7-5). It is evident in Figure 7-5 that the negative feedback effect was not maximal until two hours after the injection. This was 90 minutes after the peak of the change in plasma corticosterone produced by the exogenous corticosterone. The inhibition of the phasic neural system persisted for many hours. Beaven et al. (1964) demonstrated a similar latency in suppression of cortisol secretion rates from transplanted adrenal glands in sheep following an intravenous injection of dexamethasone. These data seem to describe a controller system with low sensitivity to negative feedback. This sensitivity is too low to

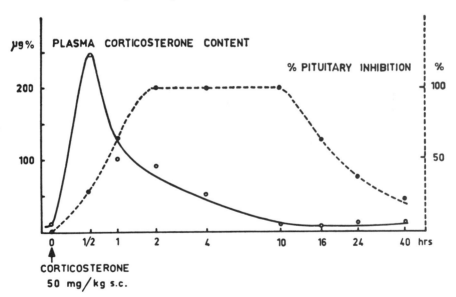

FIGURE 7-5. Temporal dissociation of the changes in plasma corticosterone and pituitary inhibition following a subcutaneous injection of corticosterone. (Smelik, P. G., *Proc. Soc. Exp. Biol. and Med.* 113:616, 1963b.)

satisfy the requirements imposed by the reset hypothesis. Yet the sensitivity was high in the injection experiments of Yates et al. (1961) and of McHugh and Smith (1964). One of the impenetrable aspects of the subject at this time is how the sensitivity of the controller system can vary over such a wide range.

Further evidence against the reset hypothesis has been produced by Hodges and Jones (1963) in rats and by Liddle and his co-workers in human patients undergoing surgery (Estep et al., 1963). Thus we still lack crucial evidence that there is a momentary negative feedback by plasma corticoid changes in the physiological range. Until this point is settled the existence of regulation of the plasma corticoid levels by the system remains in doubt.

Acknowledgments

My intellectual debt to Dr. F. Eugene Yates is obvious, and I thank him for discussing his ideas with me on several occasions. I am indebted to him and to Dr. S. M. McCann for permitting me to see their most recent manuscripts on this subject before publication. I also thank Dr. Walle J. H. Nauta for suggesting the analogy between the neural control of movement and the neural control of the adrenocortical system. Finally, I thank Dr. Paul R. McHugh for persuading me that there are interesting problems of neurohumoral control in regions of the body outside the gastrointestinal tract.

This review was prepared while the author was a Scholar of the Pennsylvania Plan to Develop Scientists in Medical Research. Experiments referred to in the text were performed while the author was a member of the Department of Neuroendocrinology, Walter Reed Army Institute of Research, Washington, D. C.

Chapter 8

GONADOTROPHIC FUNCTIONS
OF THE ADENOHYPOPHYSIS

SAMUEL M. McCANN

By comparison with the detailed information available on the regulation of temperature and respiration, our knowledge of the complex mechanisms which govern the secretion of the gonadotrophins is still in a relatively primitive state. This is in large measure due to the absence of sufficiently precise, sensitive, and convenient methods of measuring blood levels of the gonadotrophins. It is difficult to study any regulation unless what is being regulated can be measured easily and accurately. Yet, in spite of these methodological limitations, we can now present a relatively clear outline of the regulation of the gonadotrophins, as it has come from the combined efforts of many investigators over a period of some thirty years. The following discussion will first delineate some of the hormonal and environmental influences which can alter the secretion of gonadotrophins, and will then present the evidence that these hormonal and environmental influences act upon a hypothalamic regulatory area which modulates the secretion of gonadotrophins.

It is necessary to deal with not one, but at least two, and perhaps three, hormones when discussing gonadotrophins. The follicle-stimulating hormone (FSH) promotes the growth of the ovarian follicle beyond the stage of early antrum formation. By itself, at least in the hypophysectomized rat, FSH evokes no hormonal secretion by the ovary. A second gonadotrophin, the luteinizing hormone (LH), can act in the presence of some FSH to evoke further growth of follicles, ovulation, and formation of the corpus luteum. LH evokes estrogen secretion by the follicle, which rises to a peak just prior to ovulation. A third gonadotrophin, luteotrophin (LTH), is required for the maintenance of functional corpora lutea and for the secretion of progesterone in the rat and in other small rodents. In these species LTH appears to be identical with prolactin. In women, convincing evidence for the identity of LTH with prolactin is yet to be forthcoming. In fact, it has been shown that LH itself will increase progesterone secretion by human corpora lutea in vitro. So the question of the nature of LTH in man is still unsettled; however, for the purposes of this discussion, the terms LTH and prolactin are used synonymously.

One of the primary goals of study of the regulation of the gonadotrophins is to explain the fluctuating hormonal secretion during the mammalian estrous cycle or the primate menstrual cycle. Although the following discussion will give a partial explanation for these cyclic changes, a complete explanation is still beyond our reach. We shall consider all three of the gonadotrophins—i.e., FSH, LH, and LTH; however, since our knowledge of the regulation of LH is more complete than that for FSH and LTH, we shall have much more to say about the first.

METHODOLOGY

Before proceeding further, it is perhaps worthwhile to say a little about the various assays for gonadotrophins which are currently used.

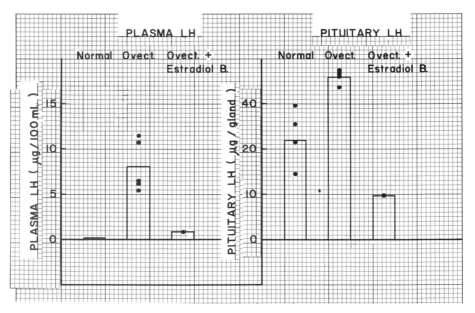

FIGURE 8-1. Plasma and pituitary LH levels in normal and ovariectomized rats; the effect of estrogen on plasma and pituitary LH. Ovect. = ovariectomized; Estradiol B. = estradiol benzoate. Dots represent values from individual assays; the height of the bar gives the mean value. Values for LH are expressed in terms of the N.I.H. ovine S-1 standard.

We are in the best position with respect to LH. Parlow (1961) has developed a highly sensitive, specific, and reasonably precise assay for LH: the ovarian ascorbic acid depletion technique. This test is performed by injecting the solution to be assayed into immature, female rats which have been pretreated with large doses of gonadotrophins. The pretreatment is essential to enhance the ovarian sensitivity to the effect of LH. If the material being tested contains LH, the ascorbic acid concentration in the ovaries of the test animal decreases in proportion to the log-dose of LH injected.

The currently accepted assay for FSH is the ovarian weight method of Steelman and Pohley (1953). Immature rats are treated concurrently with the FSH preparation to be assayed and with a large dose of human chorionic gonadotrophin (HCG). This treatment with HCG augments the response to FSH and simultaneously blocks the response to any LH which may contaminate the sample. In the presence of FSH, the ovarian weight is increased. The method appears to be specific for FSH and reasonably precise, but suffers from a low sensitivity, so that very large amounts of plasma are required to obtain detectable levels of the hormone. Recently, in our laboratory, Igarashi has developed a mouse uterine weight method which is much more sensitive than the Steelman-Pohley assay, but which is not completely specific for FSH. It does appear to measure the FSH activity in plasma in which the quantities of LH present are too low to interfere with the results obtained. In this method,

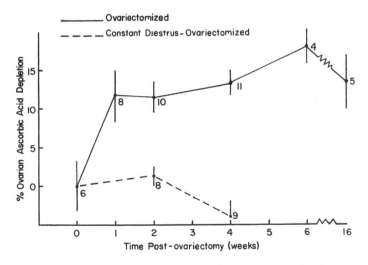

FIGURE 8-2. Plasma LH (per cent ovarian ascorbic acid depletion) after ovariectomy of normal rats and those with median eminence lesions (constant diestrus). In this and subsequent figures vertical lines indicate the SEM; the numbers give the number of test rats used. (Taleisnik, S., and McCann, S. M., *Endocrinology* 68:263, 1961.)

HCG is also administered concurrently with the test sample to enhance the sensitivity to FSH and to block the response to LH, but the dose of HCG is much lower than that used in the Steelman-Pohley assay.

LTH must be assayed by either the systemic or local pigeon crop sac method. The sample is introduced intradermally over the crop in the local technique, and the area of the underlying crop gland which is stimulated is measured (Grosvenor and Turner, 1958). This method is sufficiently sensitive for assay of pituitary LTH, but is probably not sensitive enough to detect blood levels of this hormone.

The methods described have been used in most of the work to be discussed here.

FEEDBACK EFFECTS OF GONADAL STEROIDS ON THE SECRETION OF GONADOTROPHINS

In the intact animal, the secretion of FSH and LH by the adenohypophysis is held in abeyance by the feedback action of gonadal steroids. That such feedback effects are operative was perhaps first recognized by Moore and Price in 1932. In early work, alterations in the output of pituitary hormones in response to changes in the gonadal hormone titers

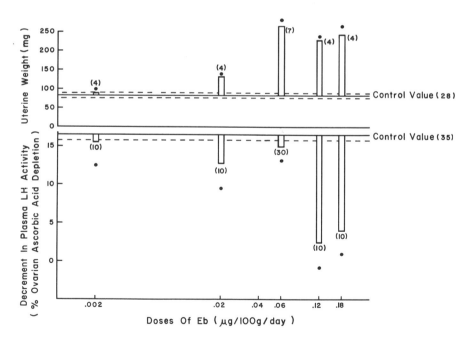

FIGURE 8-3. Decrement in plasma LH and the increase in uterine weight produced by daily injections of estradiol benzoate (Eb) in ovariectomized females. (McCann, S. M., and Ramirez, V. D., *Rec. Progr. Hormone Res.* 20:131, 1964.)

was inferred from changes in the target gland size and structure. More recently it has been possible to examine gonadal steroid feedback by measuring the plasma levels of the gonadotrophins. When the ovaries or testes are removed, both the plasma and pituitary levels of LH show a marked rise, the former from an undetectable level (Fig. 8-1). The plasma level is elevated within a week of gonadectomy, but appears to reach a plateau with no further rise over a period of months (Fig. 8-2). For unexplained reasons, the hypophysial LH content continues to rise even after the plasma level has reached the plateau. Similarly, the plasma levels of FSH are markedly elevated in the castrated rat of either sex.

If one administers estradiol to the spayed female, beginning on the day of operation, the rise in plasma and pituitary LH can be prevented (Fig. 8-1); however, the quantity of estrogen required to accomplish this result appears to be slightly higher than that which is present physiologically, since it evokes vaginal estrus and a marked uterine enlargement (Fig. 8-3). This suggests that normally the inhibition caused by the estrogen is aided by another ovarian hormone which also plays a role in inhibiting LH. The logical candidate, of course, is progesterone. Pro-

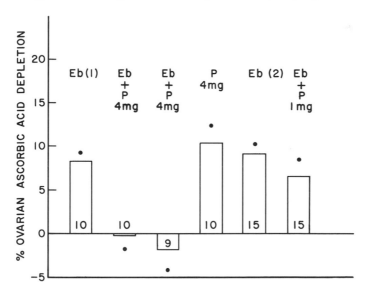

FIGURE 8-4. Effect of progesterone (P) on plasma LH activity of ovariectomized rats pretreated with estrogen. Reading from left to right, each pair of columns represents the results of an experiment in which the LH activity of plasma from donors subjected to two different treatments was compared. The first two columns compare the plasma LH activity of donors pretreated with estradiol benzoate (Eb) only with that of another group treated with Eb followed by 4 mg of P. The next two columns compare the plasma LH activity of another set of donors given Eb followed by 4 mg of P with that of a group given only 4 mg of P. In the last two columns on the right, the LH activity in plasma of another group of donors given only Eb, as in the first experiment, is compared with that from a final group given Eb followed by 1 mg of P. (McCann, S. M., *Am. J. Physiol.* 202:601, 1962.)

gesterone alone has only a feeble inhibitory effect on LH release, and supraphysiological doses are required to lower plasma LH in the spayed rat. Pretreatment of an ovariectomized female with small doses of estrogen sensitizes the animal to the inhibitory effect of progesterone; then under these conditions, doses of progesterone which appear to lie close to the physiological range have an inhibitory effect (Fig. 8-4). So we conclude that in the normal female the synthesis and release of LH are held in check by the combined actions of both estrogen and progesterone.

As might be predicted, the testicular androgen, testosterone, is capable of preventing the rise in plasma LH which follows castration in males. Here again, however, we have been puzzled by the observation that the dose required is greater than that required to return the weight of the sex accessories to normal. Possibly, small amounts of estrogen produced by the testes may synergize with testosterone in the normal animal.

The responses of castrates to administered gonadal steroids provide one of the pieces of physiological evidence for the belief that FSH and LH are in reality two discrete pituitary hormones and not merely artifacts provided to us by the biochemists. For example, Parlow (1964a) has shown that much larger doses of estrogen are required to lower pituitary FSH than to lower pituitary LH. Furthermore, the administration of testosterone, although lowering plasma FSH, produces a paradoxical elevation in pituitary FSH (Greep, 1961). The explanation for these peculiar FSH responses is not yet at hand, and much more work must be done before a coherent picture of the effects of gonadal steroids on FSH secretion can be presented.

Still less is known about control of the secretion of LTH. For example, the effect of castration on the release of prolactin is unknown. One relevant fact which has been learned is that estrogen, given in rather high doses to intact rats, can elicit a state like pseudopregnancy. This is characterized by the presence of an enlarged pituitary gland, by enlarged functional corpora lutea, and by mammary development and secretion, which almost certainly indicate that the estrogen has evoked the secretion of LTH with its lactogenic action. Progesterone administration in normal female rats fails to cause reduction in size or other signs of involution of the corpora lutea, and has even been shown to evoke pseudopregnancy when administered in large doses. These findings have led Rothchild (1960) to postulate that progesterone has a positive feedback effect on LTH secretion.

What happens when one combines treatment with estrogen and progesterone is apparently untested. In the normal animal, it is possible that the positive effects of gonadal steroids may enhance LTH secretion to aid in the maintenance of the corpora lutea during the last half of the estrous cycle. The problem then is that we know that gonadal hormones increase LTH release, but we know of nothing which inhibits it. Thus, the factors responsible for luteolysis at the end of the cycle are yet to be

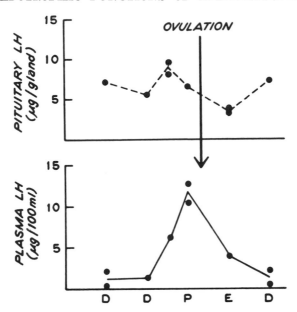

FIGURE 8-5. Fluctuations in the hypophysial and plasma LH during the estrous cycle of the rat. The data for pituitary LH changes are from studies by Dr. Neena B. Schwartz at the University of Illinois. The data for plasma are those obtained in this laboratory. The letters on the abscissa refer to the days of the estrous cycle in the rat: D = diestrus; P = proestrus; E = estrus. (McCann, S. M., *Physiol. Physicians* 1:1, 1963.)

elucidated. It is interesting that hysterectomy leads to a prolongation of the life of the corpora in several species, and it has even been suggested that a luteolytic substance from the uterus may be responsible for terminating the life of the corpus luteum (Anderson et al., 1963).

Another point which must be emphasized in any construction of hormonal relationships is that gonadotrophin secretion is cyclic in the normal human female and in many other mammals, such as the rat. Although plasma LH can not be measured during most of the estrous cycle in the rat, it rises to a readily measurable peak on the afternoon of proestrus. The value obtained is as high as that found in ovariectomized animals. This cyclic burst of LH secretion which just precedes ovulation and is accompanied by a fall in pituitary LH (Fig. 8-5) has been attributed to a positive feedback of estrogen and progesterone operative at this stage of the cycle in the normal female. The estrogen and progesterone presumably arise from the maturing ovarian follicles.

Everett (1964) has succeeded in advancing the time of ovulation in rats by the administration of either estrogen or progesterone. While the doses of estrogen which he used were much larger than the amounts present in the normal state of the animal, he achieved effects with relatively low doses of progesterone. Presumably then, the ovulatory surge of LH secretion may be caused by the stimulating effect of gonadal steroids. Why a positive feedback on LH is seen under these conditions, whereas negative feedback has always been observed in the gonadectomized animal, is an unsolved mystery. Perhaps the rapidly rising titer of the steroids which occurs at this time is an important clue. We will return later to a further consideration of this phenomenon.

Environmental Factors Which Modify Gonadotrophin Secretion

Superimposed upon the influence of gonadal steroids on gonadotrophin secretion are the effects of environmental stimuli. Copulation induces ovulation in the rabbit, cat, and ferret presumably by a reflex activation of the hypothalamus. Another example of a neural influence on gonadotrophin secretion is the induction of pseudopregnancy in the rat following stimulation of the cervix or mating with a vasectomized male. Here a prolonged inhibition of FSH and LH secretion is coupled with enhanced LTH secretion. This provides one example of the frequently observed reciprocal effects of various stimuli on FSH-LH and LTH secretion.

Constant exposure to light, apparently acting via afferent fibers of the optic nerves, can abolish the estrous cycle in rats and induce a state of constant estrus. Acute and chronic stresses have also been reported to produce changes in gonadal and accessory organ weight which have been attributed to a reduction in the gonadotrophin output. We have been unable to find any effects of stress on LH secretion, but such effects may still occur although they may be limited to a reduction in FSH output.

Lactation, presumably put in operation via the stimulus of nursing, has a profound influence on gonadotrophin secretion. It is capable of lowering pituitary stores of both FSH and LH (Parlow, 1964b), and also of lowering the plasma levels of these trophins in the spayed female (these animals continue to lactate after ovariectomy). Since the gonadal steroids are absent in this situation, the effect is almost certainly a neurally mediated one. By contrast with its inhibitory effect on FSH and LH secretion, lactation clearly demands an augmented prolactin release. This is illustrated by the high requirements of the hypophysectomized, lactating animal for LTH and by the acute drop in pituitary LTH content which occurs in intact rats following suckling (Grosvenor and Turner, 1957). This is another example of the reciprocal effects of various stimuli on the secretion of FSH and LH on the one hand and LTH on the other.

Hypothalamic Control of Gonadotrophin Secretion

The observation that environmental changes could influence gonadotrophin secretion gave rise to the concept of central nervous control over the secretion of anterior pituitary hormones. More recently, the belief has grown that neural control of gonadotrophin release is not limited to the mediation of environmental influences, but also mediates the alterations in gonadotrophin secretion which occur during the menstrual or estrous cycle. Thus opinions have swung away from the early conclusion that the pituitary is the master gland, and the view now prevails that it is itself under the control of the hypothalamus. Several lines of evidence substantiate this concept of hypothalamic control of gonadotrophin secretion.

First, electrical stimulation of the basal tuberal region will evoke ovulation in rabbits and rats (Everett, 1964). This clearly indicates the existence of a pathway from the hypothalamus to the pituitary which can cause a release of gonadotrophins, but it does not prove that this is a normal pathway of hypophysial activation.

Concrete evidence for the importance of the hypothalamus has come mainly from studies of the effects of hypothalamic lesions in man and lower forms—in particular, the rat. Hypothalamic lesions can induce precocious puberty in both man and rat, although there is considerable difference of opinion over the hypothalamic areas involved.

In adult rats, either of two syndromes can be produced. Rostral lesions in the region just over the optic chiasm evoke vaginal changes typical of continual or constant estrus. In this situation, the animal appears poised on the brink of ovulation, but never goes beyond this point, and ovulation does not occur. The ovaries are filled with large follicles, while the estrous vaginal smear shows that these follicles are actively secreting estrogen. Apparently, in this situation the animal secretes relatively constant amounts of FSH and LH, which are responsible for follicular development and estrogen secretion, but the cyclic burst of LH secretion which produces ovulation is missing.

A much more profound interference with gonadotrophin secretion follows upon destruction of the basal tuberal hypothalamus. The key region here appears to be the median eminence of the tuber cinereum, the site of union of the hypothalamus and the pituitary stalk. If the median eminence is destroyed, the animal remains in a state of diestrus with a lack of follicular development. Apparently FSH and LH secretion has been turned off. This conclusion has been verified by placing lesions in the median eminence in ovariectomized rats. Two days after the placement of such lesions, the plasma FSH and LH are markedly lowered from the high levels induced by the ovariectomy. Also, if the operations are done in reverse order, the rise in plasma LH which normally follows ovariectomy fails to occur when animals with median eminence lesions (Fig. 8-2) are ovariectomized.

Female rats with lesions in the median eminence show one change which stands in marked contrast to the atrophic condition of the ovarian follicles. It concerns the corpora lutea, which are large and appear to be functional. Proof of their function is provided by the fact that a decidu-

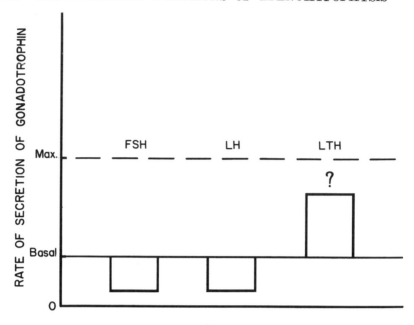

FIGURE 8-6. Effect of hypothalamic lesions on the secretion of the three gonadotrophins. Max = maximum. (McCann, S. M., *Physiol. Physicians* 1:1, 1963.)

oma will form after uterine trauma. This decidual reaction can occur only when progesterone is present. It appears, therefore, that the animal with a median eminence lesion secretes LTH which stimulates progesterone secretion by the corpora. Moreover, there is evidence that this enhanced secretion of LTH stimulates the mammary glands; rats with these lesions do not show the mammary involution which normally occurs when a litter of infant rats is taken away from the mother. Even in male rats, lesions of the median eminence will induce the secretion of milk. This situation is clearly another example of the reciprocal effects on FSH and LH on the one hand, and LTH on the other. Apparently, when hypothalamic influences are removed by median eminence lesions, the secretion of FSH and LH is severely curtailed, whereas the secretion of LTH is enhanced (Fig. 8-6). Hypothalamic influences appear to stimulate FSH and LH secretion in the normal animal while holding the secretion of LTH in abeyance.

HYPOTHALAMIC RELEASING FACTORS

One of the most striking recent advances in neuroendocrinology has been the discovery that nerve cells of the hypothalamus secrete hormones which control the secretion of the several trophic hormones of the pituitary gland. There is general agreement among neuroanatomists that the anterior lobe has no secretory nerve supply; but there is a means of communication between the hypothalamus and the anterior lobe. It is via the hypophysial portal veins, which provide a pathway for selective transmission of neurohumoral agents to the anterior pituitary. Although in the original descriptions the blood was said to flow upward from the pituitary to the hypothalamus, in vivo observations of the direction of blood flow by Houssay and Biasotti (1935) in amphibia, and by Green and Harris and by Worthington (1960) in mammals, have clearly established the direction as downward from the hypothalamus to the pituitary. Since the vessels arise from capillaries in the median eminence and pituitary stalk, substances secreted in this region clearly have preferential access to the pituitary gland. More recently, additional vessels which arise from the neural lobe proper have been shown also to perfuse the anterior lobe. Consequently, capillary blood from the entire neurohypophysis appears to pass to the anterior lobe.

LH Release

When it was shown that either vasopressin (a hormone of the posterior lobe), or a hypothalamic corticotrophin-releasing factor (CRF) extractable from the stalk–median eminence region could trigger the release of ACTH from the anterior lobe, attention was focused on the possibility that similar releasing factors might trigger the release of gonadotrophins. Such evidence was soon forthcoming from two laboratories, our own and Harris's, which began work on the problem independ-

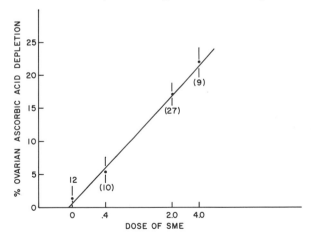

Figure 8-7. Effect of varying doses of stalk–median eminence extract (SME) on ovarian ascorbic acid depletion. Numbers on the abscissa refer to the number of hypothalami from which the extract was derived. Except for the control (0), the doses are on a log scale. (McCann, S. M., *Am. J. Physiol.* 202:395, 1962.)

ently at about the same time. We were able to show that the same crude, acidic extracts of the rat stalk–median eminence which released ACTH were capable of depleting ovarian ascorbic acid when injected into immature test rats prepared as for the assay of LH (McCann and Ramirez, 1964). A linear log-dose response curve was obtained (Fig. 8-7). Similarly prepared extracts from rat cerebral cortex had none of the activity found in the hypothalamic extracts. Furthermore, the activity found in the hypothalamic extracts could not be accounted for by the presence of other physiologically active substances such as epinephrine, Substance P, histamine, serotonin, or, in particular, vasopressin or oxytocin.

The extracts of the stalk–median eminence region were shown to have no effect upon the ovarian ascorbic acid levels of hypophysectomized test rats. This, plus the observation that heating for 10 minutes in a boiling water bath inactivated rat LH while leaving unchanged the activity of the hypothalamic extract, clearly indicated that the material in question was not LH. At this point, the unknown active substance in the extract was termed the LH-releasing factor (LH-RF).

Studies with extracts prepared from various hypothalamic areas revealed that the major activity was concentrated in the stalk–median eminence region. Minimal activity was obtained with those from the overlying ventral hypothalamus and from the suprachiasmatic region. It is still too early to hazard a guess as to the hypothalamic location of the neurosecretory cells which presumably secrete the LH-RF, but they could be in this region which overlies the median eminence.

In all the experiments performed up to this time, the extract had been injected directly into the immature test rats, which had been pretreated with large doses of gonadotrophins, and ovarian ascorbic acid depletion had been measured. Consequently, it was important to see if

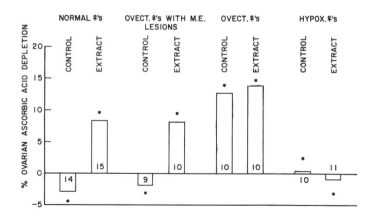

FIGURE 8-8. Effect of stalk–median eminence extract on plasma LH (per cent ovarian ascorbic acid depletion) in several types of donor rat. (McCann, S. M., *Am. J. Physiol.* 202:395, 1962.)

hypothalamic extracts could elevate plasma LH titers in other circumstances. In these experiments the extract was administered intravenously, and the rats were bled 10 minutes later. After centrifugation of the blood sample, the plasma was injected into the immature test rats to measure ovarian ascorbic acid depletion. The ascorbic acid depletion served as a measure of the plasma LH. An elevation in plasma LH occurred within 10 minutes of the intravenous injection of the extract into normal females (Fig. 8-8), and it was also effective in ovariectomized rats in which the release of LH had been inhibited by lesions in the median eminence. This latter observation is an important one, because it indicates that the extract does not act via the nervous system, but presumably directly stimulates the pituitary.

No effect was obtained in similar experiments in which ovariectomized rats were used. We do not know the explanation for its ineffectiveness in this situation, but we hypothesize that the ovariectomized animal with its high level of LH secretion is already responding maximally to endogenously secreted LH-RF and cannot respond further to the exogenously administered releasing factor. The factor failed to raise the plasma LH in hypophysectomized rats, another indication that it is not LH.

The quantity of LH-RF in the hypothalamus is small. (This is, perhaps, not surprising, since the amounts needed are probably minute because the factor has preferential access to the pituitary via the portal vessels.) Consequently, the very sensitive ovarian ascorbic acid depletion assay is necessary to demonstrate activity after systemic injection of the hypothalamic extract. A different approach to this problem has been made in Harris's laboratory. They have implanted cannulae into the pituitary and injected minute amounts of hypothalamic extract into the gland. Campbell et al. (1964), using rabbits, and Nikitovitch-Winer (1962), using rats, were able to show that stalk–median eminence extracts would evoke ovulation in either species and that extracts from other brain areas were inactive. Vasopressin and oxytocin, as well as epinephrine, Substance P, histamine, and serotonin were inactive in their system (as in ours). Ovulation could not be induced by the systemic administration of stalk–median eminence extracts unless very large amounts were used. This rules out LH contamination of the extract as a factor in their results. These important experiments clearly point to a hypophysial site of action of hypothalamic extract. The ovulation induced is presumably caused by LH; but since, as tested in hypophysectomized animals, ovulation is produced optimally by injections of mixtures of FSH and LH, it is apparent that their results can not be taken by themselves as proof of the release of LH alone. Release of FSH might have been responsible, at least in part.

FSH RELEASE

Igarashi (1964), in our laboratory, has made a systematic study of the possible FSH-releasing action of hypothalamic extracts. He has used principally the mouse uterine weight augmentation method for estimating FSH, but has also confirmed his results with the ovarian weight augmentation method of Steelman and Pohley. An increase in plasma FSH was observed 10 minutes after the intravenous administration of stalk–median eminence extract. For the test animals he used ovariectomized rats in which the release of FSH had been inhibited, either by median eminence lesions or by the injection of large doses of estrogen and progesterone. Cerebral cortical extracts and the neurohypophysial polypeptides, vasopressin and oxytocin, were inactive. Kuroshima, Ishida, Bowers, and Schally have recently confirmed these observations.

Schally's group and Mittler and Meites, employing pituitaries incubated in vitro, have also found evidence for the FSH-releasing activity of hypothalamic extracts. Schally and Bowers have observed the occurrence of LH release in the same circumstances. So the FSH- and LH-releasing activity of stalk–median eminence extract has been demonstrated by both in vivo and in vitro tests.

LTH RELEASE

Since the hypothalamus exerts an inhibitory influence over the secretion of LTH or prolactin, one might expect to find a prolactin-inhibiting factor (PIF) in hypothalamic extracts. Pasteels in Belgium, and Talwalker, Ratner, and Meites in this country, have recently reported the discovery of PIF as follows: they found that pituitaries incubated in vitro released large amounts of prolactin into the medium. This release of prolactin was inhibited by the addition of crude, acidic extracts of rat hypothalamus. Extracts of cerebral cortex were inactive, as were other physiologically active substances found in the extract. In collaboration with Grosvenor at the University of Tennessee, we have sought in vivo evidence for a PIF. Grosvenor and Turner (1957) observed that, in lactating rats, suckling for 30 minutes induced an abrupt decrease in hypophysial prolactin levels. We have observed that the intraperitoneal injection of crude, acidic extracts of rat or beef hypothalamus into lactating females one to two minutes prior to the beginning of a suckling period can prevent the suckling-induced decline in pituitary prolactin. Cerebral cortical extracts were ineffective. Consequently, it appears likely that, in addition to the LH-RF and FSH-RF which stimulate gonadotrophin secretion, there is also a neurohumoral inhibitor of prolactin release.

NATURE AND SIGNIFICANCE OF RELEASING FACTORS

To establish the physiological significance of the LH-RF, it is desirable to show that the quantity of the releasing factor stored in the median eminence fluctuates in situations associated with altered LH secretion. In our laboratory, Chowers has recently observed a lowering in the hypothalamic content of LH-RF at proestrus, and similar findings have been obtained independently by Ramirez and Sawyer. Nallar has even observed that plasma from long-term (more than six weeks) hypophysectomized rats will deplete ovarian ascorbic acid in the immature rats of the Parlow test. Plasma from recently hypophysectomized animals was without effect. If lesions were placed in the median eminence of the long-term hypophysectomized animals, the plasma lost its capacity to deplete ovarian ascorbic acid. These results suggest that the hypothalamus of the hypophysectomized rat in time secretes a large amount of LH-RF. The secretion is abolished by hypothalamic lesions. Thus, although much remains to be verified, it seems highly likely that the LH-RF is the physiological mediator for LH release.

What is the chemical nature of the hypothalamic factors which affect gonadotrophin secretion? To try to answer this question, sheep and beef hypothalami have been used, since they provide larger quantities of the releasing factors than can be obtained from rats. The LH-RF is inactivated by the proteolytic enzymes, pepsin and trypsin, which suggests that it may be a polypeptide. It is not inactivated by reduction with thioglycolate, a treatment which eliminates the biological activity of the neurohypophysial hormones, vasopressin and oxytocin, by splitting the disulfide group joining the five-membered amino acid ring in their molecules. This result again shows that the LH-RF is distinct from vasopressin and oxytocin. The technique of gel filtration through a column of Sephadex G-25 has been used in several laboratories to purify the LH-RF. This procedure separates molecules primarily according to their molecular size. When the absorbed materials are eluted by washing the column, LH-RF is eluted just prior to vasopressin, a result which suggests that its molecular weight is slightly greater than that of vasopressin. Further purification has been achieved by ion-exchange chromatography on carboxymethylcellulose (CMC). The LH-RF emerges from such columns as one increases the ionic strength and pH of the eluting buffers. These results support the view that it is a basic polypeptide. A preliminary listing of the amino-acid composition of LH-RF has just been reported by Schally and Bowers.

The next question which should be answered is whether the factors, FSH-RF and PIF, are identical with the LH-RF. In our initial experiments employing gel filtration on small columns of Sephadex G-25, we were unable to separate the FSH-RF from the LH-RF. With the use of longer columns and different conditions of elution, Dhariwal and Nallar have succeeded in separating the FSH-RF from the LH-RF. It is important to point out that the gonadotrophin-releasing factors have no CRF activity. We are currently evaluating the prolactin-inhibiting activity of fractions obtained after gel filtration through Sephadex; consequently, the identity or lack of indentity of PIF with the other factors is still an open question.

It appears highly likely that the hypothalamus exerts its control over gonadotrophin secretion by secreting appropriate releasing and inhibiting factors into the hypophysial portal system to affect the secretion of each adenohypophysial hormone. This method of control appears to extend to all the anterior lobe hormones. We have alluded to evidence for a CRF, and similar evidence has already accrued indicating the probable presence of a thyrotrophin-releasing factor and a growth hormone-releasing factor in the hypothalamus.

Site of Action of the Feedback of Gonadal Steroids on Gonadotrophin Secretion

There are two obvious loci at which gonadal steroids might exert their feedback effects to inhibit the secretion of trophic hormones. Inhibition on either the hypophysis or the hypothalamic mechanism has been postulated. Several experimental approaches have been applied to the problem. The first of these is to implant minute amounts of steroid directly into the site under investigation. Rose and Nelson (1957), the first to do this, reported a direct inhibitory effect of estrogen within the pituitary gland. Later Lisk (1960), and Sawyer and his colleagues, reported effects from placement of estrogen in the hypothalamus. They found little or no effect from the hypophysial placements of gonadal steroids and concluded that the principal site of the feedback was on the hypothalamus itself. Subsequently, Bogdanove (1963) found that implants of estrogen directly into the anterior lobe would prevent a cellular change which occurs in the anterior lobe following removal of the ovaries. This change appears to be correlated with the hormonal changes already described. In agreement with Rose and Nelson, he concluded that the physiological site of the inhibitory effect of estrogen is located at the pituitary level, and argued that the effects obtained with hypothalamic placements of the steroid were caused by the absorption of the steroid into the portal vessels, which then distributed it to the pituitary gland, where the effect was actually exerted.

At about the same time, Ramirez and Abrams, in our laboratory, were able to show that implants of estrogen into either the median eminence region or into the pituitary gland would inhibit LH secretion and enhance the secretion of LTH. In ovariectomized rats with these estrogen implants, the plasma LH levels were lowered. In intact animals, LTH secretion was induced, as evidenced by the development of a pseudopregnancy syndrome characterized by persistent diestrus, enlarged corpora lutea, mucification of the vagina, and lobuloalveolar development and secretion in the mammary gland. Kanematsu and Sawyer, working with ovariectomized rabbits, also found that hypothalamic implants of

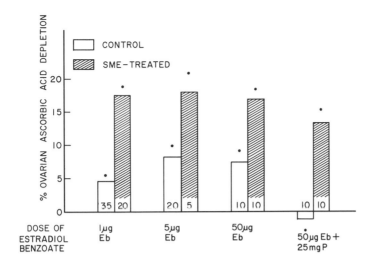

FIGURE 8-9. Failure of ovarian steroids to block the LH-releasing action of stalk–median eminence extract. The doses of ovarian steroids given are indicated on the abscissa. Eb = estradiol benzoate; P = progesterone. (McCann, S. M., *Am. J. Physiol.* 202:395, 1962.)

estrogen lowered plasma LH levels, but they found increased storage without enhanced release of LTH in intact animals with hypothalamic implants of estrogen, whereas pituitary placements of the steroid caused LTH release accompanied by low hypophysial content. The cause of this difference in behavior of LTH in the two species remains obscure. Meites and his associates have found that estrogen applied to rat pituitaries incubated in vitro will augment LTH release, which agrees with our in vivo findings. I believe that it is safe to conclude from this discussion that estrogen can act directly at the pituitary level. Whether there is also an action at the hypothalamic level cannot be decided from these data, in view of possible absorption into the portal vessels with an action on the pituitary.

Another line of evidence, however, has recently led to the conclusion that these steroids must also act at a site other than the pituitary gland, presumably on the hypothalamus. If the inhibition were exerted only at the pituitary level, one would expect that treatment of ovariectomized rats with large doses of gonadal steroids would block the response to LH-RF or FSH-RF given by injection. This is not the case. Varying doses of estrogen, which lowered the plasma LH in these ovariectomized rats, were incapable of preventing the elevation of plasma LH that followed intravenous administration of LH-RF (Fig. 8-9). Even when progesterone (25 mg) was administered along with estradiol benzoate (50 μg), the LH-RF was still effective. In fact, the ovariectomized, estrogen-progesterone-blocked rat is supersensitive to the action of LH-RF. Such animals respond to the extract made from as little as 0.015 of one rat stalk–median eminence by showing LH release, whereas injection of 0.4 of one stalk–median eminence is required for LH secretion in the immature rats which are routinely used for this test. It does appear likely, then, that there are two sites of action of gonadal steroids on gonadotrophin secretion, and that the effects are mediated both at the hypothalamic and hypophysial levels. As a working hypothesis we assume that the hypothalamic site is the more sensitive and is the one involved in mediating the physiological effects of these steroids.

MECHANISM OF PUBERTY

One of the baffling problems in studies of the control of the endocrinology of reproduction has been the mechanism by which puberty is induced. What prevents premature secretion of gonadotrophin and resultant precocious puberty at a time when the organism is ill equipped for assuming reproductive functions? Now that we have examined the mechanisms which regulate gonadotrophin secretion, we can approach this mystery.

Certain facts appear to be well established. First, gonadotrophin excretion is low in infancy and rises at puberty. Yet the immature gonad

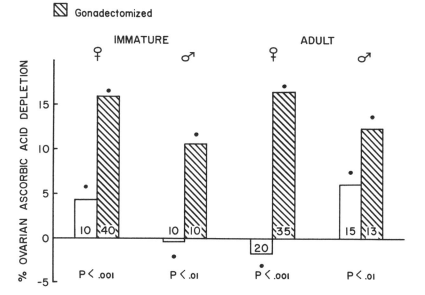

FIGURE 8-10. Plasma LH (per cent ovarian ascorbic acid depletion) in immature and adult rats. (Ramirez, V. D., and McCann, S. M., *Endocrinology* 72:452, 1963.)

(at least after a very early unresponsive period) is quite capable of responding to gonadotrophins and of developing to maturity. Moreover, the immature hypophysis is similarly capable of premature secretion of gonadotrophins, for when Harris and Jacobsohn hypophysectomized female rats and placed the pituitaries of their infant offspring in the sella turcica, the adult females again began to show estrous cycles after only a short delay. Finally, hypothalamic lesions in the infantile human of either sex or in infantile rats have been shown to evoke precocious puberty. These facts indicate that the immature gonad and pituitary are capable of assuming adult levels of function. Presumably, during the prepuberal period, the pituitary is not being stimulated by the hypothalamus.

Ramirez has studied this problem in our laboratory in regard to LH secretion. First, he found that castration of both male and female infantile rats at the tenth postpartum day was followed within two weeks by elevated levels of plasma LH which were as high as those found in adult castrates (Fig. 8-10). This occurred at an age (24 days) when an intact animal would still be sexually infantile. Apparently castration removed a brake on gonadotrophin secretion in these immature animals. Castration was also followed by a definite, though small, reduction in weight of the infantile accessory sex organs, which indicated that

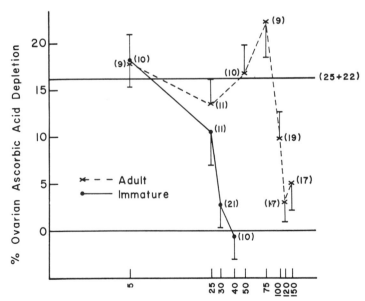

FIGURE 8-11. Decrement in plasma LH activity (per cent ovarian ascorbic acid depletion) of castrated males treated with testosterone propionate. (McCann, S. M., and Ramirez, V. D., *Rec. Progr. Hormone Res.* 20:131, 1964.)

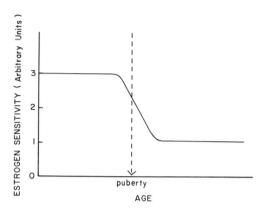

FIGURE 8-12. Schematic diagram illustrating change in hypothalamic sensitivity to estrogen at puberty. (Ramirez, V. D., and McCann, S. M., *Endocrinology* 72:452, 1963.)

the infantile gonads must have been secreting small amounts of gonadal steroids.

He then compared the sensitivity of adult and immature castrates to the inhibitory action of gonadal steroids on LH secretion, and found that the sensitivity to estradiol was two to three times greater in the immature females than in the adults. For testosterone in males, the difference in sensitivity was three- to fourfold (Fig. 8-11). In these experiments the dosage of hormonal replacement therapy was expressed in terms of body weight to correct for the great disparity in weight between the adult and the immature animals.

These experimental results can be explained by postulating that at puberty the hypothalamus becomes less sensitive to the negative feedback by gonadal steroids. At this time the low levels of steroid secreted by the immature gonad are no longer adequate to hold the gonadotrophin secretion in check. Augmented gonadotrophin secretion follows, which is stabilized at the adult level when the gonadal steroid titers reach the higher adult value, which is once again capable of inhibiting further gonadotrophin release. This hypothesis implies that a change in the hypothalamic sensitivity to gonadal steroids is a major factor in the development of puberty. Unfortunately, we are unable to measure in blood the levels of gonadal steroids which are actually produced by these injections, and this constitutes a possible loophole in the foregoing argument. The results could be explained equally well by postulating an altered metabolism of gonadal steroids at puberty. If the rate of degradation of gonadal steroids were increased at puberty, an apparent decrease in sensitivity to the feedback action of the steroids would also be seen. Even if the reset hypothesis is correct, it still leaves us completely ignorant of the factors which bring about this altered hypothalamic sensitivity at puberty. That the gonadal steroids themselves may induce the lowered sensitivity which occurs at puberty is suggested by some recent experiments of Ramirez, who has observed that minute doses of estrogen administered to infantile female rats will advance puberty.

AREA RESPONSIBLE FOR CYCLIC RELEASE OF GONADOTROPHIN

Although a cyclic release of LH appears to induce ovulation periodically in spontaneously ovulating females of such species as man and the rat, it is generally believed that the secretion of gonadotrophin by males is relatively constant. This can be shown by grafting ovaries into a castrated male; although the ovaries become filled with large follicles, ovulation fails to occur. If vaginal epithelium has also been grafted into such a male, it will be observed to show persistent cornification. By contrast, similar ovarian and vaginal grafts in spayed females show evidence of cyclic activity. What is the reason for this basic difference in the behavior of the gonadotrophin-regulating mechanism in males and females? Pfeiffer, in the early 1930's, gained the first insight into the mechanism when he found, using male rats castrated at birth, that ovarian grafts would later show evidence of cyclic behavior. Conversely, if testes were transplanted into females at birth, the females bearing both their own ovaries and engrafted testes failed to show ovarian cycles when adult, and instead showed constant vaginal estrus and follicles without corpora lutea in their ovaries.

On the other hand, if females were ovariectomized at birth, ovaries grafted into them when they had reached adulthood gave evidence of cyclic behavior. These findings have been confirmed and extended by later workers, in particular by Yazaki, by Gorski, and by Harris and Levine (Harris, 1964). They suggest that an intrinsic cyclic mechanism is inherent in females, but that male hormone from the testis can convert this to the male, acyclic pattern. Barraclough and others have indeed shown that injection of a single minute dose of testosterone into newborn females during the first 10 days of life is sufficient to block cyclic gonadotrophin secretion, so that the adults show persistent vaginal estrus and the presence of only follicles in the ovary. Since later injections of an androgen fail to evoke this syndrome, the concept has developed that there is a critical period during the first 10 days of life, in which androgen can induce the acyclic pattern of gonadotrophin secretion.

One might argue that the defect in these so-called androgenized females is located in the ovary or in the hypophysis; however, this argument is refuted, since grafting of either the ovary or the pituitary of such an androgenized female into a normal recipient with removal of its own corresponding gland is followed by a return to normal function of the graft in question. By exclusion, the most likely locus for the defect in the androgenized female, therefore, is in the hypothalamus. It will be remembered that lesions in the suprachiasmatic region evoke a similar state characterized by persistent vaginal estrus. Barraclough has shown that electrical stimulation of this same suprachiasmatic region fails to evoke ovulation in these androgenized female rats, although it readily does so in normal females. The androgenized rat, however, can be made to ovulate by electrical stimulation of the basal tuberal region, so apparently only the rostral region is refractory to such stimulation. Barraclough and Gorski postulate that neonatal androgen from the infantile testis alters the nature of this suprachiasmatic region, so that the cyclic feminine pattern of gonadotrophin secretion is changed to the male acyclic pattern. Thus, in addition to feeding back on the hypothalamus to inhibit gonadotrophin secretion, androgens appear to have an inductive influence on the infantile hypothalamus which can alter its subsequent pattern of behavior. Along with the alterations in gonadotrophin secretion which we have outlined here, androgen which is present neonatally also eliminates female sexual behavior, but this interesting aspect of the story is beyond the scope of this discussion.

CONCLUDING REMARKS

It can be seen from the preceeding discussion that a complex interplay of humoral and neural factors influences the hypothalamic mechanism which governs the secretion of hypophysial gonadotrophins. However, this complex interplay in a three-tiered hierarchy of endocrinological control begins to reveal the complexity necessary to account for the varieties of temporal organization in the sexual functions. The system must account for the acyclic pattern characteristic of males, the various forms of periodic female cycles, the problems of the appearance of puberty, and in the end, the acyclic pattern of the senescent female. The tale is still fragmentary, but these fragments reveal that the hypothalamus plays the important role. Beginning in the neonatal period, androgen from the infantile testis may act upon the rostral hypothalamus to convert a female, cyclic hypothalamic pattern into a male, acyclic one. As the animal matures, a declining hypothalamic sensitivity to the negative feedback action of gonadal steroids may set the stage for the augmented gonadotrophin secretion that initiates puberty. In the adult animal, there is a negative feedback of gonadal steroids on the release of FSH and LH. Under some circumstances, however, a positive feedback of estrogen and progesterone on LH secretion can also be demonstrated; it is thought to evoke the ovulatory burst of LH secretion at mid-cycle. Both estrogen and progesterone appear to have mainly a stimulatory or positive feedback action on LTH secretion. In most of the circumstances thus far analyzed, there is a reciprocal relationship between the secretion of FSH and LH on the one hand and LTH on the other. This appears to be a consequence of the fact that both hormonal and environmental influences act upon a hypothalamic mechanism, so as to stimulate secretion of FSH and LH, and to inhibit secretion of LTH, or vice versa. The hypothalamic control over secretion is mediated by means of neurohumoral agents FSH-RF, LH-RF and PIF, which upon secretion into the hypophysial portal vessels, are responsible for the observed alterations in output of gonadotrophins from the adenohypophysis.

Notes about References

The concepts which have been reviewed here are the culmination of work from a number of laboratories. The physiological significance of the hypophysial portal vessels was indicated clearly by experiments performed in the laboratory of Harris (see, for example, Harris, 1960). Everett (1964) is responsible for many of the ideas expressed here and in particular deserves credit for discovering the inhibitory influence of the hypothalamus on the secretion of prolactin. The regulation of gonadotrophin secretion in the rabbit, which is an excellent example of those species that ovulate as a reflex response to sexual stimulation, has been studied extensively in Sawyer's laboratory (see, for example, Kanematsu and Sawyer, 1964). The work of Meites and his associates on the PIF has been of great interest (Talwalker, Ratner, and Meites, 1963). Barraclough and his associates have made a careful analysis of the endocrine abnormalities in the androgenized female rat (Barraclough and Gorski, 1961). Lastly, in addition to the work from this laboratory, important work on the chemistry of the releasing factors has originated from Schally's (Schally and Bowers, 1964) and Guillemin's laboratories (1964). There are many others whose experiments should also be referred to; some of these are identified by reference to particular papers in the text, but it should be emphasized that this is by no means a complete listing.

Research in the author's laboratory is supported by the following awards: Contract AF-AFOSR-62-133 from the Air Force Office of Scientific Research, Grant AM-01236 from the National Institutes of Health, Population Council Grant M-6461, and a grant from the Ford Foundation.

Chapter 9

CARBON DIOXIDE BALANCE AND THE CONTROL OF BREATHING

M. W. EDWARDS, JR., AND W. S. YAMAMOTO

Carbon dioxide, the end product of animal metabolism, has a dual importance because of the necessity of getting rid of the large quantities produced, and because it forms the acid component of the most important buffer system of the body. The production and excretion of CO_2 achieve a balance such that the CO_2 concentration in the tissues and body fluids is relatively constant. Nevertheless, deviations from this constancy seem to be important in causing bodily responses which tend to restore the constancy. Thus, when the arterial P_{CO_2} is raised by breathing a gas containing CO_2, the body responds by increasing ventilation in an effort to remove the excess. Or when CO_2 production increases during exercise, the ventilation increases and more CO_2 is excreted.

The physiological mechanisms involved in these changes in ventilation are subsumed under the title, "the regulation of respiration," and have been discussed and reviewed many times. Some of the history of this work and the current trends in thinking are discussed in three recent volumes (Nahas, 1963; Cunningham and Lloyd, 1963; Fenn and Rahn, 1964). One may gain a somewhat different insight, however, by taking the slightly perverse attitude that respiration, i.e., pulmonary ventilation, is not regulated at all, but that it is varied to suit other bodily needs. In this view the regulated variable is really not certain. For example, ventilation controls CO_2 exchange, but it also controls oxygen exchange and plays an essential part in regulation of body pH, so these processes cannot easily be separated from a discussion of CO_2 exchange. For purposes of argument, however, some of this uncertainty and interdependence can be set aside.

CARBON DIOXIDE BALANCE

The concept of CO_2 balance is not always considered in discussions of control of ventilation, yet it is a basic idea. If production of CO_2 and its elimination are unequal in an animal, the CO_2 content must increase or decrease, and this will cause a change in concentration unless the volume of distribution changes. Because most animals have a fairly constant mass, any change is generally one of concentration. The precision with which the balance of CO_2 is regulated is unknown, but it certainly must be accurate in the long run, because a small imbalance, if allowed to accumulate, would eventually prove disastrous. Similarly, momentary variations too far above or below the average value would destroy the animal, since one expects that there is only a limited zone of P_{CO_2}* which is best suited for cellular function. These limits of acceptable cellular P_{CO_2} are not well known, and they may well vary according to the species and type of cell. Blood P_{CO_2} values are better known. In man, the usual arterial P_{CO_2} is 40 mm Hg, and the mixed venous about 45, while the venous drainage of certain regions may have a P_{CO_2} 10 or 15 mm Hg higher. In the acute phase of altitude acclimatization the alveolar P_{CO_2} may be as low as 15 mm Hg, whereas in severe pulmonary insufficiency the arterial P_{CO_2} may be as high as 80 mm or more. Although the tissue P_{CO_2} is naturally somewhat higher than the venous, is is difficult to know exactly what the ranges of tolerance are for the tissues themselves.

The body's balance of CO_2 depends on production, excretion, and storage. There are relatively few ways in which these can occur in mammals. The *sources* of carbon dioxide are the body's metabolic processes, the oxidation of carbohydrates, fats, and proteins, the decarboxylation of acids, the breakdown of purines and pyrimidines, and other reactions. Occasionally CO_2 may be inhaled from an abnormal atmosphere, as when one rebreathes air in a closed space or in an experimental situation.

The *sink* for CO_2 is primarily its elimination by the lungs. Besides this a very small amount is fixed by biochemical reactions whose products are usually excreted, such as urea and uric acid; or CO_2 may be transformed further into succinate and other products. CO_2 is added to β-hydroxy-isovaleryl-Coenzyme A to form β-hydroxy-β-methylglutaryl-Coenzyme A, and thus becomes acetyl-CoA which enters into many reactions, but the net CO_2 fixation by this reaction is probably very small in mammals. An "average man" may produce at least 200 to 200 ml (STPD) of CO_2 per minute, 13 to 20 moles every 24 hours. The renal excretion of urea, $CO(NH_2)_2$, is about 0.5 mole in this period, and that

---

* That is, the P_{CO_2} of the cell or its immediate surroundings, the animal's internal environment.

of uric acid a few millimoles. At normal plasma bicarbonate concentrations little or no bicarbonate is excreted in the urine, and although the P_{CO_2} of the urine may be that of the blood, the CO_2 content of a day's urine is negligible. The amount of CO_2 lost through the mammalian skin is also very small. Thus the lungs account for over 95 per cent of the CO_2 excretion, and it is with this function that we shall concern ourselves in the present discussion.

Storage of molecular CO_2 takes place by an increase in its concentration or its volume of distribution. Some animals have organs, *e.g.*, the swim bladders of fish or the air sacs of birds, which store gas that may contain CO_2. Mammals of course do not have such organs, and they can store gaseous CO_2 only in their lungs, when the breath is held. This has a small, short-term effect as during diving, but can have no great importance in CO_2 balance. If the volume of blood or extracellular fluid increases, as it does during growth or pregnancy, the body's CO_2 content rises with no change in concentration. Finally, a relatively large amount of CO_2 exists as bone salts, but these are so much less mobile than the other pools that we shall not consider them here.

A 70 kg adult with an "average" blood volume of 6 liters, an interstitial fluid volume of 14 liters, and a total CO_2 concentration in these fluids of 550 ml/L would be able to store an excess of 1100 ml of CO_2 with a 10 per cent increase in these fluid volumes. In an actual individual we might find more storage because of the corresponding increase in volume of intracellular fluid which also contains CO_2. These capacities for storage are, however, rather small compared to the hourly production of CO_2, and emphasize the importance of a close relationship between production and excretion. Furthermore, the principal manifestation of change in storage is change in CO_2 concentration rather than in the volume of these fluids.

Our hypothetical man of 70 kg might contain, by a generous estimate, 25 liters of CO_2 in some 46 liters of body water. Doubling the concentration of CO_2 in this volume would absorb the CO_2 produced in only 100 minutes of resting metabolism. Certainly changes of concentration of this magnitude are extremely unusual and occur only under very pathological conditions. Most measurements of arterial CO_2 tension show it to be constant within a few millimeters of mercury, under conditions in which 1 mm Hg corresponds to 5 or 10 ml of CO_2 per liter of blood. Thus there seems to be little detectable fluctuation in the body's CO_2 content under normal conditions.

As noted above, CO_2 is related to the acid-base balance of the body. Carbon dioxide in aqueous solution forms an acid, and its concentration is intimately linked to the pH of body fluids by reactions of the bicarbonate–carbonic acid buffer system. Carbon dioxide balance is a necessary condition for the stability of the pH of these fluids.

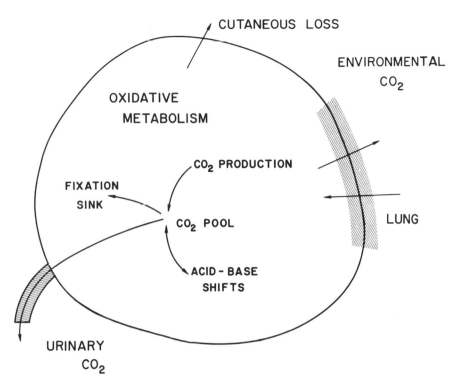

FIGURE 9-1. External and internal carbon dioxide exchanges. This diagram shows the principal features of the material balance of CO_2. Arrows indicate the direction of movement of the CO_2 molecules.

MODELS OF CARBON DIOXIDE EXCHANGE

Having listed the sources and sinks of carbon dioxide we can quite readily make a diagram to represent the organism's exchanges of this substance. These are shown in Figure 9-1.

In order to introduce physiological mechanisms as we now understand them, let us assign particular mechanisms to the paths of material flux. Production of CO_2 is a relatively independent process since it is governed by other priorities in the life of the organism, such as the need to keep warm or to move for flight or pursuit of food; moreover, this production does not occur homogeneóusly throughout the body mass but in many tissues and at many rates. The internal distribution of CO_2 is managed by diffusion between compartments and by convection via the circulation of blood. Interposed between the veins and the arteries in this convection stream is a special mechanism, the lung, which forms the effective exchange surface between the organism and its external environment. Usually this exchange is a sink, but since the physicochemical

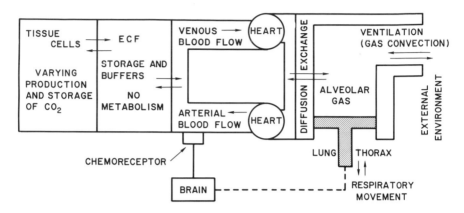

FIGURE 9-2. External and internal CO_2 exchanges diagramed in terms of some of the physiological processes known to be involved.

process involved is diffusion, the organism also becomes subject to the vagaries of the external environment with respect to CO_2; thus the lung is a facultative sink but may also be a non-facultative source. ("Facultative," in this instance, means at rates controlled by physiological mechanisms.) Pulmonary ventilation is a process in which environmental gas is moved across the blood stream allowing a diffusion exchange to occur. The organism is able to modify the gaseous convection (the pulmonary ventilation) and hence to vary the rate of flux of CO_2. These facultative changes of ventilation are controlled by neural mechanisms consisting apparently of sensors (located both peripherally and centrally) and of evaluators of need which include priority-determining and instruction-formulating processes—a way of saying that the central nervous system produces an overall respiratory response which is a function of the body's needs.

To include these physiological features in the diagram is a straightforward enough process and leads to a second level model like Figure 9-2. One can make a succession of similar diagrams to include ever more detail about this system—details such as partition of the respired gases into alveolar and dead spaces, or the peculiar properties of blood with respect to the solubility of CO_2, and so forth. These details, although making the model more accurate, would begin to conceal for our present purposes the overall picture. The diagrams are physiological diagrams; the parts are interconnected in an anatomical sense and the arrows indicate the sequence produced by convection and diffusion.

Investigation of the physiological system represented by this scheme has been extensive, including many attempts to "break into the loop" to study the behavior of isolated parts. The reviews and symposia cited above indicate to a large measure the present state of such studies. As in many fields of investigation the number of questions exceeds the number of firm assertions one may make and the assertions often appear to be unrelated. To indicate the type of problems that emerge from this type of traditional, physiological inquiry, some of the "facts" and questions concerning different parts of this system may be listed here.

1. Production of CO_2 occurs in all tissues at a rate dependent upon their metabolic activity. This CO_2 diffuses out to the capillaries according to the difference in its partial pressure. Are there any signals from the tissues to the central nervous system describing changes in rate of energy metabolism? Of CO_2 production?

2. Increased venous CO_2 concentration and increased cardiac output reflect the rate of CO_2 production. Changes in pH are minimized by the blood buffers, particularly hemoglobin. Are these changes in CO_2 concentration, Pco_2, or pH detected anywhere?

3. The relative rates with which the lung is perfused with blood and ventilated with air help determine the dynamics of gas exchange. Segments of the lung that are underventilated are relatively less effective in removing CO_2 from the blood. Does the body "know" what is happening in the lungs?

4. The action of respiratory muscles, the state of the airways, and the elasticity of the lungs and chest interact to produce the mechanical motion of gas. The work of breathing is a function of the pressure-volume relationships, and probably affects the rate and depth of breathing. Is the work detected somehow? What is the role of pulmonary stretch receptors, and muscle and tendon organs of the diaphragm and intercostals?

5. Chemoreceptors of the aortic and carotid bodies are sensitive to pH, Pco_2, and Po_2. When they are removed and denervated, hypoxia causes depression rather than stimulation of breathing. Yet increased arterial Pco_2 or acidity can still stimulate breathing. What is the role of these "peripheral" chemoreceptors in normal breathing and in exercise?

6. There is evidence that increased acidity and Pco_2 in the blood perfusing the brain will stimulate breathing. Are there specific chemoreceptors there, or is this an effect on some or all of the respiratory neurons? What role does the composition of the cerebrospinal fluid play?

7. What effect do H^+ and CO_2 have on responses of the peripheral effectors—the anterior horn cells, myoneural junctions, and their reflex arcs?

8. Where does rhythmicity arise? Are there indeed "centers" that interact and are reciprocally inhibited to produce rhythmic breathing? Emotion, voluntary action, and reflexes such as swallowing can alter the pattern of ventilation. How are these pathways involved in normal respiration?

9. Most experiments which purport to measure the effect of CO_2 on breathing involve the administration of CO_2 in the inspired air. Then the changes in arterial Pco_2 or pH are observed, and some mathematical

relationship is calculated. But when animals exercise, the ventilation and CO_2 excretion rise to 5 or 10 times the resting rate, and the arterial Pco_2 and pH do not change as would be predicted from the results of CO_2-breathing experiments. Is this because there are other stimuli to breathing? What are they? What determines the relative priorities for their controlling action?

These, then, are some of the unanswered questions about parts of the CO_2-regulating system. In a large measure they are oriented toward more and more detail, and we sometimes feel that while investigating them we lose sight of the overall problem. Very often we are not sure whether results of individual experiments are consistent with our model of the system as a whole. It would be helpful if we could make models that would point out consistencies and inconsistencies in the interpretation of experimental results and even indicate whether some of the questions are irrelevant.

The material flux shown in Figure 9-1 is still clearly marked in Figure 9-2. The next step in abstraction is the orderly translation of physiological processes into control systems abstractions. This will allow us to make mathematical statements about the different parts of the system and quantitatively to examine its behavior. In these terms Figure 9-2 is a multicompartmented diffusional system with convection streams interposed between source and sink, and between sink and source, representing respectively the systemic venous drainage and the systemic arterial flow. Since material flux is measured in moles per unit time, and since the pulmonary mechanism is one of diffusion exchange with a flowing stream of gas, the material flux is computable as the arithmetic product of the gas concentration and the air flow (the latter being the facultatively controlled process). One expects, therefore, that when the abstractions are formally stated the equations which result will be simultaneous, first-order, differential difference equations, at least one of which will be non-linear because it will include the product, concentration $\times$ volume. Further, if one makes any but the most general guess about the proper mathematical function to represent the central nervous mechanism, the equations become promptly and substantially more complex.

Mathematically rigorous diagrams suitable to this terminology can be completed, provided one is willing to accept substantial approximations (Grodins, 1963; Defares, 1963). In any event, each relationship shown on the detailed physiological diagram needs to be translated into mathematical symbolism of some unambiguous sort. These equations when represented in graphical form yield diagrams which may be called signal-flow graphs, block diagrams, or machine diagrams (if the mathematical solution is to be achieved upon an analog computer), or flow charts (if the solution is to be achieved upon a digital computer). The particular form one selects is not too important at this juncture. For the sort of representation we have described, let us show, for demonstrative purposes

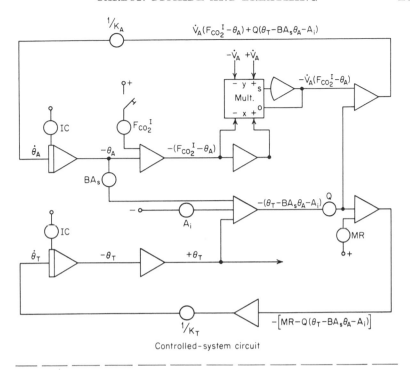

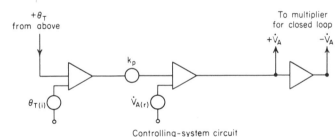

FIGURE 9-3. An analog computer circuit for the solution of dynamic respiratory chemostat equations. This diagram exemplifies the translation of the information from Figure 9-2 into an unequivocal mathematical form and thus is the third level of abstraction discussed in the text. (Grodins, F. S., *Control Theory and Biological Systems*. Columbia University Press, New York, 1963).

only, the diagram evolved by Grodins for a model of respiration (Fig. 9-3). We need not explain the symbols employed in this diagram since they are not germane to the thrust of this argument. It suffices to note that this diagram, in contrast to the diagram shown in Figure 9-2, is mathematically unambiguous and may now be solved. Moreover, it can be solved to produce numbers which imitate experimental situations. There are two ways of evaluating a model like this one. First, it should produce solutions that are the same as those observed by conventional physiological means; second, it should have some heuristic value to investigators. If the model behaves like the animal under a variety of different stresses, it is remarkably successful in the former sense. But to be worth the effort, it should also suggest new solutions or at least new questions for those who are investigating the physiological system it represents.

Models more complex than the one in Figure 9-3 can be proposed and evaluated, but one basic fact is always evident—no model permits us to inquire in detail about structures which do not appear in the scheme. To make such an inquiry we have to change the model and make it include the new structure. As an example, suppose we construct a model which controls ventilation by an equation similar to Gray's multiple factor theory (Gray, 1950):

$$VR = aP_{CO_2} + b[H^+] + cP_{O_2} + d \qquad\qquad 9.1$$

where VR is the ventilation ratio, a, b, c, and d are fixed operators, and $[H^+]$, P_{CO_2}, and P_{O_2} refer to the arterial blood. The other half of the model consists of equations describing P_{CO_2}, P_{O_2}, and $[H^+]$ as functions of VR and other constants. We can simulate the system with a computer, and let us say that the results of several experiments in simulation agree with the results of similar experiments in man: we have constructed a valid model. We can simulate further experiments with it, and the results will be consistent with the model. But such a model, however valid, does not tell us anything, for example, about the *mechanisms* through which arterial P_{O_2} acts to change VR. What happens when the carotid body chemoreceptors are denervated? We cannot answer this question from the model. We must either do the experiment in vivo, or alter the model to include a "carotid body."

Thus, to agree with a variety of experiments, the model must repeatedly be made more complex, and we begin to lose the simplicity with which we started. Yet perhaps we have gained something else. Most physiologists have within their minds a conceptual set of reflexes, anatomical connections, and other correlations—a model, if you will—that describes for them how the body or one of its systems works. When one reads of a new experimental observation, he compares it to this model to see if it is consistent with what he already knows. If it fits he accepts it, if not he may either alter his model or question the author's experimental technique. Our mathematical model, growing more and more complex, is similar to the physiologist's, but is not subject to the vagaries of memory. Would it be possible to use this model, then, as a mnemonic against which experimental results and interpretations of function could be checked? If this model could be programmed on a computer it would give us ready access to the current state of knowledge of the system, and any future experiments inconsistent with it would have to lead to a reexamination of part of the theory. If a revision were made, all previous experiments would have to be repeated in simulation to see if they would remain consistent with the revised theory. More important, the model could eventually get to be so complex that finding exactly where to revise it would be complicated. Simulating this model would be difficult, but it would be instructive to have such an aid to our memory.

MINUTE VOLUME: AN EXAMPLE OF THE UTILITY OF MODELS

Sometimes modeling leads to a critical examination of terminology. An example is the concept, ventilation, which is expressed as respiratory minute volume (RMV) or gas flow per minute (V). This is the primary response variable in most experiments and models dealing with respiratory control. It is easy to measure by collecting gas or metering it out from a container over a known interval of time. This is its operational definition; mathematically it is a time average of the unidirectional gas flow in the respiratory airway. The implication in both the experiments and the model is that the body mechanisms, whatever they are, produce an entity which this variable measures. Because control of ventilation is contingent upon the body's "knowledge" of its current state, one may well ask whether the minute volume is "knowable," since it has no reasonable definition over periods of time shorter than a breath cycle. In fact, one can argue that the single breath, not the minute volume, is the physiological unit of action. Each breath has two properties, duration and depth, which the body can conceivably measure. Whether the body can use them to compute the minute volume through some neural algorithm is a moot question. Thus, although a model like that of Figure 9-3 which uses minute volume as the mode of feedback control may have value, it prevents any analysis of rate or depth because these variables do not even occur in it.

Let us look for a variable which relates minute volume to rate and depth of breathing. Suppose we assume that the central nervous system controls breathing by generating not an "average breath," but a succession of individual breaths of variable duration and depth. We then have a waveform, the chest position, whose frequency and amplitude are modified, or modulated, by some central mechanism. The physical properties of the chest and lungs transform this waveform into another waveform, the airflow, which we can measure, for example, with a pneumotachograph. At this point we have a variable which is both measurable in the laboratory and also relatable to respiratory rate, depth, and minute volume. From the points where flow equals zero (twice per cycle) we obtain the rate, and from the integral over a complete inspiration or expiration we obtain the depth. To obtain the ventilation we rectify, excluding values above (or below) zero, integrate the resulting half-wave over a sufficiently long time and then divide by that time. The record of airflow thus contains more information than the magnitude of minute volume alone.

If control is accomplished by modulation of an oscillating mechanism, then to relate the effects of a change such as altered metabolism merely to ventilation is a relatively inefficient approach, since it neglects all features of modulation except the short term averages of amplitude. Using ventilation to study this modulation is like trying to extract the message of an amplitude-modulated radio signal by rectification and passage through a too-narrow band, low pass filter. Much of the information in the original signal is lost, even though the result may be recognizable. Similarly, some of the information regarding respiratory drive can be obtained from the minute volume. Were this not so, the concept would not have become as useful or as entrenched as it is. Radio signals carried by frequency modulation cannot be analyzed by rectification and filtering. If the central nervous system's control signals are analogous to FM, then minute volume is inadequate for analyzing them. Perhaps this explains why the phenomena of rate and depth, which fascinate respiratory neurophysiologists, are so difficult to relate to the usual variables of control theory.

CHANNELS OF INFORMATION: ANOTHER EXAMPLE

If one accepts the premise that respiratory mechanisms in the brain stem have an intrinsic rhythmicity, ventilation becomes a modulated oscillation, and one can then proceed into the further analysis of the control of ventilation using concepts of signal analysis. Let us consider one interesting aspect of the assumption that CO_2 balance is regulated by control of pulmonary ventilation through informative feedback. Because the amount of pulmonary ventilation, however it is measured, is produced through the output of the nervous system via the respiratory motor neurons, then we must in some very concrete sense find a connection between the existing state of CO_2 balance and the central nervous system. This connection must be some message-carrying channel or channels that can inform the nervous system about the current state of the success or failure of its performance.

Even a casual scrutiny of the anatomical and physiological evidence now available indicates that there are many such paths; certainly they include at least the pulmonary stretch afferents passing over the vagus, the musculotendinous proprioceptors of the thorax, abdomen, and diaphragm, sensory receptors in the upper airway, peripheral chemoreceptors lying adjacent to the blood stream, and the arterial blood itself arriving in the central nervous system. Experimental evidence of their participation exists for all the foregoing and for some others. Any or all of these paths might be channels for messages in languages we do not understand, written with symbols whose identity is uncertain. It is in fact this *plénitude de richesse* that complicates the serious study of the relationship of ventilation to CO_2 balance. Since we cannot, like Mark Twain's famous horseman, mount our steed and charge madly in all directions, we must elect to examine one of these at a time. Even this turns out to require of us an almost perverse persistence.

Let us look at the arterial blood as a message form and see, in the first place, if it can serve as an informative channel; secondly, if it does, if mechanisms exist that might conceivably encode and decode messages in the language of the channel so as to lead to appropriate responses at the next stage. It should be emphasized at the outset and all through this discourse that we do not regard the arterial Pco_2 as the sole and unique message which plays upon the decision-making processes, although it would be remiss not to admit that at present this seems to be a message of primary importance. If the arterial blood contains a message, what is the message about? How might the message be constituted?

Since we know neither the code (or language) in which the message is written nor the priority with which the nervous system may consider it, we have a very challenging problem in cryptography. Let us approach it by stages, through a succession of hypotheses and assumptions. In the first place let us compare ventilation (as measured by the minute volume, whose limitations we have already discussed) with the composition of blood drawn from an artery (in particular its CO_2 tension). We are then comparing values averaged over a short time interval. As different conditions are prescribed, we find different relationships between our variables. For example, during the steady states achieved by breathing gases of different CO_2 concentrations, there is a linear relationship between arterial blood Pco_2 and the respiratory minute volume. Yet if we measure arterial Pco_2 during exercise of varying severity, we find that ventilation and arterial CO_2 tension are virtually independent. Finally, if we combine the two situations or impose some other condition like a simultaneous hypoxia upon the organism, a still different form of dependence can be discovered. The problem that the hyperpnea of exercise poses in our present context is that if the CO_2 tension is a form of language it seems to have the characteristics that Humpty Dumpty would give it—an unpredictable and capricious denotation.

A physiologist's problem is like that of a child learning a language by associating the actions of the parent with the spoken sounds. At first he cannot tell the sounds from the words from the sentences. There are many sentences which have the same sound intensity and seem uncorrelated with any overt acts of his parent. His earliest understanding (though we really do not know this) may be his awareness of the level of sound, and his association of shouting with punishment and of cooing with affection. It is only much later that a child can distinguish between the ideas expressed by "you're agreeable" and "you're not disagreeable." By then he has penetrated many layers of exploration in the language. If we can risk for a moment the perils of a too close analogy, we would like to submit that the contemporary scheme of comparing minute volume in a steady state of ventilation with an averaged sample of blood taken over a non-commensurable interval is in much the same category as the earliest level of language exploration.

Between the most primitive level and a rational appreciation of the communication there are many intermediate levels of understanding. Advancing to these will require us to abandon the use of the minute volume and the slowly sampled arterial blood. What are some of the levels? To identify them we must first elect some medium in which the conversation might occur, just as a child becomes aware that fluctuations in sound pressure, rather than transfer of bodily heat from his mother, is the relevant communication medium. Since we are discussing CO_2 balance, which is largely reflected in changes in CO_2 concentration, let us choose different values of blood P_{CO_2} as the symbols of the language. The message contained in a sequence of changing symbols is then transmitted through the arterial system. (This analogy does not deny that there may be other, even more important channels or languages, but merely focuses upon a particular object for study.) We thus conjecture that the symbols of this language may assume any value of P_{CO_2} between zero and the highest partial pressure compatible with life, which, in certain instances of severe disease, may be as high as 100 mm of mercury.

Although we do not know for certain how this range is to be divided, plausible conjecture and some of the formalisms of information theory may be combined to yield a numerical estimate. The estimate can be made only in terms of the worst case, and requires an assumption about the size of a P_{CO_2} change that is representative of a new symbol. By assuming that the chemosensitive mechanisms can usually discriminate changes in arterial P_{CO_2} of the order of 0.1 mm of mercury, then we may estimate by techniques too involved for review here, that a P_{CO_2} symbol transmits about four bits of information (Yamamoto, 1962). This is about equivalent to four English letters in a consecutive English text. Suffice it to say that the number of symbols is probably finite and that they probably do not differ from each other by increments of equal magnitude.

Being unable to estimate the number of different symbols exactly, we may turn to another question: Is it possible to make a reasonable estimate of the number of symbols emitted in any unit of time? Here elementary physiological knowledge is of considerable value. If a flowing stream of fluid is carrying a message, and the message is not packaged in a bottle, one needs to consider the possibility that the symbols can become mixed through dilution, chemical reaction, and turbulence. If one tried to send a message down a river by throwing wooden letters into it so that they arrived in the proper sequence at some downstream point, the presence of waterfalls and whirlpools would create an alphabet soup which could only indicate that someone was attempting a communication.

Likewise, when portions of blood dissimilar with respect to CO_2 are mixed, the chemical processes of the blood tend to make it quite homogenous, with a uniform CO_2 tension. If messages of importance to the brain are to be transmitted as a series of ups and downs in the arterial CO_2 tension, it is important to know at what interval successive volumes of arriving blood are independent of each other. This knowledge would permit us to inquire whether there are segments in the sequences of symbols like sentences, phrases, paragraphs, and so forth. Such inquiry may lead to difficult, abstract problems for which there are no general solutions; but as physiologists we can proceed a considerable way in this particular case.

The symbols, a succession of P_{CO_2} values, preserve information about the CO_2 balance in the form of a changing chemical potential which can be easily translated into chemical reactions. Arterial CO_2 tension is the resultant of an operation which can be reasonably well defined mathematically as consisting of the pulmonary blood flow, the buffering functions of the blood, the pulmonary air flow, and the delivery of CO_2 from the metabolizing tissues. Even if we allow these to vary independently and randomly, there is still only a limited range over which the P_{CO_2} can vary. Since we do not wish to be bound by arbitrary choices in discriminating between one symbol and another, or in deciding how brief an instant a symbol can occupy, we can begin by merely asserting that these P_{CO_2} values are bounded in size. Nevertheless, the collection of infinitesimally-separated-in-time values of arterial CO_2, even if bounded in magnitude, is infinite in number and infinite in frequency of occurrence.

For these reasons the characteristics of the arterial path from lung to brain which produce mixing are of interest. One is the beating heart. During each cardiac cycle, blood to be expelled is sequestered momentarily, first in the left atrium and pulmonary veins, and then in the left ventricle. Whether intracavitary mixing is good or poor, the net tendency is to average out any disparity in composition between the first and the last blood flowing into the cavity. There is, therefore, a tendency to obliterate fluctuations that occur close together. In fact, if mixing in the ventricle were complete and the process of sequestration very sharp, and if blood issued from each ventricular systole as a slug with sharp front and back boundaries, one would have to conclude that no matter what the lung does to blood composition, no more than one symbol per heart beat can be transmitted. For a resting man, with heart beating at 70 per minute, this is about the same as the symbol emission rate of the human hand writing letters of the alphabet.

Actually, it is possible to make a somewhat more realistic estimate than the one symbol per beat, if one recognizes that the two chambers of the heart are cascaded (i.e., follow each other in sequence) and that each chamber has a residual volume; both of these effects extend the mixing influence of the heart over many heart cycles. Consequently, the blood produced in the lung during a particular short interval is mixed in decreasing proportions with many subsequent portions of blood. Thus it becomes impossible to emit even as many as one independent symbol per beat. In fact, the mixing of successive portions of blood in the residual volume of the chambers causes a loss of about 50 per cent of the information which each heart beat might pass on as a particular value of P_{CO_2}. Thus the heart alone loses about half the information which the lung might generate.

If other mixing processes like the parabolic flow of blood in the arteries and diffusional lags at the blood-chemoreceptor interface are considered, the information loss is even greater, and correspondingly the amount of information which the ventilatory mechanisms receive concerning the CO_2 balance is smaller and perhaps of a more reasonable magnitude. The amount of information which remains may be illustrated as follows: As far as CO_2 balance is concerned, the regulation of respiration can be conducted by telling the brain either "increase" or "decrease" once every two or three seconds.

SPECTRAL ANALYSIS

While it is nicely colloquial and anthropomorphically satisfying to speak of the arterial P_{CO_2} in terms of symbols and information measure, the potent concepts for an experimental approach are the same ideas expressed in the terminology of continuous wave radio transmission. The P_{CO_2} in time is more like the fluctuating voltages of radio transmission than like the written word. Ventilation may be regarded as a process which modulates the CO_2 tension in the blood. In turn the CO_2 tension, through central nervous mechanisms, modulates the oscillatory motion of the chest.

One way in which continuous waves can be examined is in terms of their frequency components. Physiologists acknowledge that there are such waves or cycles when they perform or report experiments which are characterized as being in the "steady state" or in a "transient" changing state. The procedures by which a complex wave can be dissected and reconstituted in terms of its frequency components lead naturally to the representation of messages by spectra, which are plots of amplitude or power vs frequency. Thus, pulmonary air flow, which is an oscillation, has a spectrum, as do the mixed venous P_{CO_2} and the arterial P_{CO_2}. It is the variations in the spectra that describe both the messages that occur and the transformations that physiological mechanisms produce during message handling.

For example, the spectrum of air flow has a characteristic frequency called the respiratory rate; but since both rate and depth vary with the state of the animal, a full description contains many frequencies extending down to very slow frequencies. Therefore, the problem of measuring the information handling in the arterial Pco_2-ventilation system can be rephrased in terms of spectra. Here again certain elementary physiological facts are of primary usefulness. The inertia of the chest, the airway resistance, and the speed of movement of the chest musculature eliminate spectral components above a few tens of cycles per second. The beating heart, previously described as a sort of information confounding gate, becomes now a filtering process acting upon the continuous message wave produced by pulmonary ventilation. Since the heart can produce only one estimate of Pco_2 amplitude per beat, the maximum spectral frequency which passes the heart is about one half the existing heart rate.* In general the heart beats about three or four times per breath, so that at best, the arterial message can contain frequencies only up to the respiratory rate. Any relevant control message must lie in parts of the signal spectrum between zero cycle per second and something like one cycle per three seconds in a man at rest. In severe exercise the spectrum may extend to as high as one cycle per second. One can examine the spectra of ventilation and arterial Pco_2 for changes resulting from exercise, CO_2 breathing, and other maneuvers. These changes can be quantified so that differences in the amount of information transferred and the processing of this information should be detectable. But before going on to outline some work in this line of thinking, it should be noted that the two views of information transmission, as exemplified (a) in the discussion of symbols and probabilistic measure of information and (b) in the discussion of continuous waves and spectra, merge in the subject of generalized harmonic analysis.

One of the important difficulties in using control and communication theory in physiology is the need for a long sequence of hypotheses before an interpretable experiment is performed. In most cases, as in this one, this sequence cannot be examined piecemeal. An analogy is found in bridge making: Before one begins to build a bridge he must ask whether there is suitable ground at each end for the foundation. The piers of our bridge are the propositions that pulmonary ventilatory control of CO_2 balance occurs through informative feedback, and that the orderly study of the information-handling parts is expressible as a sequence of spectra at physiologically meaningful sites.

---

* Calculated by Nyquist's theorem, which states that sampling with a frequency of 2f will yield data adequately describing oscillations of frequency up to f.

One experimental approach to the laboratory study of the message-decoding problem has the following basis. It has been observed that when a pair of electrodes is placed so that one is in the circulating blood and the other is in some point within the mass of tissue of the central nervous system, they show a potential difference related to the difference between the average pH or Pco_2 of the blood and the brain (Tschirgi and Taylor, 1958). Although it is not clear why this potential exists or what are the exact tissue layers intervening in the electrical pathway, the potential is experimentally reproducible and consistent. For slow changes, i.e., those lasting over several minutes, there is a predictable relationship between the logarithm of CO_2 tension and the blood-brain potential difference. This type of potential is seen only between the blood and the brain. We conjecture that the electromotive force may reflect some process which occurs in the natural sequence of message transfer from blood to brain. If so, the "blood-brain barrier" can be regarded as a transducer, with the potential as its output.

Thus, the "conversation" which ventilation carries on with the brain can be "intercepted" as a potential at this point in the anatomical sequence. The pulmonary ventilation and the blood-brain potential become the time series relevant for the study of messages, a study undertaken by means of statistical communication theory. This statistical approach is required since, in addition to ventilation, the time series representing the present state of the brain, the external environment, metabolism, and blood flow are party to the conversation. First of all it is desired to identify the part of the blood-brain potential coherent with the ventilatory gas flow; we can regard this as a problem in signal-noise separation, although a careful scrutiny will indicate that this assumption is almost as large a conjectural leap as any already taken.

Figure 9-4 shows a record of air flow in the trachea of a rat and

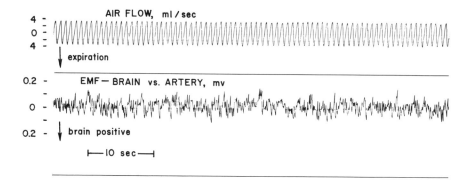

FIGURE 9-4. Simultaneous recording of tracheal air flow and brain-artery potential difference in a lightly anesthetized rat. (Raub, W. F., doctoral dissertation, University of Pennsylvania, 1965.)

the concurrently recorded blood-brain potential difference. The animal is lightly anesthetized, breathing 100 per cent oxygen. The blood-brain potential difference has been filtered (with a low pass filter with a gain that is reduced to one half the amplitude at 15 cycles per second) so that only frequencies in the vicinity of 10 cps and below are faithfully recorded (Raub, 1965). (Thus the record is considerably different in frequency content from the usual EEG record.) When the skull is not damaged badly there is very little cerebral displacement during respiration, and movement of the tissue relative to the electrode can be eliminated. The importance of this physical stability can be seen by observing the potential during the occasional gasp which the rat gives spontaneously. If a movement occurs, the cerebral potential deviates promptly, but the chemical consequences of the gasp, if observable at all, do not occur until after the lapse of a circulatory delay of several seconds. Usually the pattern of respiratory movement is regular and shows some changes in respiratory depth from cycle to cycle. One can regard these small changes in ventilatory depth as uncontrolled random variations in the mechanism of the chest, or as responses of a corrective nature upon the current state of CO_2 balance, or as a mixture of both. Whatever their origin, to a first approximation the airway record can be regarded as an amplitude modulated wave. Both ventilation and potential difference can be observed in a variety of situations in which a ventilatory response occurs and the animal's CO_2 balance is affected.

To see if there is a relationship between the two records one calculates approximately the auto- and cross-correlation functions of these records, and from them the power spectra. These techniques are somewhat formidable mathematically, and it is not appropriate to try to represent their derivation here. It is enough to state that the autocorrelation function measures the relationship of a time-varying process to itself at any specified epoch in the past. The numerical magnitude of the function has a connotation like that of the usual correlation coefficient computed for the relationship of two variables. A positive value indicates that the two selected values of the time series are related in the sense that an increase in one is accompanied by an increase in the other; a negative value indicates a converse relationship. The cross-correlation function, on the other hand, in a similar fashion relates two *different* time series, measuring relationships between the present value of one of the functions and a past value of the other. If one considers all time intervals of separation, the value of the correlation function is a function of the time interval. This means that such a function can be analyzed into frequency components; one such analysis is a Fourier series representation of the correlation function, and it is related to the spectrum of the original processes.

These functions, then, are the tools by which one seeks to discover the regular portions of an irregular record, or the relation between

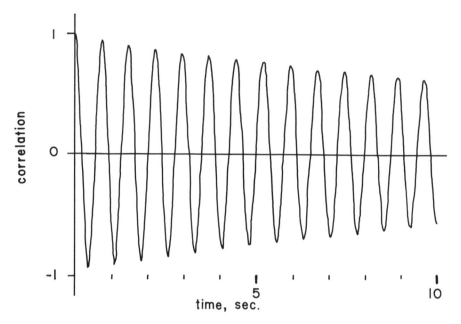

FIGURE 9-5. Autocorrelation of the tracheal air flow in Figure 9-4; a tracing of a graph generated by a computer from the data of Figure 9-4. (Raub, W. F., doctoral dissertation, University of Pennsylvania, 1965.)

regular portions of two irregular but presumably linearly related records. The autocorrelation function of the tracheal air flow is shown in Figure 9-5. With a rat at rest breathing oxygen, one observes that the function is strongly oscillatory and that the frequency of its oscillation is the frequency of respiratory movement. The envelope of the oscillation converges slowly over many breath cycles, indicating probably a very long-range relationship; that is to say, that in most instances any given breath shares common influences with predecessors as far as 12 or 13 breath cycles previous. With the data currently available, however, the full extent of this range cannot be asserted; the present data are based upon computations using over 3000 data points spaced 0.5 second apart, and yet they permit the examination of events having a maximum separation of only 10 seconds, if the reliability of the analysis is not to exceed the 5 per cent confidence band. The long term smoothing of the breathing pattern noted here is in accord with the obvious physiological fact that when breathing is interrupted in various ways (such as in voluntary apnea, yawning, sneezing, or coughing), there are no immediately observable changes in the subsequent breath pattern. This analysis shows that whereas the top frequency that might possibly get through the heart is of the order of the respiratory frequency, important message frequencies are probably much lower than this.

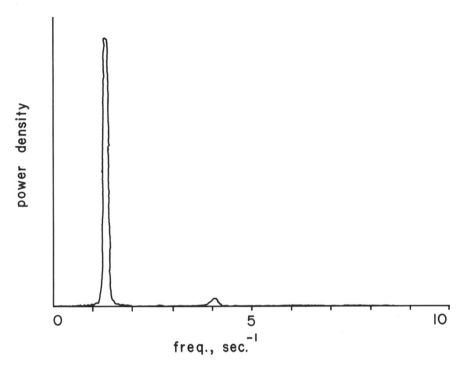

FIGURE 9-6. Power density spectrum of the tracheal air flow shown in Figure 6-4; a tracing of a computer plot. (Raub, W. F., doctoral dissertation, University of Pennsylvania, 1965.)

In the spectrum of the air flow shown in Figure 9-6, the signal power is concentrated in a narrow band about the respiratory rate, 1.3 cycles per second, with some images of this at the second harmonic. Since in the rat we accepted the respiratory movement as a "carrier" oscillation of about 1.5 cycles per second, with this carrier modulated by control information ranging perhaps 0.3 to 0.001 cycle per second, the spectrum illustrated is, as expected, relatively sharply peaked with the side bands clustered close to the carrier.

Since the heart limits the higher frequencies as previously discussed, we would most certainly expect the sort of spectrum shown in Figure 9-6 to be highly distorted by passing through the heart. This would mean that the amplitude modulation hypothesis could not be correct, were it not for one additional fact: The amplitude modulated signal is at least partially demodulated (detected) before the heart is reached. This is accomplished by diffusion exchange between the blood and the gases in the lungs, and the relationship is an inverse one. If $F(t)$ is the air flow and $B(t)$ is the P_{CO_2} of the pulmonary venous blood, the relation is of the general type $B(t) = \dfrac{a}{b + F(t)}$ in which a and b are arbitrary variables. However, b is always larger than the absolute value of $F(t)$ and is related to the cardiac output. Diffusion in the lung has the effect of mathematically transforming air flow into a second function.

If one performs abstractly the manipulation of taking an amplitude modulated wave $F(t)$ and substituting for it the corresponding wave $B(t)$, one sees a redistribution of signal power such that it is spread over the complete spectrum. The important feature of this transformation is that the modulating frequencies, whether relevant signals or noise, are separated from the carrier and appear at frequency positions near zero. The information which constitutes modification of respiratory depth is thus extracted from the carrier, and demodulation has been effected. The transformation also preserves at diminished magnitude the basic respiratory frequency and its modulating side bands in the usual manner. Other images of the spectrum are duplicated at higher frequencies along with the added appearance of cross product frequencies, even though the filtering which the recording instrumentation imposes quite excludes these higher frequency components.

With this background one can examine the autocorrelation function of the blood-brain potential difference which corresponds to the simultaneously recorded ventilation record. The autocorrelation function in Figure 9-7 shows some long range correlation, but less than in the auto-

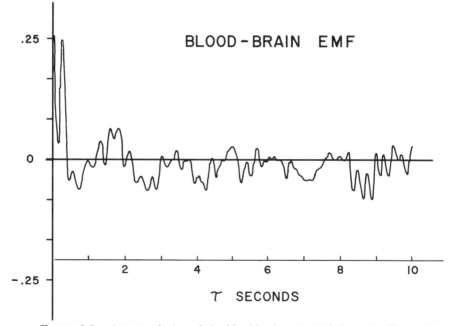

FIGURE 9-7. Autocorrelation of the blood-brain potential shown in Figure 9-4; a tracing of a computer plot. (Raub, W. F., doctoral dissertation, University of Pennsylvania, 1965.)

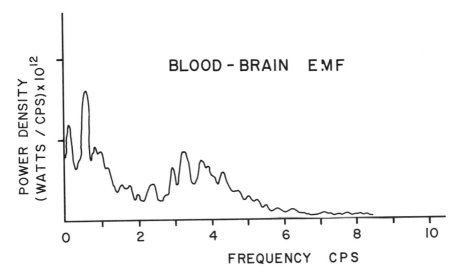

FIGURE 9-8. Power density spectrum of the blood-brain potential shown in Figure 6-4; tracing of a computer plot. (Raub, W. F., doctoral dissertation, University of Pennsylvania, 1965.)

correlation function of ventilation. For the potential there are no obvious periodic components. One sees a rapidly decaying peak at zero representing the random components of the wave, and then some oscillations about the line of zero correlation which reflect the relationship of the blood-brain potential to itself. The spectrum of this blood-brain potential (Fig. 9-8) shows a small amount of power about the respiratory frequency and a broad smeared-out spectrum throughout lower frequency ranges. Since these spectra were computed after the least squares line was removed from the data (thus correcting for mean and trend), the lowest portion of the scale where one would expect to find the modulating signals is deficient by the power belonging to the very lowest frequencies (below 0.1 cycle per second).

The evidence suggests that at an intracerebral site an electrical potential can be recorded which has a spectrum compatible with the notion that the change of respiratory air flow can be transmitted to the brain as changes in the chemical composition of the blood. To clarify this conclusion, one must note that the blood-brain potential, to a first approximation, is a logarithmic transformation of the difference between blood P_{CO_2} and brain P_{CO_2}. Thus, the idea that the potential difference is a transducer output introduces a second serious alteration of the signal configuration, in that the observed spectrum contains not only the effects of logarithmic transformation but also the regular components of a second set of processes, those occurring within the brain.

The number of intermediate steps between air flow as one message form and blood-brain potential difference as another is very great. The attendent number of mental reservations that one needs to carry along with each step is greater still. However, the thrust of this approach is exciting and probably valid. A control systems orientation toward the study of the balance of CO_2 exchanges in the organism leads to the formulation of highly effective models which simulate many of the salient features of the control of CO_2 balance and the patterns of pulmonary ventilation seen in certain natural states. These models indicate several points where physiologists need to re-examine traditional basic definitions which have been both operationally and conceptually useful to them in the past. We have attempted to exemplify this need in the small sector of the system which is concerned with the problem of how the brain is informed about ventilation and CO_2 balance via the blood. Developments in the theory of communication provide both new viewpoints and experimental designs which tremendously increase the possibilities of extracting information from physiological observations. The prospect is that the understanding of the processes by which our breathing is matched to our CO_2 balance will be improved, even if many of these present attempts turn out to be erroneously conceived.

SUMMARY

Beginning with the concept of CO_2 balance, we have shown how models of CO_2 control are developed. All models require certain simplifications which limit their generality while making them simple enough to work with, but even simple models can make us question some of our assumptions and our usual way of looking at things. The minute volume, one of the most common measures of respiratory drive, has serious limitations when treated as a continuous variable. For many purposes the tracheal air flow is more useful, because it is truly continuous and contains information about the rate and depth of individual breaths, which are probably the units of action of the central nervous system–pulmonary complex. The idea that the arterial P_{CO_2} may be a message from the lung to the brain has been examined, and some speculations made about ways of decoding this message. The control systems approach to respiration and CO_2 balance raises many interesting new questions.

This chapter was prepared during Dr. Edwards' tenure as Samuel Lloyd Irving Scholar of the Pennsylvania Plan to Develop Scientists in Medical Research. The research was supported in part by Grant G-18483 from the National Science Foundation.

Chapter 10

REGULATION OF
ACID-BASE EQUILIBRIUM

ROBERT W. WINTERS and RALPH B. DELL
College of Physicians and Surgeons, Columbia University

Mammalian organisms appear to regulate the pH of their extra-cellular fluid with great precision. Thus in health the mean normal blood pH is about 7.40, variations around this value being approximately ± 0.05 pH unit. In disease, the blood pH may vary more widely than this, the usually quoted values for the extremes compatible with life being about 7.00 and 7.80. This chapter is devoted to a detailed discussion of the mechanisms by which normal values for blood pH are maintained in health and abnormal values are countered in disease.

In considering this complex topic, it is useful to recognize two basically different types of mechanisms. The first type, which we shall refer to as physicochemical, is concerned with the various buffer systems in body fluids. Essential to an understanding of the operation of these buffers is an understanding of certain basic physicochemical principles. Accordingly these principles are presented first, followed by an examination of the blood buffers and an application of these principles to this complex fluid. The second set of mechanisms, referred to as physiological mechanisms, is concerned principally with the functions of the lungs and the kidneys and their effects upon the blood buffers. In considering this set of mechanisms, we have attempted to present a reasonable summary of what is known concerning the mechanisms by which the lungs and kidneys are able to alter the acid-base relationship in blood, as well as a brief consideration of the factors which are believed to control these actions.

181

PHYSICAL CHEMISTRY OF BLOOD

Buffers and Buffer Systems

DEFINITION OF BUFFERS

Buffers are substances which impart to a solution the ability to resist a change in pH when acid or base is added.* A buffer consists of an acid-base pair, such as acetic acid and acetate or ammonium and ammonia. Such an acid-base pair is often termed a conjugate pair since members of the pair possess a common ion. For example, acetate is the conjugate base of acetic acid with the common radical of the pair being the acetate ion, while ammonia is the conjugate base of ammonium with the common radical being ammonia. The components of acid-base pairs may be either a weak acid with a strong base or a strong acid with a weak base. Most buffers of biological interest are composed of a weak acid and its conjugate base.

When a strong acid is added to a buffered solution it reacts with the base to form the weak acid. Since the dissociation of the weak acid is slight, most of the added H^+ is bound and thus not free to cause a change in pH. This may be illustrated with the reaction for acetate-acetic acid buffer:

$$Na^+ Ac^- + H^+ Cl^- \rightarrow HAc + Na^+ Cl^-$$
$$\Updownarrow \qquad\qquad\qquad 10.1$$
$$H^+ + Ac^-$$

Thus, as the result of the buffer reaction, the free hydrogen ion concentration will be low compared to the concentration of free hydrogen ions that would exist if no buffer were present. The situation is similar upon the addition of a strong base:

$$H^+ Ac^- + Na^+ OH^- \longrightarrow Na^+ Ac^- + H_2O$$
$$\Updownarrow \qquad\qquad\qquad 10.2$$
$$H^+ + OH^-$$

Again the effect is to reduce the concentration of free hydroxyl ions, thus "buffering" the rise in pH which would have occurred from dissociation of the strong base in the absence of the weak acid.

---

* Throughout this chapter, the Brønsted-Lowry terminology for acids and bases is employed. Thus an acid is any substance which can act as a proton donor, and a base is any substance which can act as a proton acceptor. The symbol H^+ is used to represent the proton, instead of the more cumbersome but technically more accurate symbol H_3O^+. The reader should note, however, that only one of every 10^{190} protons exists as H^+, the remaining being hydrated with one or more molecules of water.

DERIVATION OF THE HENDERSON-HASSELBALCH EQUATION

The dissociation reactions for members of the buffer pair are as follows:

$$\text{Weak acid (Ha): Ha} \rightleftharpoons H^+ + a^- \qquad \qquad 10.3$$

$$\text{Conjugate base (Ba): Ba} \rightarrow B^+ + a^- \qquad \qquad 10.4$$

It is important to recognize that, in a mixture of these two, nearly all of the a^- ions present come from the conjugate base and very few from the dissociation of the weak acid. Applying the Law of Mass Action to the dissociation of the weak acid, one obtains:

$$K'_A = \frac{[H^+]\,[a^-]}{[Ha]} \qquad \qquad 10.5$$

in which [   ] indicates concentration and K'_A is the equilibrium constant of the acid. Rearranging the equation to solve for $[H^+]$ gives Henderson's equation:

$$[H^+] = \frac{K'_A\,[Ha]}{[a^-]} \qquad \qquad 10.6$$

Thus the $[H^+]$ in the buffered solution will be determined by the equilibrium constant, K'_A, and the ratio of the concentration of the weak acid to the concentration of the conjugate base (since nearly all of the a^- comes from the dissociation of the conjugate base).

Since pH can be defined as the negative logarithm of the hydrogen ion concentration:

$$pH \equiv - \log [H^+] \qquad \qquad 10.7$$

Hasselbalch rewrote Henderson's equation in a more convenient logarithmic form:

$$pH = - \log K'_A - \log \frac{[Ha]}{[a^-]} \qquad \qquad 10.8$$

By analogy with the expression for pH, Hasselbalch introduced the term pK'_A in which

$$pK'_A \equiv - \log K'_A \qquad \qquad 10.9$$

Substituting this into Equation 10.8, one obtains the Henderson-Hasselbalch equation:

$$pH = pK'_A + \log \frac{[a^-]}{[Ha]} = pK'_A + \log \frac{[\text{conjugate base}]}{[\text{weak acid}]} \quad\quad 10.10$$

ACTIVITY AND ACTIVITY COEFFICIENTS

Ions in solution cannot be regarded as being completely free in respect to their chemical activity. Rather they are hindered to some extent by the electrostatic effects of other ions in the solution. Thus the *effective* concentration of any given ion is less than its true concentration and decreases as the density of the charge contributed by other ions increases. The effective concentration or activity of the ion (a) is thus dependent upon the number and charge of *all* ions present in the solution, and this charge density is called the *ionic strength* (μ). The relationship between the activity of an ion and its concentration is given by:

$$a_{ion} = \gamma[\text{ion}] \quad\quad 10.11$$

where γ is the activity coefficient of the particular ion and is dependent upon the ionic strength of the solution.

DERIVATION OF THE HENDERSON-HASSELBALCH EQUATION USING ACTIVITIES

The Henderson-Hasselbalch equation may be written using the activities of the substances involved and defining pH as the negative logarithm of the activity (rather than the concentration) of H^+:

$$pH = pK_A + \log \frac{a_B}{a_A} \quad\quad 10.12$$

in which a_B is the activity of the conjugate base and a_A the activity of the weak acid. When activities are used, the equilibrium constant, K_A, becomes a function of temperature only, and hence pK_A is constant at any given temperature. Ordinarily, however, the Henderson-Hasselbalch equation is written using concentrations instead of activities (see Eq. 10.10). This means that the pK ordinarily used becomes a function not only of temperature but also of ionic strength. This may be understood by deriving the Henderson-Hasselbalch equation as it is ordinarily written from the form using activities:

$$a_B = \gamma_B [B] \text{ and } a_A = \gamma_A [A] \qu\quad 10.13$$

$$pH = pK_A + \log \frac{\gamma_B [B]}{\gamma_A [A]} = pK_A + \log \frac{[B]}{[A]} + \log \frac{\gamma_B}{\gamma_A} \quad\quad 10.14$$

This form of the equation is equivalent to the form usually given:

$$pH = pK_A + \log \frac{[B]}{[A]} + \log \frac{\gamma_B}{\gamma_A} = pK'_A + \log \frac{[B]}{[A]} \qquad 10.15$$

Because of this relationship, pK'_A can be defined in terms of pK_A by eliminating the $\log \dfrac{[B]}{[A]}$ from the last two parts of the equation:

$$pK'_A = pK_A + \log \frac{\gamma_B}{\gamma_A} \qquad 10.16$$

Thus the superscript prime on pK'_A denotes that the pK as it is ordinarily written is dependent upon the logarithm of the ratio of the activity coefficients of the base and acid form of the buffer. Since the activity coefficients are in turn dependent on the charge density or ionic strength, pK'_A is also dependent on ionic strength. The pK'_A, also called the

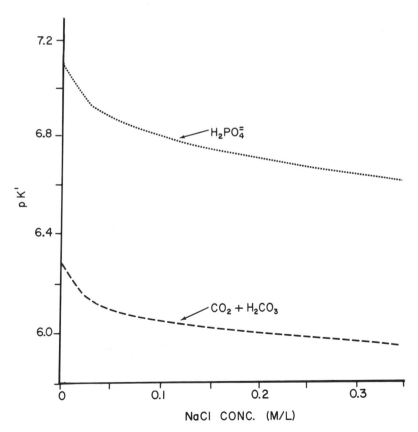

FIGURE 10-1. Effect of the variation of ionic strength by varying the sodium chloride concentration upon pK'_2 of $H_2PO_4^-$ and pK'_1 of $CO_2 + H_2CO_3$.

apparent pK_A, is thus dependent on both temperature and ionic strength (Fig. 10-1). The ionic strength of blood plasma is about 167 mEq/L, and if one is dealing with buffers in blood, the pK'_A valid for this ionic strength must be used.

DEFINITION OF BUFFER CAPACITY

If one inspects the titration curve of a buffer, it can be seen that it takes more base to cause a given change in pH at the center of the curve than at either end (Fig. 10-2). At the center of the curve one half of

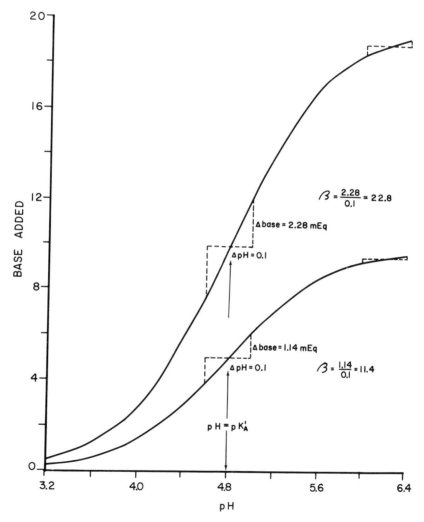

FIGURE 10-2.   Titration curve for acetic acid–acetate buffer system. The upper curve represents twice the concentration of buffer as the lower curve. Note that the slope of the curve is steepest at the point where pK' = pH, but varies with varying concentration of buffer.

the weak acid has been converted to the base, and the ratio of concentration of acid to base is 1. At this point, the Henderson-Hasselbalch equation becomes:

$$pH = pK'_A + \log \frac{[B]}{[A]} = pK'_A + \log 1 \qquad 10.17$$

Since the log of 1 is 0, the pH $= pK'_A$ and the pH at the center of the titration curve is equal to the pK'_A. Thus when the pH $= pK'_A$ the buffer is most efficient in preventing pH changes, since it takes more base to produce a given change in pH at this point than at any other point on the titration curve. The ratio of the amount of base added to the pH change caused by the addition of the base is called the buffer capacity* (β) and is defined as:

$$\beta \equiv \frac{d\ base}{d\ pH} \cong \frac{\triangle\ base}{\triangle\ pH} \qquad 10.18$$

The buffer capacity is a measure of the ability of the buffer to prevent changes in pH when acid or base is added. It is related to the slope of the titration curve in that the steeper the slope the greater the buffer capacity. The slope of the titration curve is steepest in its center portion and hence has the maximum buffer capacity at this point, while at the ends of the curve the slope is nearly 0 and there is very little buffering ability.

The effect of the changing buffer concentration upon buffering action is also shown in Figure 10-2. As the concentration of the buffer increases, the buffer capacity increases again because the slope increases. Hence the buffer capacity is related to how far the actual pH of the system is from the pK'_A of the buffer pair and to the concentration of the buffer in the system.

Buffers in Body Fluids

THE BICARBONATE–CARBONIC ACID BUFFER SYSTEM

The bicarbonate–carbonic acid buffer pair is composed of a weak acid, carbonic acid, and its conjugate base, bicarbonate. This system is one of the most important buffers of the body because the weak acid member of the pair, carbonic acid, can undergo certain unique reactions. These will now be examined in some detail.

Ed. Note: For a given molar strength, this ratio is sometimes referred to as buffer efficiency, and the term buffer capacity is reserved for the change in this ratio with the molar concentration of the buffer.

Carbonic acid can rapidly dissociate into bicarbonate and hydrogen ions, the reaction being:

$$H_2CO_3 \rightleftharpoons HCO_3^- + H^+ \qquad\qquad 10.19$$

The Henderson-Hasselbalch equation for this buffer pair takes the form:

$$pH = 3.80 + \log \frac{[HCO_3^-]}{[H_2CO_3]} \qquad\qquad 10.20$$

The $pK'_{H_2CO_3}$ of 3.80 at 38°C in H_2O makes H_2CO_3 a stronger acid than acetic acid (pK'_{HAc} of 4.80), and would at first glance seem to make it useless as a buffer in the physiological range of pH. However, H_2CO_3 is also in equilibrium with the CO_2 dissolved in the solution by the following hydration reaction:

$$CO_{2\,(d)} + H_2O \rightleftharpoons H_2CO_3 \qquad\qquad 10.21$$

where $CO_{2\,(d)}$ denotes the dissolved CO_2. Unlike many hydration reactions, this one proceeds relatively slowly, perhaps because CO_2, a collinear molecule, must be rearranged to form carbonic acid, a coplanar molecule having a new C-O bond and different bond angles and lengths. At 38°C, Reaction 10.21, uncatalyzed, takes about 40 seconds to go to 99 per cent completion. In blood, however, a specific catalyst for the reaction occurs in the form of carbonic anhydrase, a zinc-containing enzyme present in abundance in the erythrocyte. In the presence of this enzyme, the rate of Reaction 10.21 is increased several thousandfold.

Applying the Law of Mass Action, the equilibrium constant for the hydration reaction, K_h, is as follows:

$$K_h = \frac{H_2CO_3}{CO_{2\,(d)}} = 3.2 \times 10^{-3} \text{ (at 38° C in water)} \qquad 10.22$$

The small value for the equilibrium constant shows that the equilibrium lies far over to the side of $CO_{2\,(d)}$: Solving the foregoing equation for $CO_{2\,(d)}$ gives:

$$CO_{2(d)} = 317\ [H_2CO_3] \qquad\qquad 10.23$$

The pK for the hydration reaction is:

$$pK_h = -\log K_h = 2.50 \qquad\qquad 10.24$$

The interrelationships between the dissociation of carbonic acid (Reaction

10.19) and the hydration reaction (Reaction 10.21) are shown in the following overall reaction sequence:

$$CO_{2(d)} + H_2O \rightleftharpoons H_2CO_3 \rightleftharpoons H^+ + HCO_3^- \qquad 10.25$$

The overall reaction constant (K_1) may be derived:

$$K_1 = \frac{[H^+]\,[HCO_3^-]}{[CO_{2(d)}] + [H_2CO_3]} \qquad 10.26$$

or in the form of the Henderson-Hasselbalch equation.

$$pH = pK_1 + \log \frac{[HCO_3^-]}{[CO_{2(d)}] + [H_2CO_3]} \qquad 10.27$$

Since $pK_{H_2CO_3}$ has a value of 3.80,

$$pK_1 = 3.80 + 2.50 = 6.30 \text{ (at } 38°\text{ C in } H_2O)^* \qquad 10.28$$

The value of this pK_1 is close to the physiological range of pH. Thus carbonic acid in the presence of dissolved CO_2 under conditions in which the rate of the hydration reaction is not limiting (i.e., in the presence of carbonic anhydrase) behaves as a much weaker acid than at first appears.

Sometimes the Henderson-Hasselbalch equation is written with H_2CO_3 only as the denominator of the logarithmic ratio term instead of $CO_{2(d)} + H_2CO_3$. This is incorrect since the H_2CO_3 is in rapid equilibrium with the dissolved CO_2 in the presence of carbonic anhydrase, and the overall reaction must be considered when dealing with this buffer system.

The dissolved CO_2 in the plasma is, in turn, in equilibrium with the CO_2 in the gas phase in the lungs:

$$CO_{2(g)} \rightleftharpoons CO_{2\,(d)} \qquad 10.29$$

where $CO_{2(g)}$ denotes CO_2 in the gas phase. According to Henry's law, the concentration of $CO_{2(d)}$ in the system is dependent on the partial pressure of CO_2 in the gas phase (P_{CO_2}):

$$CO_{2(d)} = S \times P_{CO_2} \qquad 10.30$$

in which S is the solubility factor of CO_2; its dimensions are mM of CO_2 dissolved/L solution/mm Hg P_{CO_2}. The relationship between S and α, the Bunsen coefficient, is as follows:

$$S = \alpha \left[\frac{1000 \text{ ml/L}}{(760 \text{ mm Hg/atm}) (22.26 \text{ ml of } CO_2/\text{mM})} \right] = 0.0591\alpha$$

$$10.31$$

---

* In plasma the effect of ionic strength would be to lower pK_1 to a pK'_1 of 6.10, since both $pK_{H_2CO_3}$ and pK_h would be lower in plasma than in water.

S depends on both temperature and the solute concentration, since the solubility of gases is affected by both factors. S is 0.0322 mM of CO_2/L/mm Hg for water at 38°C, while it is 0.0301 mM of CO_2/L/mm Hg for plasma at 38°C. Since at equilibrium, a minute amount of dissolved CO_2 is hydrated, the foregoing values for S include this portion as well, and the full equation is:

$$CO_{2(d)} + H_2CO_3 = S \times PCO_2 \qquad\qquad 10.32$$

Because of this expression, $S \times PCO_2$ can be incorporated as the denominator term of the Henderson-Hasselbalch equation:

$$pH = pK'_1 + \log \frac{[HCO_3^-]}{S \times PCO_2} \qquad\qquad 10.33$$

SOLUTION OF THE HENDERSON-HASSELBALCH EQUATION FOR THE BICARBONATE–CARBON DIOXIDE SYSTEM

The Henderson-Hasselbalch equation contains three unknowns, pH, $[HCO_3^-]$ and $[CO_{2(d)} + H_2CO_3]$, and two constants, pK'_1 and S. In order to solve the equation, two of the three unknowns must be determined and the appropriate values for the constants must be used.

With plasma or urine, the pH is most often measured electro-

TABLE 10-1.   EQUATIONS RELATING THE VARIOUS COMPONENTS OF THE BICARBONATE–CO_2 BUFFER SYSTEM

---

(1) $[CO_{2(d)} + H_2CO_3] = S \times PCO_2$

(2) $[HCO_3^-] = [CO_2]_{(T)} - [CO_{2(d)} + H_2CO_3]$

(3) $pH = pK'_1 + \log \dfrac{[CO_2]_{(T)} - (S \times PCO_2)}{S \times PCO_2}$

(4) $PCO_2 = \dfrac{[CO_2]_{(T)}}{S \times (10^{pH-pK'_1} + 1)}$

(5) $[HCO_3^-] = \dfrac{[CO_2]_{(T)}}{1 + 10^{pK'_1 - pH}}$

(6) $[CO_2]_{(T)} = S \times PCO_2 (10^{pH-pK'_1} + 1)$

---

metrically and the total CO_2 content is determined in the Van Slyke apparatus. The total CO_2 content ($[CO_2]_T$) is defined as follows:

$$[CO_2]_T = [CO_{2(d)} + H_2CO_3] + [HCO_3^-] \qquad 10.34$$

By using Equation 3 in Table 10-1, the Henderson-Hasselbalch equation can then be solved. Other equations in Table 10-1 show other combinations of two measured unknowns and the specific form of the equations used to calculate the third.

VARIATION OF pK′ WITH pH

Many investigators have found that the pK'_1 is dependent not only upon the temperature and ionic strength but also upon the pH itself. This is because the "bicarbonate" fraction of the total CO_2 content is composed of two compounds in addition to bicarbonate: $CO_3^=$ and $NaCO_3^-$. Carbonate originates from the dissociation of bicarbonate as follows:

$$HCO_3^- \rightleftharpoons CO_3^= + H^+ \qquad 10.35$$

and one can write a Henderson-Hasselbalch equation for this reaction:

$$pH = pK'_2 + \log \frac{[CO_3^=]}{[HCO_3^-]} \qquad 10.36$$

in which $CO_3^=$ is the conjugate base and HCO_3^- is the weak acid. The value for pK′ ($38°C$, $\mu = 0.167$ Eq/L) is 9.8.

The $NaCO_3^-$ ion originates as follows:

$$CO_3^= + Na^+ \rightleftharpoons NaCO_3^- \qquad 10.37$$

The equilibrium constant for this reaction is about 0.075, so that appreciable quantities of $NaCO_3^-$ are formed when carbonate is formed, especially in plasma in which the sodium concentration is high (about 140 mEq/L). The concentrations of both carbonate and $NaCO_3^-$ are thus quite dependent upon pH, both increasing as the pH rises. This means that the apparent bicarbonate fraction of the total CO_2 content will contain more of these ions and less true bicarbonate at an alkaline than at an acid pH. In order to correct for this effect in calculating the true bicarbonate from total CO_2 content and pH, pK'_1 must be decreased as pH rises. Thus at pH 7.0, pK'_1 is 6.089. The difference is small but it should be taken into account for accurate work.

PROTEINS AS BUFFERS

Proteins may act as buffers since they can reversibly combine with
H^+ or OH^- and thereby prevent large changes in pH. Unlike simpler
buffers, proteins do not display sharp inflections in their titration curves,
but rather exhibit some buffer action over almost the whole of the pH
scale. This is because proteins are composed of several different kinds
of amino acids, each of which has its own pK_A, the values for which are
distributed across the pH scale. In other words, titrating a protein is
roughly analogous to titrating a solution of amino acids with values
for pK_A over the range of 1 to 12. If each of the amino acids were present
in equal concentration and if the pK values were spaced 1.0 unit apart,
then as the buffer capacity of one constituent fell off to 50 per cent of its

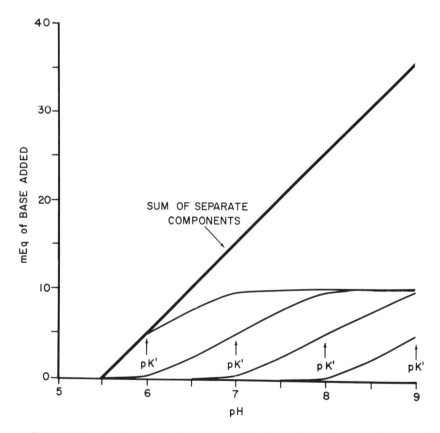

FIGURE 10-3. A hypothetical example to illustrate the production of a straight
titration curve for proteins. The protein is conceived to be made up of specific individual components whose separate titration curves are shown at the bottom of the
figure. The pK' values for these components are spaced at intervals of one pH unit.
Over the range shown, the straight resultant titration curve represents the sum of
the heights of all curves of the components at every value for pH.

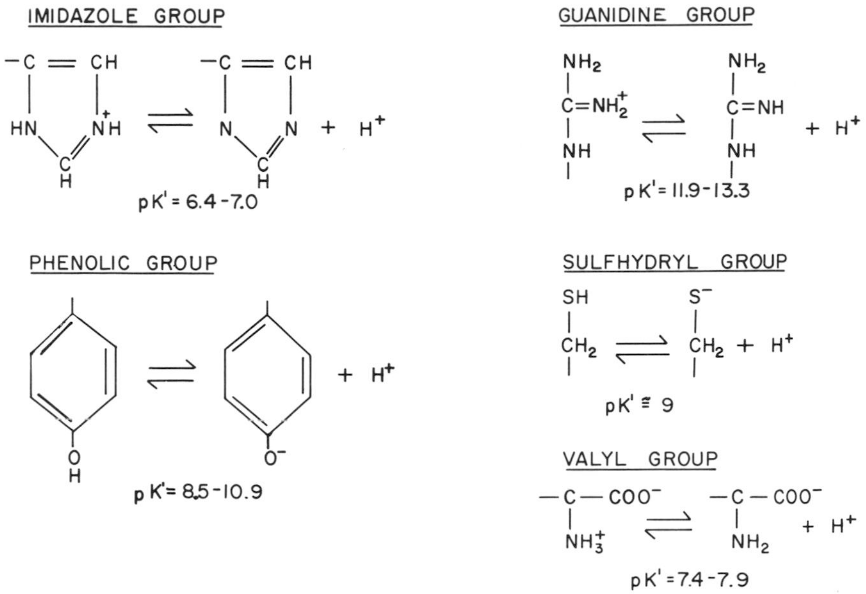

IMIDAZOLE GROUP

$$-C = CH \qquad -C = CH$$

$$HN \quad \overset{+}{N}H \rightleftharpoons N \quad N + H^+$$

$$\underset{H}{C} \qquad \qquad \underset{H}{C}$$

pK' = 6.4-7.0

GUANIDINE GROUP

$$NH_2 \qquad NH_2$$

$$C=NH_2^+ \rightleftharpoons C=NH + H^+$$

$$NH \qquad NH$$

pK' = 11.9-13.3

PHENOLIC GROUP

$$\rightleftharpoons \quad + H^+$$

OH   O$^-$

p K'= 8.5-10.9

SULFHYDRYL GROUP

$$SH \qquad S^-$$

$$CH_2 \rightleftharpoons CH_2 + H^+$$

pK' ≅ 9

VALYL GROUP

$$-C-COO^- \qquad -C-COO^-$$

$$NH_3^+ \rightleftharpoons NH_2 \quad + H^+$$

pK' = 7.4-7.9

FIGURE 10-4. Titratable groups which may occur in proteins and their approximate values for pK'$_{int}$.

peak, the next would have risen to 50 per cent of its peak value. A straight titration curve would thus be generated for the mixture (Fig. 10-3).

Not all the amino acids in a protein are titratable, since unless they are in an end position, the amino and carboxyl groups are combined in peptide linkages. However, the dicarboxylic and the diamino amino acids have free titratable groups and the imidazole group of histidine, the phenolic group of tyrosine, the guanidine group of arginine, the sulfhydryl group of cysteine and the valyl group are also titratable (Fig. 10-4). Because of electrostatic effects from their neighbors, the pK values for the titratable residues of a protein are not necessarily the same as those for the same groups in free solution. In fact, the pK for the same group may vary slightly from protein to protein. The particular pK$_A$ which a group possesses in a particular protein is called the pK$_{intrinsic}$ or pK$_{int}$.

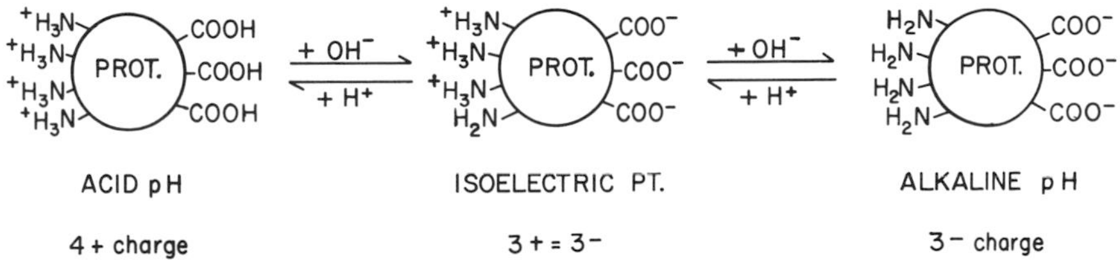

ACID pH	ISOELECTRIC PT.	ALKALINE pH
4 + charge	3 + = 3 -	3 - charge

FIGURE 10-5. Change in the net charge of proteins as a function of pH.

At an acid pH the amino groups of the protein are all positively charged and the carboxyl groups are all combined with H^+ ions and have no charge. The net charge of the protein is therefore positive (Fig. 10-5). If one now titrates this protein by adding base, the first groups to react will be the ones with the H^+ most loosely bound (i.e., the ones with the lowest values for pK_A), which are the carboxyl groups. When most of the carboxyl groups have been titrated, the pH will have risen considerably and the amino groups will begin to react. At an alkaline pH, all the carboxyl groups will be ionized, while the amino groups will have no charge and the net charge of the protein will be negative (Fig. 10-5). In changing from a positively charged molecule to a negatively charged molecule, the protein goes through a point where it has no *net* charge. The pH at which this occurs is called the isoelectric point. In the range of physiological pH, almost all proteins are above their isoelectric points

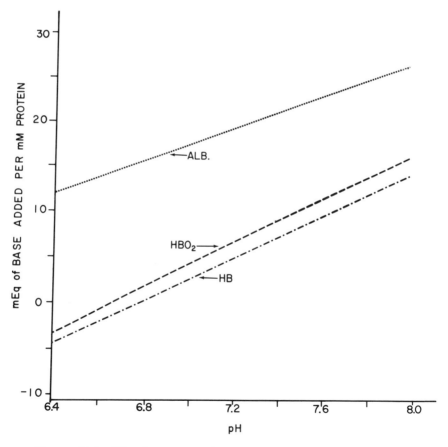

FIGURE 10-6.   Titration curves of reduced hemoglobin (Hb), oxyhemoglobin (HbO₂), and serum albumin (Alb). Note that both Hb and HbO₂ have nearly the same slope while that of albumin is less steep.

and hence they bear a net negative charge which varies with the pH. The weak acid form can be represented as HPr and the conjugate base form as Pr⁻. The buffering reaction may be written as

$$HPr \rightleftharpoons H^+ + Pr^- \qquad\qquad 10.38$$

In blood, hemoglobin is the protein of greatest importance since it occurs in such high concentration. One hemoglobin molecule contains about 540 amino acids, approximately 171 of which are titratable. It is of particular interest that about 36 of these histidyl groups which have a pK'_{int} of about 6.5 to 7.0. This makes hemoglobin an excellent buffer in the physiological range of pH. In addition, there are four end valyl-NH_2 groups with a pK'_{int} of 7.8. The combination of these two groups makes the titration curve for hemoglobin a nearly straight line in the physiological range of pH (Fig. 10-6). The slope of this line gives the buffer capacity which is 12.1 mM of H^+ per mole of hemoglobin per pH unit. There is an "oxylabile" group in hemoglobin whose pK_{int} changes considerably when the hemoglobin is reduced. The pK'_{int} of this group, which is probably a histidine group, changes from approximately 6.7 in

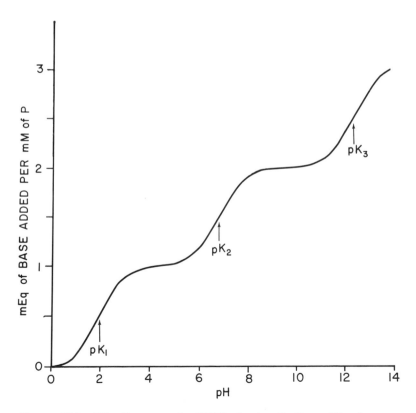

FIGURE 10-7.   Titration curve for H_3PO_4 showing its three pK' values.

oxyhemoglobin to approximately 7.9 in deoxygenated hemoglobin. This change does not change the buffer capacity appreciably, since the lines for oxygenated and deoxygenated hemoglobin are nearly parallel, but deoxygenation does shift the curve to the right (Fig. 10-6). Thus, at any given pH, deoxygenated hemoglobin will combine more avidly with H^+ and act as a weaker acid. This property is of special importance in the buffering of the CO_2 produced by tissues as oxygen is unloaded.

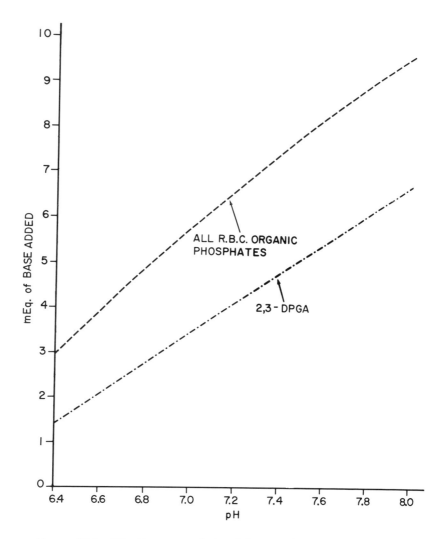

FIGURE 10-8.   Titration curve calculated for organic phosphates of erythrocytes. (Data on concentration of the separate organic phosphates are taken from Bartlett, G.R., *J. Biol. Chem.* 234:449, 1959; data on values for pK' of these components are taken from Johnson, M.J., Chapter 21 in Vol. 3 of *The Enzymes*, edited by Boyer, P.D., Landy, H., and Myrback, K., Academic Press, N.Y., 1960.) The curve for 2,3-diphosphoglyceric acid (2,3-DPGA) is shown separately, since it is the largest single component of the organic phosphates group in the erythrocyte.

Plasma proteins act as buffers in the same fashion as hemoglobin, except they are stronger acids (their titration curve is shifted to the left in Figure 10-6) and their buffer capacity is less, being about 8.8 mEq H^+/mole/pH unit for albumin and about 10 mEq/mole for the globulins. Also plasma proteins are present in much lower concentration per liter of whole blood than is hemoglobin. These two factors make plasma proteins far less significant buffers than hemoglobin in blood.

INORGANIC AND ORGANIC PHOSPHATES AS BUFFERS

The titration curve for phosphoric acid is shown in Figure 10-7. This acid has three ionizable hydrogens, but as the curve shows, only one of these occurs in the range of physiological interest:

$$H_2PO_4^- \rightleftharpoons HPO_4^= + H^+ \qquad\qquad 10.39$$

The pK'_2 of phosphoric acid is approximately 6.75 (38°C and $\mu = 0.16$ Eq/L) which is close to the physiological range of pH. This would make inorganic phosphate a good buffer were it not for the fact that its concentration in plasma is so small (2 to 3 mM/L). In urine and possibly in intracellular fluid, in which its concentration is higher, inorganic phosphate plays an important role as a buffer.

Organic phosphates are probably significant buffers in intracellular fluid. They are a heterogeneous group made up principally of 2,3-diphosphoglyceric acid, adenosine triphosphate, and hexose phosphates. The pK'_A values for these compounds are in the range of 6.0 to 8.1. Figure 10-8 shows a titration curve of these organic phosphates as derived from analytical data on the erythrocyte. The curve for 2,3-diphosphoglycerate, the largest component of the organic phosphate compounds, is also shown separately in Figure 10-8.

Whole Blood Buffers

COMPONENTS OF THE WHOLE BLOOD BUFFERS

The components of the buffer systems in whole blood have already been discussed separately. They will now be viewed as a whole and their interactions will be discussed.

Blood is a two-phase system composed of erythrocyte water and plasma water. Hemoglobin, organic phosphates, and some of the blood bicarbonate are in the erythrocytes, while the rest of the buffers occur in the plasma. Under usual conditions, the pH of erythrocyte water is lower than the pH of plasma* by about 0.15 to 0.20 unit, but the two

---

* When blood pH is measured, the measurement is essentially the pH of the plasma and not the intraerythrocyte pH.

phases are not independent of each other with respect to pH because of the ability of bicarbonate as well as dissolved CO_2 to cross the erythrocyte membrane. Thus hemoglobin, organic phosphates, and the erythrocyte bicarbonate exist in an environment which has a different pH from that to which the plasma components of the whole blood buffers are exposed.

It is convenient in considering the blood buffers to divide them into two groups: the non-bicarbonate system—i.e., hemoglobin, plasma proteins, organic and inorganic phosphates—and the bicarbonate system. The rationale for this division will become apparent later.

TABLE 10-2. COMPONENTS OF THE NON-BICARBONATE BUFFERS OF WHOLE BLOOD

BUFFER SYSTEM	TOTAL CONC. OF CONJUGATE BASE + WEAK ACID IN WHOLE BLOOD		CONC. OF CONJUGATE BASE	MOLAR BUFFER[a] CAPACITY	BUFFER[b] CAPACITY IN 1 L OF WHOLE BLOOD	% OF TOTAL NONBICARB. BUFFER CAPACITY
	GM%	mM/L	mEq/L WHOLE BLOOD	mEq BASE/ ΔpH/MOLE	mEq BASE/ ΔpH/L	
In erythrocytes: (pH = 7.25)						
Hemoglobin	15.0	2.21	15.8	12.1	21.90[c]	78
Organic phosphates	0.107	3.64	3.38	0.65[d]	1.94	6
In plasma: (pH = 7.40)						
Albumin	2.48	0.358	7.5	8.84	3.16	12
Globulins	1.38	0.107[e]	2.7	10.0[e]	1.07	4
Inorganic phosphate	6.02 mg%	0.621	0.5	0.402	0.25	1
					Total 28.32	

a. The pH used here is a change of 1 unit in the pH of the phase containing the buffer.

b. The pH used here is a change of 1 unit in the pH of the plasma.

c. The buffer capacity in 1 liter of whole blood for hemoglobin and organic phosphate is not the concentration multiplied by the molar buffer capacity, since intracellular pH changes only 0.82 unit when the plasma pH changes 1 unit. Therefore, the calculation for hemoglobin is $2.21 \times 12.1 \times 0.82 = 21.9$, and for organic phosphate the calculation is $3.64 \times 0.65 \times 0.82 = 1.94$.

d. This is a virtual molar buffer capacity, since organic phosphates are made up of several different chemical species, each having its own molar buffer capacity. The buffer capacity shown above is a weighted average of the buffer capacities of the separate components.

e. The concentration of globulins in mM/L and their molar buffer capacity are both given only to indicate order of magnitude, since they are not a single chemical entity but are made up of molecules ranging in weight from 100,000 to 1,000,000. A weighted average molecular weight of 129,000 was used to calculate the mM/L from the concentration in grams per cent and to calculate the molar buffer capacity from the buffer capacity per gram (0.0778).

THE NON-BICARBONATE BUFFERS

Table 10-2 shows the details of the non-bicarbonate buffers of normal whole blood. Of all the non-bicarbonate buffers, hemoglobin not only has the highest molar buffer capacity but is also present in high concentration. These two factors account for its being the most important component of the group of non-bicarbonate buffers (78 per cent of the total). Plasma proteins, albumin, and the globulins have the next highest buffer capacities, but because they are present in much lower concentration than hemoglobin, they contribute only 16 per cent of the total buffer capacity of the non-bicarbonate buffers. Organic phosphates of the erythrocyte and inorganic phosphate of the plasma collectively contribute the remaining 6 per cent.

Figure 10-9 shows the titration curves of each component of the

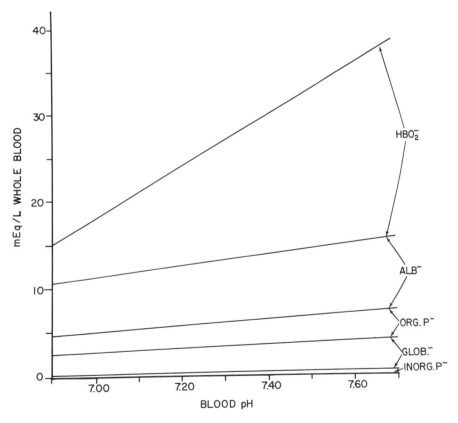

FIGURE 10-9. The titration curves for each component of the non-bicarbonate buffers in normal blood. The figure is constructed in such a manner that each line represents the sum of all components below it. The slope of the highest line is therefore a measure of the buffer capacity of the entire group of non-bicarbonate buffers.

non-bicarbonate buffers in which each line represents the sum of all the buffers below it. The total buffer capacity of the entire group is 28.3 mEq base added/$\triangle$pH/L of whole blood. Figure 10-9 demonstrates the important point that it is the *change* of concentration rather than the *absolute* concentration of buffer anion which is the most important determinant of the ability of any given buffer to prevent a change in pH with a given addition of acid or base. For example, at pH 6.9, the concentration of the anionic form of albumin (Alb⁻) is 5.9 mEq/L and is greater than that of the anionic form of hemoglobin (HbO₂⁻) which is 4.4 mEq/L. Yet the addition of base sufficient to change the pH to 7.0 increases the concentration of Alb⁻ by only 0.3 mEq (to 6.2 mEq/L), while the HbO₂⁻ concentration increases by 2.4 mEq/L (to 6.8 mEq/L). Thus the change in HbO₂⁻ is eight times that of albumin.

In passing it should also be noted that the concentrations of the non-bicarbonate buffer pairs do not appear to be subject to physiological regulation to the specific end of the maintenance of acid-base equilibrium. That is to say, there is no evidence to indicate that the body regulates the concentrations of hemoglobin, of plasma proteins, or of organic or inorganic phosphate in a rapid and precise way in response to the acid-base status. Rather in any one individual, the total concentration (i.e., weak acid + conjugate base) of each is relatively constant and the part

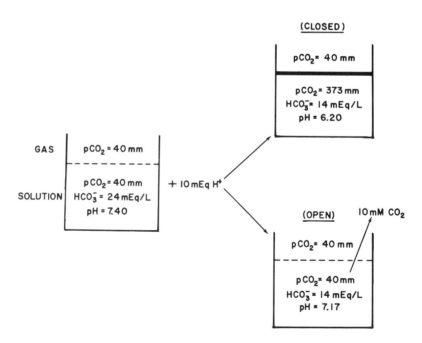

FIGURE 10-10.   The effect of the addition of strong acid to a bicarbonate-CO₂ buffer in a closed system and in an open system in which the Pco₂ of the gas phase is held at 40 mm Hg.

of the total which is present as the anion is solely dependent upon the pH.

This is in marked contrast to the situation with the bicarbonate system in which physiological mechanisms do exist which can rapidly change the concentrations of both the conjugate base and weak acid members of the buffer pair in response to changes in the acid-base status of the organism. This difference between the bicarbonate and the non-bicarbonate systems is the justification for the separate treatment of these two classes of blood buffers.

THE BICARBONATE BUFFER SYSTEM

The bicarbonate buffer system of both the plasma and erythrocytes occupies a special role in the regulation of blood pH, since it is the direct target of physiological regulation. Despite the fact that the pK' of this system is 6.10 and is therefore 1.3 pH units away from the normal blood pH of 7.40, the bicarbonate-CO_2 system is a very good buffer in vivo. This is because the weak acid member of the buffer pair—i.e., H_2CO_3—is, on the one hand, volatile as CO_2 (Eq. 10.21 and 10.29) while on the other, it is available in very large amounts as a product of oxidative metabolism.

The importance of these factors is illustrated in Figures 10-10 and 10-11 which demonstrate the differences between the operation of a bi-

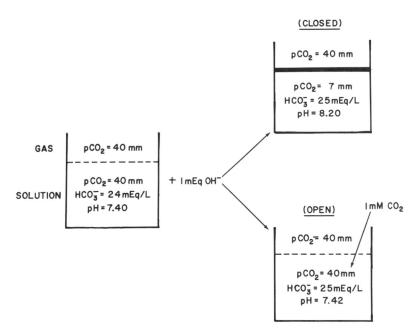

FIGURE 10-11. The effect of the addition of a strong base to a bicarbonate-CO_2 buffer in a closed system and in a system equipped with a large supply of CO_2.

carbonate-CO_2 system as a closed system and its operation in a system equipped with the ability to excrete CO_2 or to produce CO_2. The addition of strong acid without a loss of CO_2 would produce a very low pH and a very high P_{CO_2}. If the non-metabolic CO_2 produced by buffering, however, can be expelled, the pH change is considerably less. The bicarbonate-CO_2 system is even less efficient in the buffering of a strong base, since in the absence of CO_2 production, only 1.2 mM/L of weak acid is available. With CO_2 production, however, a virtually infinite supply of weak acid is available from metabolism, and buffering is considerably more efficient.

In the body, the buffering from this system is made even more efficient than is shown in Figures 10-10 and 10-11 because of the ability of the respiratory system to make such additional adjustments of P_{CO_2} that pH shifts are further ameliorated. Thus, in loading with strong acid, the new steady-state value for P_{CO_2} is less than the normal of 40 mm Hg, while with strong base loading it is likely to be greater than normal. The physiological mechanisms underlying these compensatory changes in P_{CO_2} will be dealt with later.

Figure 10-12 shows a synthesis of all the buffers in whole blood in which the pH is varied by the addition of strong acid or base, and there is an additional readjustment of P_{CO_2}. The figure demonstrates the contributions of each system to the total buffer capacity under physiological circumstances and, quite clearly, the major contribution which bicarbonate makes under these conditions.

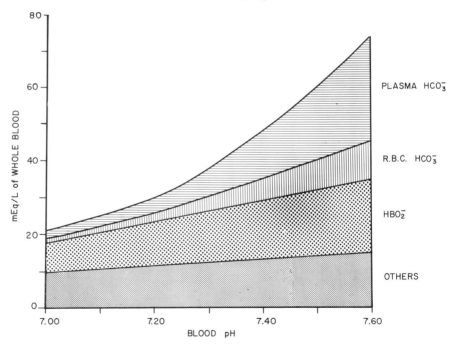

FIGURE 10-12.   The titration curves for each of the major components of whole blood. Each curve is the sum of the ones below it so that the top line represents the titration curve of whole blood. In constructing the bicarbonate lines the P_{CO_2} resulting from the average respiratory compensation was used (see Fig. 10-17).

RESPONSE OF BUFFERS TO ACID-BASE DISTURBANCES

Types of Acid-Base Disturbances

Before proceeding it is useful to define the types of disturbances which can occur in normal acid-base equilibrium. Traditionally, four such disorders are recognized, and they are classified in terms of the nature of the primary distortion as it is reflected in the ratio of the Henderson-Hasselbalch equation. For purposes of classification, the numerator of the ratio ($[HCO_3^-]$) is referred to as the metabolic component, while the denominator, $S \times P_{CO_2}$, is referred to as the respiratory component. Disorders which *primarily* affect the concentration of bicarbonate are therefore called metabolic disturbances* while those which primarily affect P_{CO_2} are called respiratory disturbances. With each type of disturbance there are two possible effects—the specific component (i.e., metabolic or respiratory) may either increase or decrease. These pos-

---

* The term *metabolic* is rather ambiguous as used in this connection, since in a broad sense, all changes in acid-base equilibrium (including respiratory ones) are due to metabolic causes. To avoid ambiguity, some authors use the terms *non-respiratory* (metabolic) and *respiratory* to describe the two types of acid-base disturbances.

TABLE 10-3.   CLASSIFICATION OF ACID-BASE DISORDERS

$$pH = pK' + \log \frac{[HCO_3^-]}{S \times P_{CO_2}}$$

DISTURBANCE	EFFECT OF PRIMARY FACTOR	COMPENSATION	COMMON ETIOLOGICAL FACTORS
Metabolic acidosis	Reduction in $[HCO_3^-]$	Reduction in P_{CO_2}	*Gain of strong acid:* diabetes, uremia, lactic acidosis, NH_4Cl *Loss of bicarbonate:* diarrhea, renal tubular disease
Metabolic alkalosis	Increase in $[HCO_3^-]$	Increase in P_{CO_2}	*Gain of strong base:* loss of HCl, K depletion *Gain of bicarbonate:* sodium bicarbonate loading
Respiratory acidosis	Increase in P_{CO_2}	Increase in $[HCO_3^-]$	*Gain of P_{CO_2}:* disease of lung, chest wall; depression of respiratory center
Respiratory alkalosis	Decrease in P_{CO_2}	Decrease in $[HCO_3^-]$	*Loss of P_{CO_2}:* altitude, emotional hyperventilation, certain drugs

sibilities along with their attendant effects upon the blood pH are listed in Table 10-3 along with some of the factors which are responsible for the primary distortions.

Once a primary distortion in the ratio has occurred, physiological mechanisms are activated which tend to compensate for the deviation in pH. Thus in metabolic disturbances, there is an adjustment in respiration whereby a new level of P_{CO_2} is established such that the pH tends to be restored toward normal. In respiratory disturbances, on the other hand, there tends to be a compensatory change in the metabolic component, effected in this case through renal and extrarenal mechanisms, so that

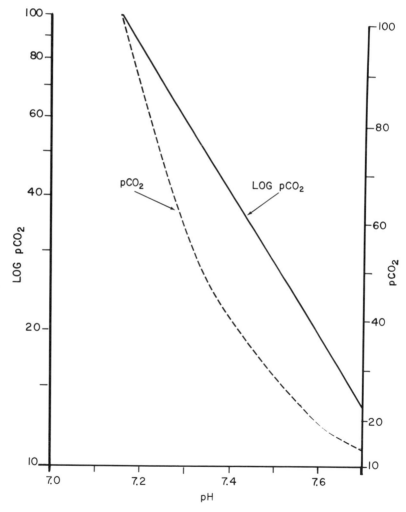

FIGURE 10-13.   The CO_2 titration curve for whole blood. The broken line represents a linear plot (right-hand scale), and the solid line represents a logarithmic plot (left-hand scale).

the pH is restored toward normal. The term *compensation* as used in this context should be carefully distinguished from the term *correction*. Correction is used to connote the mechanisms by which the initial disturbance is completely corrected—i.e., the body fluids are brought back to normal. For example, in metabolic acidosis caused by loading with strong acid, compensation for the low pH is effected by respiration, but correction requires that the kidneys excrete the acid and restore all the constituents of the body fluids to normal. Similarly, in respiratory disturbances, the kidneys may compensate for the alteration in the pH, but only removal of the primary respiratory factor will ultimately correct the disorder.

Response of Blood Buffers

Respiratory Disturbances

Buffers in blood provide the first line of defense against acid-base disturbances. In respiratory disturbances the primary etiological event is an alteration between the rate of CO_2 production by the tissues and the rate at which CO_2 is excreted by the lung so that a new steady state for Pco_2 is maintained at either a higher (respiratory acidosis) or lower (respiratory alkalosis) value than the normal of 40 mm Hg. Such disturbances may come about because of disease of the lungs, of the chest wall, or of the neural mechanisms which control respiration.

Regardless of how it is produced, the alteration in Pco_2 is rapidly reflected by an alteration in the concentration of dissolved CO_2 and carbonic acid in the arterial blood and hence a change in pH (Eq. 10.33). Changes in the blood pH produced by respiratory disturbances are buffered exclusively by the non-bicarbonate buffers, since bicarbonate, the conjugate base of carbonic acid, cannot itself buffer carbonic acid.

Figure 10-13 shows the general relationship between change in Pco_2 and change in pH for normal whole blood. In effect this is a titration curve of blood in which carbonic acid is the titrating acid. Changes in the titrating acid are produced by changing the Pco_2 to which the blood is exposed. Such a curve is therefore called a CO_2 titration curve. The curves shown in Figures 10-13 and 10-14 demonstrate several important points. *First*, plotting the logarithm of the Pco_2 against the pH yields a straight line, an effect which has been ascribed to the fortuitous spacing of the pK' values and the concentrations of the various buffer groups of the non-bicarbonate buffers. *Second*, the slope of the curve (i.e., $\triangle \log Pco_2 / \triangle pH$) is a function of the concentration of non-bicarbonate buffers of which hemoglobin is the most important (as previously stated). *Third*, the position of the curve, and to a lesser extent its slope, is affected by the degree of oxygenation of the hemoglobin. The shift in position is due to the "oxylabile" groups in the hemoglobin molecule.

If the total non-bicarbonate buffers of blood are represented by the symbols, HBuf (weak acid forms) and $B^+ Buf^-$ (conjugate base forms), the buffer reaction for respiratory disturbances can be written:

$$H_2CO_3 + B^+ Buf^- \underset{b}{\overset{a}{\rightleftharpoons}} HBuf + B^+ HCO_3^-$$ 10.40

The reaction from left to right (Reaction 10.40 a) represents the direction from which an increase in Pco_2 (respiratory acidosis) would be buffered, while the opposite reaction (Reaction 10.40 b) occurs in respiratory alkalosis. In either case, it should be noted that there is a reciprocal relationship between change in Buf^- and change in HCO_3^- and thus the concentration of the buffer anion remains unchanged. Thus, in respiratory acidosis, Buf^- accepts H^+ from carbonic acid and in the process generates new HCO_3^-. The overall effect is to minimize the fall in pH produced by the addition of H_2CO_3 as compared to the greater fall which would occur if the same amount of H_2CO_3 were added to a bicarbonate solution lacking non-bicarbonate buffers.

This reaction sequence in buffering in respiratory acidosis is exactly similar to that which occurs normally as arterial blood passes through the

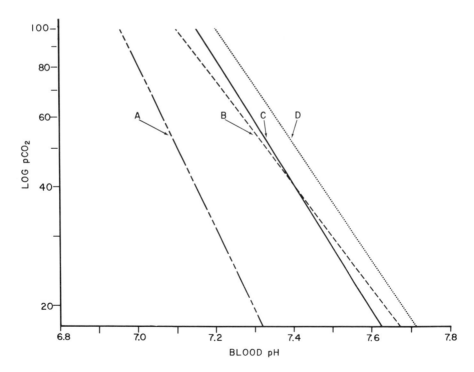

FIGURE 10-14. Factors affecting the position and slope of the CO_2 titration curve. Curve C is normal, oxygenated blood; B is blood with a low hemoglobin concentration; D is reduced blood; and A is blood to which strong acid has been added.

tissues and gains CO_2.* Venous blood is in a sense in a state of mild respiratory acidosis compared to arterial blood, and the buffer reaction we have shown allows most of the CO_2 produced by the tissues to be carried as bicarbonate.

In respiratory alkalosis, the non-bicarbonate buffers operate in a fashion opposite to that which occurs in respiratory acidosis, but similar to that which occurs in normal venous blood as it passes through the lungs. In both situations, the concentration of carbonic acid is lowered, HBuf contributes H^+ to HCO_3^-, thus generating more H_2CO_3, which is lost into the alveolar gas while concomitantly reducing the bicarbonate concentration so that the rise in pH caused by the loss of H_2CO_3 is minimized.

METABOLIC DISTURBANCES

The prototype of metabolic acidosis is usually considered to be the loading of the blood with strong acid ($H^+ A^-$). In this case, both the bicarbonate and non-bicarbonate buffer systems provide the immediate defense of the blood pH:

$$H^+ A^- + B^+ HCO_3^- \rightarrow B^+ A^- + H_2CO_3 \qquad 10.41$$

$$H^+ A^- + B^+ Buf^- \rightarrow B^+ A^- + HBuf \qquad 10.42$$

As already discussed, such buffer reactions carried out in a closed system have quite different quantitative effects from those which are carried out in the presence of a functioning respiratory system. In the body, loading with a strong acid stimulates the respiratory system, and an increase in alveolar ventilation results. The resultant effect is that the H_2CO_3 produced through buffering is "blown off" in the lungs as CO_2. The close integration between the buffer reaction and the physiological response minimizes the degree of fall in pH compared to that which would occur if there were no respiratory response. The respiratory control of acid-base equilibrium will be dealt with later.

There are a number of specific types of metabolic acidosis which conform to the general pattern of strong acid loading—i.e., diabetic acidosis (loading with ketone acids), lactic acid acidosis, and ammonium chloride acidosis (loading with hydrochloric acid). On the other hand, there is a second general type of metabolic acidosis which occurs secondary to a loss of extracellular bicarbonate—i.e., diarrhea (loss of bicarbonate-rich intestinal fluids) or renal tubular disease (loss of bicarbonate in the urine due to failure of the normal tubular reabsorptive mecha-

---

* In the course of normal CO_2 transport, buffering by hemoglobin is made more efficient by the deoxygenation of hemoglobin as oxygen is delivered to the tissues.

nisms). Metabolic acidosis due to loss of bicarbonate differs in some respects from that due to loading with a strong acid, since in the former case, a member of a buffer pair (HCO_3^-) is being lost directly while in the latter case it is being converted to the weak acid member of the same pair. With direct bicarbonate loss, only the non-bicarbonate buffers can be operative, and the only buffer reaction will be Reaction 10.40 a.

A similar analysis can be made of the buffer responses in metabolic alkalosis. Loading with a strong base produces the following reactions:

$$OH^- + H_2CO_3 \rightarrow H_2O + HCO_3^- \qquad \text{10.43}$$

$$OH^- + HBuf \rightarrow H_2O + Buf^- \qquad \text{10.44}$$

The reader should recall that Reaction 10.43 is made much more efficient by virtue of the large supplies of H_2CO_3 available as CO_2 from metabolism. The effects of both the foregoing reactions are in essence to bind the OH^- of the strong base as water, thus minimizing the rise in pH. Loading with *exogenous* strong base (such as NaOH) does not occur in clinical medicine.* However, in a sense, loss of gastric acid through vomiting is tantamount to an endogenous load of strong base, since the stomach makes H^+ of H^+Cl^- from water and for every H^+ lost as hydrochloric acid, one OH^- is left behind in the blood to be buffered. An analogous situation exists when potassium leaves muscle in exchange for H^+, thus leaving behind OH^- in the extracellular fluid. The interrelationships between potassium and acid-base equilibrium will be discussed later.

Loading with strong base differs in some respects from the metabolic alkalosis due to loading with sodium bicarbonate. Since in the latter instance we are again concerned directly with a member of the bicarbonate buffer pair, buffering must be exclusively through the non-bicarbonate buffers and proceeds by Reaction 10.40 b.

Response of Extravascular Buffers

Although the buffer properties of blood have been extremely well characterized, they constitute only a fraction of the total buffers of the body fluids and the major buffer systems in each compartment. Interstitial fluid, being an ultrafiltrate of plasma, contains virtually no protein and therefore has the bicarbonate system as its only important buffer. The buffer properties of intracellular fluid can only be guessed at, since direct,

---

* Loading with sodium carbonate is occasionally done experimentally, and its buffer reactions are analogous to loading with sodium hydroxide:

$$Na_2CO_3 + H_2CO_3 \rightarrow 2 NaHCO_3 \qquad \text{10.45}$$

$$Na_2CO_3 + HBuf \rightarrow Na + Buf^- + NaHCO_3 \qquad \text{10.46}$$

TABLE 10-4. THE MAJOR BUFFERS IN EACH COMPARTMENT OF THE
BODY FLUIDS

| | BLOOD | | INTERSTITIAL | INTRACELLULAR |
	ERYTHROCYTES	PLASMA	FLUID	FLUID
Volume (ml/kg body weight)	30	50	150	400
Major buffers	Hemoglobin	Bicarbonate	Bicarbonate	? Organic and inorganic phosphates ? Protein ? Bicarbonate

detailed study of this problem is hampered by almost insuperable technical difficulties. It is generally assumed that organic and inorganic phosphates, protein, and bicarbonate are the chief buffers of this compartment. The total body buffers are thus compartmentalized, with each compartment being separated from its neighbor by a membrane—the capillary membrane in the case of the plasma-interstitial boundary and the cell membrane in the case of the interstitial-intercellular boundary (see Table 10-4).

BUFFERING BY INTERSTITIAL FLUID

The movement of ions from plasma to interstitial fluid is practically unrestricted unless the ion is a protein, since the capillary membrane is quite permeable to all small ions, molecules, water, and dissolved CO_2. The interstitial fluid can therefore enter into buffer reactions quickly and directly. Compared to blood, interstitial fluid has a much lower buffer capacity per unit volume since it lacks hemoglobin, its only major buffer being the bicarbonate system which is present in approximately the same concentration as in plasma. Despite its reduced buffer capacity, interstitial fluid, by virtue of its large volume, can make a quantitatively important contribution to the buffering of strong acid or base.

Total Body Buffering

Despite our ignorance as to the site and mechanisms of intracellular buffering, there is a large body of evidence which strongly suggests that non-extracellular sites do actively participate in countering acid-base disturbances. Some of this evidence will now be examined.

RESPIRATORY DISTURBANCES

A number of experiments have shown clearly that in respiratory acidosis, from 50 to 90 per cent of the total CO_2 gained by the animal must be retained in extravascular sites. This implies that the CO_2 readily passes through cell membranes and is buffered within cells by the non-bicarbonate buffers as well as perhaps being converted to carbonate in bone. Conversely in respiratory alkalosis, analysis of the total loss of CO_2 from the body shows that a comparably large fraction must have come from tissues or bone.

A second line of evidence implicating non-extracellular buffering consists of the measurements of changes which occur in the total quantity of extracellular bicarbonate as the result of increasing or decreasing the P_{CO_2}. If non-extracellular buffers did not participate, one would expect that the total amount of extracellular bicarbonate would change only by the action of the non-bicarbonate buffers (Reaction 10.40) of the blood.*

If measurements of total extracellular bicarbonate show a significant discrepancy between the observed and the predicted, then a transfer of bicarbonate in one direction (or transfer of H^+ in the opposite direction) between cells or bone and the extracellular fluid has occurred. In turn, such a transfer requires a concomitant movement of anions or cations in order to maintain electroneutrality, and changes in the extracellular content of various cations and anions can be measured to account for the movements of bicarbonate (or H^+).

Induction of respiratory alkalosis by hyperventilation leads to substantially the same results in both dog and man; only about one third of the total decrease in extracellular bicarbonate can be attributed to the non-bicarbonate buffers of the blood (see Table 10-5). The remaining two thirds must therefore have been lost either by the movement of

---

* The role of the kidneys in these experiments is either eliminated by nephrectomy or accounted for by measuring changes in bicarbonate excretion. Since these are short-term experiments, the renal contribution in any case is small.

TABLE 10-5.   DISTRIBUTION OF BUFFERING IN RESPIRATORY DISTURBANCES AS PER CENT OF TOTAL

	DUE TO NON-BICARBONATE BUFFERS OF BLOOD	DUE TO EXCHANGE WITH NA°	K°	ANION°	TOTAL NON-EXTRACELLULAR BUFFERING
Respiratory alkalosis					
Man	38	+16	+4	−35	55
Dog	30	+37	−3	−36	70
Respiratory acidosis					
Man	34	−37	−14	+6	57
Dog	35	−47	−9	+9	65

* A positive value indicates a gain, and a negative value indicates a loss of the ion in question by the non-extracellular fluids.

bicarbonate into cells or bone (or by the outward movement of H^+ which yields carbonic acid and this is lost in expired air). In either case, electroneutrality is maintained by a combination of loss of extracellular cation, principally sodium, and gain of anion, principally lactate, by the extracellular fluid.

Thus, by means of these ion exchanges, non-extracellular sites accounted for two thirds of the total change in extracellular bicarbonate content. In terms of the maintenance of extracellular pH, this is an important compensatory mechanism in that it further reduces the concentration of bicarbonate so that with a new lower steady-state value for P_{CO_2}, the pH rise is less great than in the absence of such ion exchanges.

Roughly comparable data are found with the induction of respiratory acidosis (Table 10-5). In this case two thirds of the observed gain in extracellular bicarbonate comes from non-extracellular sites and is usually accompanied by a shift of sodium and smaller shifts of potassium and anion (lactate). As in the case of respiratory alkalosis, these linked transfers of ions serve to ameliorate the fall in plasma pH by providing more bicarbonate from non-extracellular sources.

METABOLIC DISTURBANCES

Experiments in which the contributions of the buffers of each fluid compartment were assessed have amply demonstrated the importance of extravascular buffers in neutralizing a load of strong acid (Table 10-6). Thus the non-bicarbonate and bicarbonate systems of the blood buffer only about one fifth of the load of acid, while the bicarbonate system of the interstitial fluid accounts for an additional one quarter. This leaves more than half of the total load to be buffered by non-extracellular sites. This buffering is accompanied by a large gain of sodium and a smaller gain of potassium by the extracellular fluid, these cations

TABLE 10-6. DISTRIBUTION OF BUFFERING IN METABOLIC DISTURBANCES
AS PER CENT OF TOTAL

	IN BLOOD	IN INTERSTITIAL FLUID	DUE TO EXCHANGE WITH NA*	K*	TOTAL NONEXTRACELLULAR BUFFERING
Metabolic acidosis					
Man	16	29	−22	−34	55
Dog	18	25	−36	−15	51
Metabolic alkalosis					
Man	26	44	+35	−15	30

* A positive value indicates a gain and a negative value a loss of the ion in question by the non-extracellular fluids.

either exchanging for extracellular H^+ or accompanying an outward movement of bicarbonate from the cells.

Similar experiments in which a load of sodium bicarbonate is administered likewise reveal the importance of non-extracellular buffers. As shown in Table 10-6, 26 per cent of the total load is buffered by non-bicarbonate buffers of blood and an additional 44 per cent is retained in the extracellular volume. The remaining 30 per cent presumably enters cells or bone or is buffered by an egress of H^+ from these sources. These exchanges are accompanied by a gain of sodium and a loss of potassium by the non-extracellular fluids.

MUSCLE ELECTROLYTES IN ACID-BASE DISORDERS

In studies of total body buffering, the intracellular space is necessarily treated as a single compartment. By weight, muscle is the most important tissue component in the intracellular compartment, and changes inferred from the study of electrolyte transfers within the whole organism might be expected to be reflected in a direct analysis of muscle tissue.

With respect to intracellular sodium, analysis of muscle of rats in general shows a fall in acidotic states and a rise in alkalotic states. This is similar to the pattern seen in the total body buffering experiments, although it is not clear whether the magnitudes of these intracellular sodium changes in muscle can account for the changes observed in the extracellular content of sodium in the total body buffering experiments. With respect to muscle potassium, the changes are generally reciprocally related to changes in sodium, so that in alkalotic states muscle potassium falls, whereas in acidotic states there is a slight rise. This pattern deviates from that observed in the total body buffering experiments (Tables 10-5 and 10-6).

It should be pointed out that the two sets of experiments are not strictly comparable. The total body buffering experiments were short-term and therefore exclude any renal contribution to the whole picture, whereas the muscle data represent results of long-term experiments with the kidneys intact. Furthermore, the former set of experiments were carried out on either dogs or men, whereas most of the muscle data represent experiments carried out on rats. Finally, as was already pointed out, it may not be justifiable to extrapolate from muscle to all the non-extracellular buffer sites, since bone sodium (and to a lesser extent, potassium) is known to fall in chronic metabolic acidosis.

RELATIONSHIP BETWEEN MUSCLE POTASSIUM AND METABOLIC ALKALOSIS

That a special relationship exists between muscle potassium and metabolic alkalosis has been recognized for many years. For example, the normal animal or man given large loads of sodium bicarbonate becomes minimally and transiently alkalotic (because of rapid renal excretion), but loads of this salt given to potassium-depleted subjects produce a marked degree of sustained alkalosis. Rats given a diet deficient in potassium along with a load of sodium (which need not be as the bicarbonate salt) show a metabolic alkalosis which is correctable by administration of potassium. That this effect is not primarily dependent upon the kidney has been shown by two sets of experiments in rats: (1) repair of potassium-deficiency alkalosis in the rat is accompanied by the excretion of excess H^+ in the urine and not by the excretion of the excess extracellular bicarbonate which would seemingly be a simple expedient, and (2) the metabolic alkalosis of potassium deficiency can be repaired in the complete absence of the kidneys by infusing potassium-containing salts.

On the basis of such observations, it has been suggested that the loss of muscle potassium during the development of potassium deficiency is accompanied by a gain of both the sodium and hydrogen ion from the extracellular fluid. Based upon the discrepancy between the amount of potassium lost and the amount of sodium gained, it has been postulated that for every three K^+ ions moving in one direction, 2 Na^+ ions and 1 H^+ ion move in the other. These movements are diagramed in Figure 10-15 which shows the further consequence of the movement of H^+—

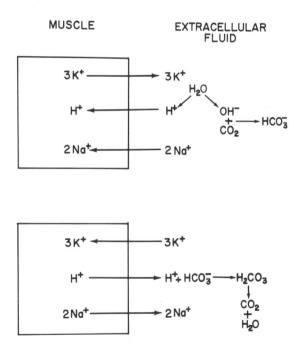

FIGURE 10-15. Interrelationships between the loss of potassium and the gain of sodium and hydrogen ion in muscle (above). Interrelationship between repletion of potassium and loss of sodium and hydrogen ion in muscle (below).

namely, that an OH^- is left in the extracellular fluid. This ion, buffered according to Reactions 10.43 and 10.44, accounts for the origin of the metabolic alkalosis in potassium deficiency on an entirely extrarenal basis, while the intracellular contents should pari passu become more acid because of the gain of H^+ which is presumably buffered by intracellular buffers. There is evidence to suggest that the potassium-deficient muscle has increased concentrations of basic amino acids (such as lysine), and these may provide additional buffering for the H^+ ion:

$$R - NH_2 + H^+ \rightarrow R - NH_3^+ \qquad\qquad 10.47$$

During repair of potassium deficiency, the sequence of electrolyte transfers outlined in Figure 10-15 presumably occurs. Muscle takes up potassium and extrudes both sodium and H^+. The later now titrates the extracellular buffers back to normal, and indeed the amount of H^+ added to the extracellular sources may be greater than this so that the excess is excreted in the urine.

This provocative hypothesis of the relationship between K^+ and H^+ has several direct corollaries: (1) the intracellular pH should be low in potassium-deficient alkalosis, (2) the alkalosis should be resistant to all measures except those which provide the potassium ion, and (3) if kidney cells resemble muscle, then they too should become more acid as potassium depletion develops. Intracellular pH and renal behavior in potassium-deficiency will be discussed later; however, it should be pointed out that recent observations in the dog and human do not support the corollary that the alkalosis is responsive to potassium and to potassium only. Rather, a specific role of chloride depletion has been implicated. It is not clear whether this represents a mechanism peculiar to the dog or whether the foregoing hypothesis may need modification.

Likewise it is difficult at present to integrate this hypothesis into the general role of body buffers. From the point of view of protection of extracellular pH, the K^+ for H^+ exchange does not play a compensatory role. It is more likely that this picture represents the sequence of events starting with potassium deficiency rather than the sequence starting with the induction of metabolic alkalosis. As we will see later, the induction of alkalosis, either metabolic or respiratory, is peculiarly likely to lead to potassium losses in the urine so that it is possible to see how the induction of alkalosis leads to potassium deficiency which in turn aggravates the alkalosis through the exchange at the muscle level.

INTRACELLULAR pH

As already indicated, the intracellular buffers are the least well characterized of any of the buffer systems of the body fluids. One reason for this has been the lack of a reliable and precise method for the measure-

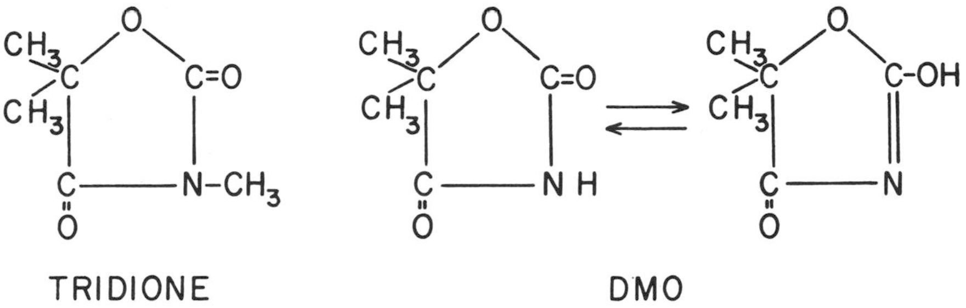

TRIDIONE   DMO

FIGURE 10-16.   Structural formula of 5,5-dimethyl-2,4-oxazolidinedione (DMO) compared to its parent compound, tridione, an antiepileptic drug.

ment of the intracellular pH in intact living tissue. This limitation may now have been partially overcome with the use of a method involving the measurement of the distribution of the weak acid, 5,5-dimethyl-2,4-oxazolidinedione (DMO) between cells and extracellular fluid. This substance exists in two forms (Fig. 10-16);

$$HDMO \rightleftharpoons H^+ + DMO^- \qquad 10.48$$

The pK' of HDMO is 6.13 at 38°C and μ of 0.16 Eq/L. The acid and its conjugate base are believed to be inert metabolically, and neither form is bound to protein. The principle of the method rests upon the reasonable assumption that HDMO readily and freely penetrates cells, where it dissociates according to the intracellular pH. The DMO$^-$ so produced is assumed not to penetrate the cell membrane in either direction. If only tracer quantities of HDMO are used, the amount of H$^+$ produced by dissociation can be disregarded. At equilibrium, the HDMO concentration in cell water is assumed to be equal to the HDMO concentration of the extracellular fluid:

$$[HDMO]_i = [HDMO]_e \qquad 10.49$$

By measuring the concentration of total DMO (i.e., HDMO + DMO$^-$) in the intracellular fluid ($[DMO]_{Ti}$), the intracellular pH (pH$_i$) may be calculated by an equation analogous to Equation 3 in Table 10-1:

$$pH_i = 6.13 + \log \frac{[DMO_{Ti} - HDMO_e]}{[HDMO]_e} \qquad 10.50$$

The concentration of the undissociated acid in the extracellular fluid can be determined from the analysis of plasma for total DMO ($[DMO]_{Te}$) and blood pH (pH_e):

$$[HDMO]_e = \frac{[DMO]_{Te}}{1 + 10^{pH_e - 6.13}} \qquad\qquad 10.51$$

In effect the pH_i calculated from the DMO method gives an "average" or aggregate intracellular pH. The method quite obviously cannot differentiate between areas within single cells or areas within the total intracellular fluid which have different values for pH. Yet the method may be of some value in assessing *differences* in the pH, in which case some of these objections may not apply with equal force.

There is at present a small amount of literature dealing with intracellular pH changes in skeletal muscle as assessed by the DMO method. In addition, there has been an attempt to apply the method to the calculation of the aggregate pH of the intracellular space of animals or men by concomitant measurement of the total body water and extracellular volume (from which intracellular volume can be calculated by difference) and by calculating the amount of $(DMO)_T$ in the intracellular compartment from the difference between the total amount administered and the amount in the extracellular volume after equilibration. The results obtained by this method generally parallel those obtained with skeletal muscle, although since this specific application of the method is subject to a greater number of assumptions and greater sources of error, they cannot be accepted with the same degree of confidence as the results from isolated tissues.

Normal resting skeletal muscle has an intracellular pH of about 7.00. In acute respiratory disturbances, the pH_i varies inversely with the P_{CO_2} as would be expected if dissolved CO_2 readily penetrates the cell membrane. With a given change in the P_{CO_2} the degree of change of the pH_i of muscle is less than that of blood. This suggests that muscle may be a very well buffered tissue, particularly if the exchanges of bicarbonate occurring in the total body buffering experiments previously discussed involve muscle as the chief site of non-extracellular buffering.

In metabolic disorders produced by the acute loading of acid or of bicarbonate, the intracellular pH varies in the same direction as the extracellular. These changes in pH_i must reflect the net interaction of the effect of change in P_{CO_2} as well as the possible participation of muscle in buffering of the load of acid or bicarbonate. As in the case of respiratory disturbances, the degree of change of pH_i is less than that of blood pH. However, since the load of acid or bicarbonate buffered by muscle is not known, conclusions cannot be drawn about the relative buffer capacities of muscle versus blood.

The mechanisms by which muscle is apparently able to maintain the pH_i in the face of major changes in pH of the extracellular fluid are obscure. Recent experiments on the in vitro behavior of the rat diaphragm suggest that active metabolism is a necessity for the maintenance of this property. Such results may imply that active extrusion or uptake of H^+ or OH^- may be involved in the maintenance of the intracellular pH and point up the fallacy of regarding the intracellular pH as the result of purely physicochemical static factors.

Studies of the intracellular pH using the DMO method of the muscle of potassium-deficient animals have shown acid values compared to controls. However, the separate contributions of the elevated P_{CO_2} (due to respiratory compensation) and of the K^+ for H^+ exchange cannot be assessed from this single observation.

RESPIRATORY REGULATION OF ACID-BASE EQUILIBRIUM

General Effects on P_{CO_2}

Changes in pulmonary ventilation exert an effect upon the acid-base equilibrium of the body fluids by virtue of the fact that the P_{CO_2} of alveolar gas is equal to that of arterial blood. If the alveolar P_{CO_2} changes through a change in the rate of alveolar ventilation, the arterial P_{CO_2} will also change rapidly and in the same direction, and this in turn will change the pH of the blood according to the Henderson-Hasselbalch equation (Eq. 10.33).

In a steady state, the P_{CO_2} of the alveolar gas is directly related to the rate of production of CO_2 by the body and inversely related to the rate of alveolar ventilation:

$$P_{CO_2} \propto \frac{\text{rate of } CO_2 \text{ production}}{\text{rate of alveolar ventilation}} \qquad 10.52$$

In respiratory disorders, there is a primary abnormality in some component of the respiratory apparatus such that the rate of alveolar ventilation is set at some new abnormal level—higher than normal in respiratory alkalosis and lower than normal in respiratory acidosis. Under these new conditions, the lungs again excrete the CO_2 produced in the tissues, but at a new abnormal steady-state value for P_{CO_2}. In metabolic disorders, alveolar ventilation is altered as the result of altered activity of the respiratory center which is presumably responsive to some chemical factor(s) in blood. The direction of alteration of alveolar ventilation in these disorders is such as to tend to restore the blood pH to normal—there being hyperventilation in metabolic acidosis and hypoventilation in metabolic alkalosis. These phenomena are collectively known as respiratory compensation.

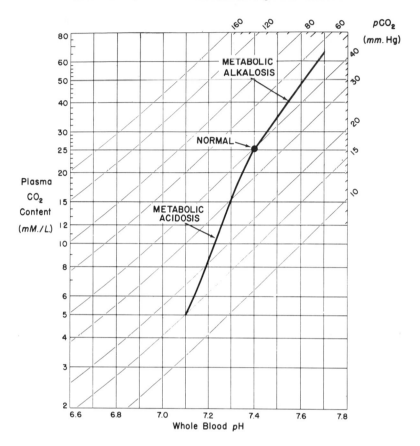

FIGURE 10-17.   General pattern of acid-base abnormality in metabolic disturbances. The grid is a graphic representation of the Henderson-Hasselbalch equation with log total CO_2 on the ordinate and pH on the abscissa. With these two variables fixed, Pco_2 is defined (scale upper right).

Figure 10-17 shows a summary of data on the general patterns of acid-base abnormalities in patients with metabolic acidosis and metabolic alkalosis. Although there is a good deal of scatter in the original data (particularly in alkalotic patients), it is quite clear that there is a general relationship between the degree of change of plasma Pco_2 and the degree of change of blood pH. In metabolic acidosis, for example, the lower the blood pH the lower the Pco_2—at least over the range of blood pH from 7.40 to 7.10. In metabolic alkalosis on the other hand, the higher the pH, the higher the Pco_2.

CHEMICAL STIMULI IN METABOLIC DISORDERS

These observations therefore demonstrate that respiratory compensation does occur and, in turn, pose the question as to the nature of the stimulus or stimuli to which the respiratory center responds in these situations. Three possible chemical stimuli are rocognized as able to cause changes in pulmonary ventilation—Pco_2, pH, and Po_2. While each of these factors may act independently of the others with respect to their effects upon the control of ventilation, two of the factors—Pco_2 and pH— are interdependent as shown by the Henderson-Hasselbalch equation (Eq. 10.33). Hence increasing Pco_2 will increase ventilation, but this effect cannot be ascribed simply to an increased Pco_2 since blood pH also falls. The infusion of strong acid, on the other hand, causes the pH to fall and alveolar ventilation to rise with a fall in the Pco_2. The effect on alveolar ventilation need not represent the pure effect of the lowered blood pH, since the stimulation due to the acid pH may be partially offset by an inhibition due to the lower Pco_2.

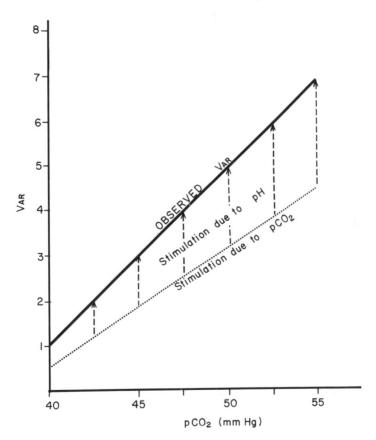

FIGURE 10-18.   Summation of stimuli in CO_2 inhalation. (After Gray, J. S., *Pulmonary Ventilation and its Physiological Regulation*. Chas. C Thomas, Springfield, Ill., 1950.)

Some years ago, Gray (1950) sought to examine these interrelationships and to quantitate the separate effects of pH, P_{CO_2}, and P_{O_2}. On the basis of a mathematical treatment of data in the literature, a single equation was derived which represented the separate independent contributions of each factor. The resulting equation is as follows:

$$V_{AR} = 0.262 \; P_{CO_2} + 0.22 \; [H^+] + \frac{105}{10^{0.038} \; P_{O_2}} - 18 \qquad 10.53$$

in which V_{AR} represents the ratio of alveolar ventilation in the experimental state over that of the control, and $[H^+]$ represents the concentration of the hydrogen ion in nanomoles per liter (10^{-9} moles/L). It should

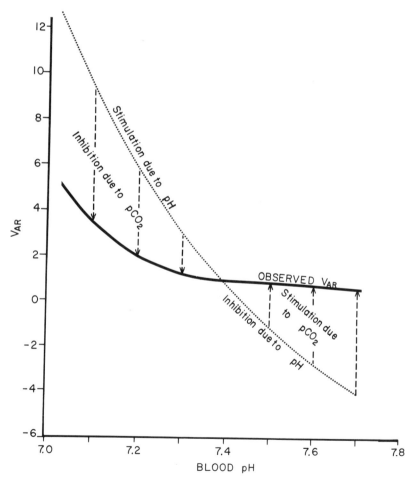

FIGURE 10-19. Summation of stimuli in metabolic acidosis and alkalosis. (After Gray, J. S., *Pulmonary Ventilation and its Physiological Regulation.* Chas. C Thomas, Springfield, Ill., 1950.)

be emphasized that the Gray equation provides no clue as to the mechanism by which any of these presumed stimuli exert their effects upon ventilation, nor was it intended to. It does provide an elegant representation of the respiratory performance of the subjects from which the data was drawn, and it contains the interesting suggestion that the chemical drive to ventilation may be represented by a summation of multiple factors. Each of the three factors is conceived to represent an independent stimulation or inhibition depending upon the direction away from normal in which it moves. Thus breathing CO_2 increases both Pco_2 and $[H^+]$, and these two effects summate to produce the increase in alveolar ventilation. This summation is illustrated in Figure 10-18. On the other hand, in metabolic disturbances of acid-base equilibrium, the final result in terms of alveolar ventilation represents the net effect of $[H^+]$ and of Pco_2, which in these cases move in opposite directions (Fig. 10-19).

From the outset, it was clear that certain phenomena could not be accounted for by the Gray equation. For example, at high altitude, Po_2 is lower than normal and would be the dominant drive to increasing ventilation, which would lower both Pco_2 and $[H^+]$ and so decrease their influence as stimuli. Yet studies of subjects during acclimatization to altitude reveal that alveolar ventilation, initially increased as would be predicted, shows a further increase over the first day or two with further falls in Pco_2 and $[H^+]$ despite the constancy of the Po_2 of the atmosphere.

There is suggestive but by no means conclusive evidence that an analogous situation occurs in metabolic acidosis in that, with the passage of time, there is an increase in alveolar ventilation compared to the acute initial value so that the initially low pH rises, and Pco_2 shows a further fall. Furthermore, studies of patients during recovery from metabolic acidosis demonstrate clearly that most if not all patients go through a period of "respiratory overshoot" in which Pco_2 is low, but pH is normal or even alkaline.

Thus, in all three of the above situations, there is evidence of a dissociation between the presumed stimuli (as predicted by Equation 10.52) and the response. Within the past few years, a new hypothesis has been formulated which may help to explain these and other findings. This hypothesis assigns a central role to the acid-base composition of the cerebrospinal fluid acting upon a medullary chemoreceptor as an additional factor in the control of pulmonary ventilation.

BLOOD–CEREBROSPINAL FLUID ACID-BASE RELATIONSHIPS

The major buffer system of cerebrospinal fluid (CSF) is the bicarbonate system, there being practically no protein and very little phosphate present. Therefore, unlike blood, the CSF bicarbonate concentration is independent of the Pco_2. In normal individuals, CSF bicarbonate con-

TABLE 10-7. RELATIONSHIPS BETWEEN ARTERIAL BLOOD AND CSF ACID-BASE STATUS IN ACUTE AND CHRONIC CONDITIONS

| | PLASMA | | | CEREBROSPINAL FLUID | | | | | |
| | | | | ACUTE | | | CHRONIC | | |
	pH	HCO₃	PCO₂	pH	HCO₃	PCO₂	pH	HCO₃	PCO₂
Metabolic acidosis	↓	↓↓	↓	↑	N°	↓	N	↓	↓
Metabolic alkalosis	↑	↑↑	↑	↓	N	↑	N	↑	↑
Respiratory acidosis	↓	↑	↑↑	↓	N	↑	N or ↓	↑	↑
Respiratory alkalosis	↑	↓	↓↓	↑	N	↓	N	↓	↓

* N = normal.

centration is approximately the same as the concentration of bicarbonate in arterial plasma water. CSF PCO_2, however, is consistently higher than arterial PCO_2 but is quite close to the PCO_2 of internal jugular venous blood. Thus compared to arterial blood, CSF has a lower pH and a higher PCO_2.

Considerable study has been devoted to the effects of alteration of the systemic acid-base equilibrium upon the acid-base status of the CSF. A summary of these results is shown in Table 10-7. In acute respiratory disturbances, changes in CSF pH follow those in blood because of rapid re-equilibration of CSF PCO_2 when the arterial (and internal jugular) PCO_2 is changed. These observations therefore suggest that the blood-CSF barrier is freely permeable to dissolved CO_2. CSF bicarbonate, however, shows little if any change from normal in short-term experiments, and therefore CSF pH is entirely a function of CSF PCO_2. Similarly in acute metabolic disturbances, CSF PCO_2 rapidly re-equilibrates, but CSF bicarbonate is unchanged so that the CSF pH varies inversely with the arterial pH.

A rather different picture emerges from studies of CSF acid-base status in more sustained disturbances. In these situations, there is evidence of readjustment of CSF bicarbonate in the direction of restoration of the CSF pH to normal. Thus at high altitude, CSF PCO_2 falls rapidly and, within a day or two, CSF bicarbonate has fallen and the CSF pH is normal. In patients with sustained metabolic acidosis or alkalosis, CSF bicarbonate is lower or higher than normal and the CSF pH is therefore normal. In respiratory acidosis, there is usually an increase in the CSF bicarbonate concentration, but complete restoration of CSF pH does not occur uniformly.

These two sets of results—acute versus sustained—suggest that some mechanism (perhaps an active transport process) exists which, given enough time, is capable of readjusting CSF bicarbonate concentration and hence CSF pH. The specific time relationships are not clearly established but they are probably of the order of a day or two for some if not most of the effect to become manifest.

CHEMORECEPTORS

Chemoreceptors sensitive to changes in the plasma P_{CO_2} and pH are known to occur in the aortic and carotid bodies. The mechanisms by which these receptors sense changes in these components are not clearly understood. The effect of P_{CO_2} could be mediated by a change in intracellular pH since, by analogy with other cells, dissolved CO_2 probably passes readily into these cells. Extracellular pH might be detected through a change in the hydrogen ion gradient between the intracellular and the extracellular fluids. To the authors' knowledge neither of these speculations has as yet been tested, primarily because of lack of appropriate methods for the measurement of intracellular pH.

It is likely that the P_{CO_2} also exerts a central effect independent of the peripheral chemoreceptors, since some but not all of the respiratory stimulation produced by inhalation of CO_2 is abolished by cutting the nerves from the carotid and aortic bodies.

There is now evidence that, in addition to the peripheral receptors, there are superficial medullary chemoreceptors which are sensitive to CSF pH but not specifically sensitive to P_{CO_2}. The action of these receptors will be discussed in more detail.

RESPIRATORY CONTROL IN METABOLIC ACIDOSIS

Figure 10-20 depicts the interplay of stimuli in the respiratory response in acute metabolic acidosis. Addition of strong acid (or loss of bicarbonate) produces a lowering of plasma bicarbonate and an acid pH.

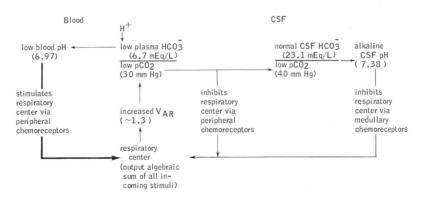

FIGURE 10-20. Development of respiratory compensation for metabolic acidosis. The numbers in this figure represent values for plasma and CSF after the addition of the acid load but before respiratory compensation has been fully developed and before the CSF bicarbonate has been readjusted.

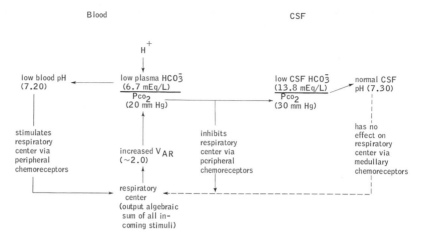

FIGURE 10-21. Steady state respiratory compensation for metabolic acidosis. The numbers represent typical values after the CSF bicarbonate has been readjusted and respiratory compensation is fully developed.

This stimulates the peripheral chemoreceptors and constitutes the main drive to the respiratory center. By virtue of the increase in alveolar ventilation, Pco_2 falls and this has two consequences: it may provide an inhibiting stimulus via the peripheral chemoreceptors, and it causes the CSF Pco_2 to fall and the CSF pH to rise. This latter effect provides another inhibitory stimulus which is mediated by the medullary chemoreceptors. Thus the net drive to ventilation in acute metabolic acidosis is conceived to be composed of a stimulation due to low blood pH, the magnitude of which more than offsets the combined inhibitions of low plasma Pco_2 and alkaline CSF pH.

With the continuing addition of strong acid (or continuing loss of bicarbonate), the acidosis becomes sustained, and the CSF bicarbonate concentration falls so that the CSF pH returns toward normal. Figure 10-21 depicts the hypothetical new steady state at which the CSF pH is completely restored to normal. This readjustment of CSF pH has the effect of lessening the inhibitory drive, and accordingly alveolar ventilation shows a further increase with a rise in the blood pH. In other words, respiratory compensation improves. The blood pH, however, is never restored completely to normal in progressive metabolic acidosis, since the low pH is the principal stimulus to the respiratory center.

The "respiratory overshoot," so characteristic of the recovery pathway of metabolic acidosis, can also be explained by the interplay of blood and CSF stimuli. This is depicted in Figure 10-22. Abrupt restoration of the plasma bicarbonate concentration to normal, either through administration of exogenous sodium bicarbonate or through renal correction of the acidosis (see the following discussion), leads to an increase in

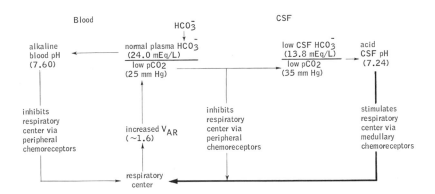

FIGURE 10-22. Respiratory alkalosis in recovery from metabolic acidosis. The numbers in this figure represent values for both plasma and CSF which might be found at a time when the plasma bicarbonate has been returned to normal but before the CFS bicarbonate has been readjusted back to normal.

the blood pH. This in turn causes some decrease in alveolar ventilation and the P_{CO_2} rises from its previous low level, although normal values are not achieved. This increase in P_{CO_2}, however, is sufficient to cause an acid shift in the CSF pH, and this in turn produces a stimulation via the medullary chemoreceptors. The final balance of stimuli is therefore the net of stimulation produced by acid CSF, inhibition produced by lowered P_{CO_2}, and inhibition produced by rising blood pH. As the CSF bicarbonate readjusts, the drive from CSF pH lessens until normal P_{CO_2} is restored.

RESPIRATORY CONTROL IN METABOLIC ALKALOSIS

Although severe metabolic alkalosis is generally characterized by some compensatory increase in P_{CO_2}, milder cases do not always appear to have detectable respiratory compensation, and in general the degree of compensation is rather less impressive than in metabolic acidosis. Presumably in severe cases, there is a balance of stimuli opposite to those in metabolic acidosis, with high blood pH tending to inhibit while low CSF pH tends to stimulate ventilation. In sustained alkalosis, the latter is presumably lessened by the increased bicarbonate concentration of CSF. Figure 10-23 depicts the presumed interaction of these stimuli. During recovery one would expect to see a transient respiratory acidosis, although this has not been documented as yet.

Two factors have been implicated as possibly limiting the respiratory compensation in metabolic alkalosis. First, it has been pointed out that

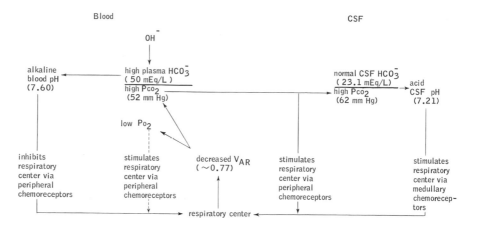

FIGURE 10-23.   Development of respiratory compensation for metabolic alkalosis. The numbers in this figure represent values which might pertain at a time after plasma HCO₃ has increased but before the CSF bicarbonate has been readjusted.

any large reduction in alveolar ventilation would lead to a fall in arterial Po_2 and hence provide another stimulus to breathe. Second, there may be a difference between subjects who have simple metabolic alkalosis and those who have metabolic alkalosis complicated by large potassium deficits, since in the latter, the K^+ for H^+ shift, known to occur in muscle, might also involve cells of the chemoreceptors or of the respiratory center, and the resultant intracellular acidosis might provide an additional stimulus for respiration. Further work is required in order to evaluate the potential contribution of each of these factors.

RENAL REGULATION OF ACID-BASE EQUILIBRIUM

General Effects of the Kidney in Acid-Base Disturbances

The kidney plays an important role in the maintenance of normal acid-base equilibrium of the body fluids in health and in compensating for or correcting acid-base disturbances in disease. A normal adult eating an average diet ingests the equivalent of about 70 mEq of strong acid per day. This is sometimes called "metabolic" hydrogen ion, since it represents acid products of the metabolism of various foodstuffs. It has been estimated that if none of the metabolic hydrogen ion were excreted by the kidney, the total body buffer would be consumed in about ten days. In fact, however, the kidney does excrete this acid, and thereby the buffers of the body fluids are maintained intact.

Reference has already been made to the compensatory and corrective roles which the kidney plays in various disorders of acid-base equilibrium (see p. 205). Thus, in respiratory disorders, there is a compensatory change in the concentration of bicarbonate in the plasma in the direction of restoration of the pH toward normal. In part these compensatory changes are a product of the non-extracellular buffers and in part they are the product of more prolonged renal action. In metabolic disorders, on the other hand, the kidney provides the ultimate means of correction. In metabolic acidosis due to loading with strong acid, the urine is the only route by which the acid can be excreted, and as will be shown, the elimination of acid occurs pari passu with the synthesis of new bicarbonate. The kidney may also play a corrective role in metabolic alkalosis, although its ability to do so is limited. In some types of metabolic alkalosis, however, the excretion of the excess bicarbonate and the restoration of bicarbonate concentration to normal proceed via the urine.

RENAL HANDLING OF BICARBONATE

The total effect which the kidney exerts upon systemic acid-base equilibrium is mediated by its ability to manipulate the bicarbonate concentration of the plasma, since, aside from small amounts of inorganic phosphate, bicarbonate is the only conjugate base of any of the buffer systems of the body fluids to which the renal tubules have direct access. From a physicochemical point of view, however, only bicarbonate need be regulated by physiological means, since by virtue of the interrelationships between various buffers in the blood (see Eq. 10.40) and between blood buffers and the remainder of the buffers in other compartments of the body fluids (see p. 211), all body buffers will be automatically regulated through physiological regulation of bicarbonate alone.

From the point of view of acid-base regulation, the kidney, under appropriate circumstances, can perform three basic operations: it can conserve existing bicarbonate stores; it can excrete strong acid, thereby making new bicarbonate available to the body; or it can excrete excess bicarbonate, thereby reducing the amount of bicarbonate in the body.

The first step in the renal manipulation of bicarbonate consists of filtration at the glomeruli. Bicarbonate, like many other small ions and molecules, is freely filterable through the glomerular membrane, and the amount of bicarbonate filtered is therefore the product of the concentration of bicarbonate in the plasma water and the glomerular filtration rate. In health the total amount of bicarbonate filtered each day by an adult is:

$$26 \text{ mEq/L} \times 180 \text{ L/d} = 4680 \text{ mEq/d}$$

(concentration   (glomerular
in plasma water) filtration rate)

10.54

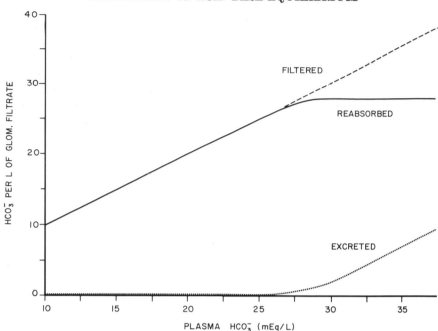

FIGURE 10-24.   The renal threshold for bicarbonate.

This is a huge amount, being several times that total amount of bicarbonate available in the whole of the extracellular fluid volume. Under these conditions, the kidney must return all the filtered bicarbonate to the body or else the body bicarbonate stores will be rapidly depleted. The process by which filtered bicarbonate is returned to the plasma is called tubular reabsorption.*

Figure 10-24 demonstrates the overall operation of the tubular reabsorptive mechanism for bicarbonate under conditions in which the plasma bicarbonate (and hence the filtered load of bicarbonate) is varied. At low plasma concentrations, no bicarbonate is excreted, but as the plasma concentration approaches the normal value, small amounts of bicarbonate appear in the voided urine. As the plasma concentration increases to supernormal levels, large quantities of bicarbonate are excreted. Bicarbonate thus behaves as a threshold substance, with the normal threshold being set at about 26 mEq/L in man and slightly lower in the dog. This renal threshold, however, is not constant under all conditions, as will be discussed later.

---

* Certain conventions of terminology should be observed with respect to renal mechanisms. *Reabsorption* refers to the process of movement of a solute or water from tubular to peritubular plasma; *secretion* refers to movement from peritubular plasma to tubular lumen; *excretion* is the appearance of a solute in the final urine regardless of the renal mechanism by which the solute is handled.

It is important to recognize the effects which this pattern of tubular reabsorption of bicarbonate can have upon plasma bicarbonate concentration. If all the filtered bicarbonate is reabsorbed (i.e., below the threshold) then the extracellular bicarbonate content is unchanged, and if there are no major changes in the volume of that compartment, the plasma bicarbonate concentration will be unchanged. On the other hand, when plasma bicarbonate is above the threshold concentration, bicarbonate is excreted and the extracellular bicarbonate content and therefore the plasma concentration will fall. In other words, the threshold behavior of the tubular reabsorptive mechanism can reduce plasma bicarbonate concentration if it is high, or it can maintain a subthreshold concentration at the same level, but it cannot by this mechanism *alone* increase the concentration.

Another renal mechanism is required in order to increase the plasma bicarbonate concentration, and this mechanism is linked to acidification of the urine. The overall process may be looked upon as follows:

$$CO_2 + H_2O \quad \rightarrow \quad H_2CO_3 \quad \rightarrow \quad H^+ \quad + \quad HCO_3^-$$

$CO_2 + H_2O$	H_2CO_3	H^+	HCO_3^-	
(from plasma)	(renal tubular cell)	(excreted in urine)	(returned to plasma)	10.55

In this process, the renal tubular cell produces H_2CO_3 and then separates H^+ and HCO_3^- ions, secreting the H^+ ion into the tubular urine while returning the HCO_3^- ion to the plasma as new bicarbonate.

Intrinsic Renal Mechanisms

BICARBONATE REABSORPTION

Much experimental work supports the scheme diagramed in Figure 10-25 as the mechanism by which filtered bicarbonate is reabsorbed by the tubules. In this scheme, CO_2 is hydrated, the reaction being speeded by carbonic anhydrase, which is present in high concentration in the renal cortex. The carbonic acid so formed dissociates, and the hydrogen ions are secreted by the tubular cell into the tubular urine in exchange for sodium which is reabsorbed. This ion exchange maintains electroneutrality at the luminal border of the cell. The hydrogen ion secreted into the tubular urine reacts with filtered bicarbonate to form carbonic acid. The carbonic acid undergoes non-catalyzed dehydration, the CO_2 so formed diffusing back into the cell. The bicarbonate formed within the cell enters the peritubular plasma; electroneutrality at this site is maintained by the entry into the peritubular plasma of the sodium which was reabsorbed. The overall result of the mechanism is the reabsorption of sodium bicarbonate, although it should be noted that the particular bicarbonate ion which is filtered is not the same one that is reabsorbed.

LUMEN   CELL   BLOOD

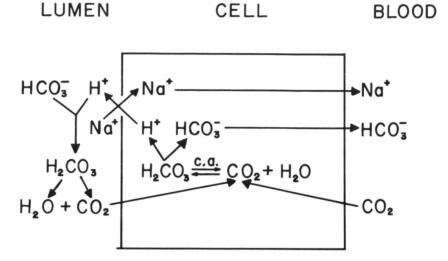

FIGURE 10-25. The renal mechanism for bicarbonate reabsorption.

It is probable that this mechanism is present in both proximal and distal portions of the nephron. In the proximal tubule, the mechanism operates to reabsorb most of the filtered bicarbonate, but it is capable of working only against low gradients for bicarbonate between tubular urine and plasma. In the more distal portions of the nephron, the remainder of the filtered bicarbonate is removed. Since the final urine can be virtually free of bicarbonate, the distal mechanism (which may be largely in the collecting duct epithelium rather than in the distal tubule) can reabsorb bicarbonate against very high gradients.

BICARBONATE EXCRETION

Under some conditions not all of the bicarbonate which is filtered is reabsorbed, and bicarbonate appears in the final urine. Electroneutrality requires that bicarbonate be excreted with a cation, and either sodium or potassium is available for this purpose. The amounts of these cations excreted with the bicarbonate are dependent largely upon the conditions. If a healthy individual ingests sodium bicarbonate, most of the bicarbonate is excreted with sodium. On the other hand, in subjects previously depleted of sodium who thus have stimuli causing the kidney to conserve sodium, bicarbonate is excreted largely with potassium.

The interrelationships between potassium, sodium, and bicarbonate are shown in Figure 10-26. One may suppose that there is a competition within the tubular cells between hydrogen ions (produced as outlined in Figure 10-26) and potassium ions in exchanging for sodium which is

LUMEN   CELL   BLOOD

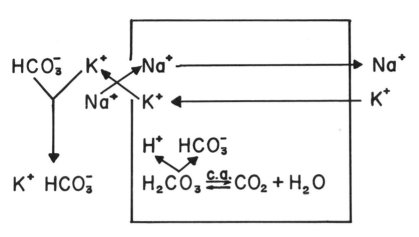

FIGURE 10-26.   The renal mechanism for bicarbonate excretion.

being reabsorbed. Under conditions in which bicarbonate reabsorption is complete, H^+ dominates K^+ for the exchange mechanism, whereas when bicarbonate is to be excreted, K^+ dominates H^+ for the sodium being reabsorbed and the final urine contains bicarbonate and potassium.

PRODUCTION OF NEW BICARBONATE (ACIDIFICATION OF THE URINE)

The third operation which the kidney can perform in the interests of maintenance of acid-base equilibrium is the production of new bicarbonate for the extracellular fluid incident to acidification of the urine. As discussed above, the source of the new bicarbonate is carbonic acid, from which the H^+ is secreted in the urine and the HCO_3^- is returned to the plasma. The kidney cannot eliminate large quantities of free H^+, since even at the lower limit of urine pH (pH about 4.0), only 0.1 mEq of free H^+ will be excreted per liter of urine. In order to eliminate large quantities of hydrogen ions within this lower limit of urine pH, the hydrogen ions must be buffered. Two renal mechanisms exist for buffering of such hydrogen ions: the formation of titratable acid and the formation of ammonia.

Figure 10-27 depicts the mechanism for the formation of titratable acid. The same mechanism operative in the reabsorption of bicarbonate (Fig. 10-25) makes H^+ available from carbonic acid. The H^+ is secreted in exchange for sodium which is reabsorbed. The H^+ reacts with dibasic phosphate present in the filtrate to form monobasic phosphate:

$$HPO_4^= \; + H^+ \rightarrow \quad H_2PO_4^-$$
$$\text{(dibasic)} \qquad\qquad \text{(monobasic)} \qquad\qquad 10.56$$

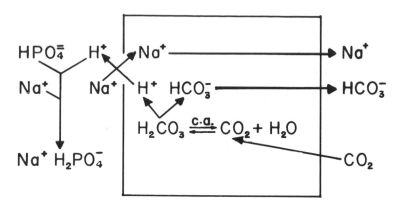

FIGURE 10-27.   The renal mechanism for the formation and excretion of titratable acid.

Thus for every milliequivalent of monobasic phosphate formed by this reaction, one H^+ secreted by the tubule is bound and carried into the final urine.*

Figure 10-28 depicts the second renal mechanism involved in acidification of the urine—namely the formation of ammonia. In this mechanism, an additional base (NH_3) is formed by deamination. Being

---

* Titratable acid is determined by back-titrating the urine with base to the pH of the glomerular filtrate (plasma):

$$H_2PO_4^- + OH^- \rightarrow HPO_4^= + H_2O \qquad\qquad 10.57$$

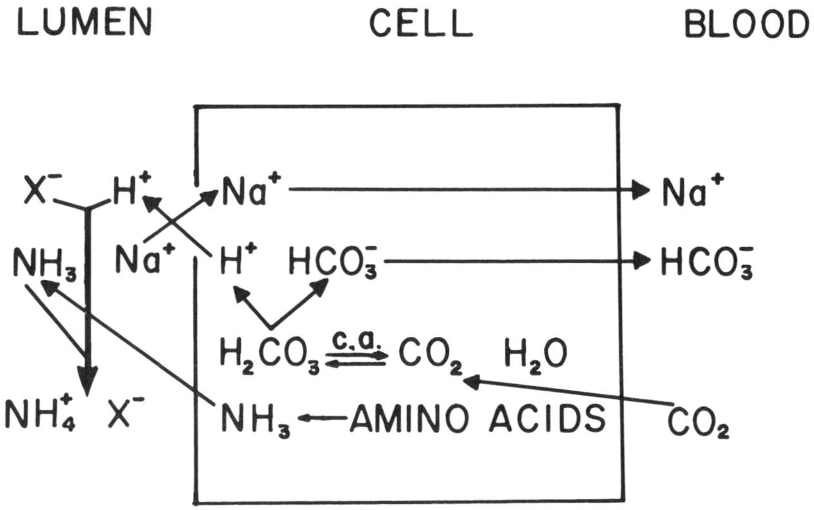

LUMEN   **CELL**   **BLOOD**

FIGURE 10-28.   The renal mechanism for the formation and excretion of ammonia.

uncharged and lipid soluble, the NH_3 diffuses into the tubular urine where it reacts with H^+ secreted by the ion exchange mechanism:

$$NH_3 + H^+ \rightarrow NH_4^+ \qquad\qquad 10.58$$

The ammonium ion so formed is charged and is not lipid soluble and is therefore trapped in the tubular urine and excreted in the final urine.

There are certain similarities and certain differences between these two mechanisms for acidification of the urine. Both provide a means of carrying H^+ from the body in a bound or buffered form. In both cases, for every mEq of H^+ so excreted, 1 mEq of new bicarbonate ion is made available to the extracellular fluid as it enters the peritubular plasma along with the sodium which has been reabsorbed. The differences between the two mechanisms concern their quantitative contributions to acidification. The amount of titratable acid excreted depends to a large extent upon the amount of phosphate buffer available in the tubular urine. Since the plasma inorganic phosphate is never very high (2 to 3 mM/L) and since some phosphate is reabsorbed by the tubules, the amount of titratable acid which can be made from the available phosphate is small. This is in contrast to the ammonia mechanism which has a very large capacity. In well-established metabolic acidosis, for example, the ammonium excretion may be five or more times that of titratable acid. In order to achieve this high capacity, however, several days are required in order to "tool up" the ammonia-producing mechanism. In contrast, titratable acid formation, being dependent only upon the secretion of H^+ and the availability of phosphate, is available immediately.

Present evidence indicates that the formation of both titratable acid and ammonia proceed both in proximal and distal portions of the nephron. In other words, acidification as well as bicarbonate reabsorption in these sites accounts for the secreted H^+. These interrelationships will be explored in the following section in the consideration of stimuli controlling the secretion of H^+.

Factors Controlling the Tubular Secretion of H^+

Components of the Total Tubular Secretion of H^+

In considering the factors controlling the tubular secretion of H^+ by the kidney, it is important to recall that the total amount of H^+ secreted is the sum of the H^+ added to HCO_3^- and then reabsorbed as bicarbonate plus the additional H^+ which is excreted in the final urine as titratable acid and ammonium ion. Of these several moieties, the reabsorption of bicarbonate requires by far the largest amount of H^+ under all conditions. This is illustrated in Table 10-8 which shows examples of the amount of H^+ used for each of these purposes in various disturbances of acid-base equilibrium.

In normal subjects ingesting an average diet, 70 mEq of metabolic H^+ is excreted as titratable acid and ammonium. This, however, represents only 1.5 per cent of the total H^+ secreted, the remaining 98.5 per cent being used to reabsorb filtered bicarbonate. Even in metabolic acidosis in which the filtered load of bicarbonate is less than half of normal, more than three fourths of the secreted H^+ is used for bicarbonate reabsorption, while in all the other situations depicted in Table 10-8, this process claims from 97 to 100 per cent of the secreted H^+. The important point about these illustrative data is that in seeking factors which regulate the acidification mechanism one must take account of the total H^+ secreted and not merely the H^+ which is excreted in the final urine.

Relationship between Total H^+ Secreted and Plasma P_{CO_2}

If the data in Table 10-8 are examined, one finds that, compared to normal, total H^+ secretion is reduced in both metabolic acidosis and respiratory alkalosis, while in respiratory acidosis and in complex forms of metabolic alkalosis, H^+ secretion is increased. In simple metabolic alkalosis due to loading with sodium bicarbonate, however, H^+ secretion

TABLE 10-8. TUBULAR SECRETION OF H^+ IN DISTURBANCES OF ACID-BASE EQUILIBRIUM*

	NORMAL	METABOLIC ACIDOSIS	RESP. ACIDOSIS	RESP. ALKALOSIS	SIMPLE METABOLIC ALKALOSIS DUE TO INGESTION OF NaHCO₃	COMPLEX METABOLIC ALKALOSIS
GFR (L/24 hr)	180	180	180	180	180	180
[HCO₃]ₚw (mEq/L)	26	10	30	20	28	35
Bicarbonate filtered (mEq/24 hr)	4680	1800	5400	3600	5040	6300
Bicarbonate excreted (mEq/24 hr)	0	0	0	25	190	0
Bicarbonate reabsorbed (mEq/24 hr)	4680	1800	5400	2575	4850	6300
T. A. + NH₄⁺ excreted (mEq/24 hr)	70	300	140	0	0	100
Total H⁺ secreted (mEq/24 hr)	4750	2100	5540	3575	4850	6400
% secreted for HCO₃	98.5%	85.8%	97.4%	100%	100%	98.5%
% secreted for acidification	1.5%	14.2%	2.6%	0%	0%	1.5%
P_{CO_2} of plasma (mm Hg)	Normal	Low	High	Low	Normal†	High†
pH of urine	Acid	Acid	Acid	Alkaline	Alkaline	Acid
pH of blood	Normal	Acid	Acid	Alkaline	Alkaline	Alkaline

* The data are idealized and chosen to make the specific points in the text. They are, however, representative of actual data in these various conditions.

† Respiratory compensation in mild metabolic alkalosis is minimal (see p. 225), whereas in severe metabolic alkalosis it is more likely to be present.

is normal. If total H⁺ secretion is the important variable, then it is apparent from the pattern of results shown in Table 10-8 that it correlates best with changes in plasma P_{CO_2}, and that it does not correlate with changes in either blood pH or urine pH.

The hypothesis of a regulatory influence of plasma P_{CO_2} upon H⁺ secretion has been directly tested by systematic variation of plasma P_{CO_2} over a wide range in the dog and measuring changes in the amount of bicarbonate reabsorbed at each value for P_{CO_2}. The smoothed curve fitted to a large number of such determinations is shown in Figure 10-29 and demonstrates the curvilinear relationship between plasma P_{CO_2} and the bicarbonate threshold.

Such results as these substantiate the view that plasma P_{CO_2} is an important regulator of the amount of H⁺ secreted by the renal tubules. Although the mechanism by which the P_{CO_2} alters the operation of the H⁺-Na⁺ ion exchange is uncertain, it is possible that it is mediated through a change in the intracellular pH. Thus when the P_{CO_2} is high, the intracellular pH falls and this leads to augmented H⁺ secretion. The fate of the H⁺ which is secreted is then dependent upon the composition of the tubular urine which it encounters. If the bicarbonate concentration is high, most of the H⁺ will be used for bicarbonate reabsorption. On the

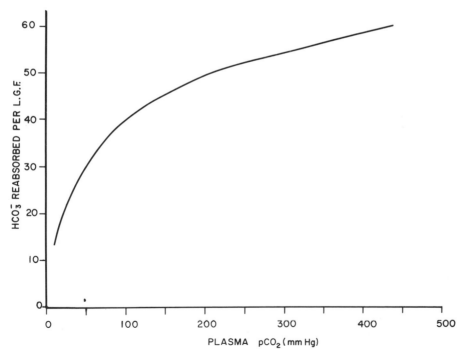

FIGURE 10-29. The effect of P_{CO_2} on bicarbonate reabsorption. (After Rector et al., *J. Clin. Invest.*, 39:1706, 1960.)

other hand if the bicarbonate concentration is not high, all the bicarbonate will be reabsorbed, and the remaining H^+ will be excreted as titratable acid or ammonium. This later effect will in turn lead to the generation of new extracellular bicarbonate as shown in Figures 10-27 and 10-28.

By contrast, in situations in which the Pco_2 is low, the absolute rate of secretion of H^+ is diminished. If, as in metabolic acidosis, the bicarbonate concentration of the filtrate is markedly reduced, all the bicarbonate can be reabsorbed even with a diminished supply of H^+, and enough H^+ remains to provide for a large excretion of titratable acid and ammonium. In acute respiratory alkalosis, on the other hand, a comparable reduction of Pco_2 in the face of a high bicarbonate load leads to only partial reabsorption of filtered bicarbonate, the remainder being excreted in an alkaline urine.

RELATIONSHIP BETWEEN H^+ SECRETION AND POTASSIUM STORES

There is a variety of experimental and clinical evidence suggesting that body potassium stores can condition the amount of H^+ secreted by the tubule. These interrelationships can be visualized in terms of a presumed competition between intracellular K^+ and intracellular H^+ of the tubular epithelium for the reabsorption of sodium (Fig. 10-26). The infusion of potassium accentuates the K^+ exchange for Na^+, and the urine becomes more alkaline because of the suppression of H^+ secretion. On the other hand, with the depletion of body potassium, K^+ in the tubular cells presumably falls, and H^+ dominates the exchange. Thus the urine is acid and contains titratable acid and ammonium (paradoxical aciduria) despite the systemic metabolic alkalosis (see p. 214).

Paradoxical aciduria in metabolic alkalosis with potassium deficiency has the consequence of failure of adequate renal correction of the disturbance. Indeed to the extent that the H^+ secretion is augmented, bicarbonate reabsorption is increased, thus perpetuating or even aggravating the alkalosis. It is possible that these effects of intracellular K^+ concentration within the tubular epithelium are mediated by a change in the intracellular pH if the K^+ loss proceeds in a manner similar to that in which it occurs in skeletal muscle (Fig. 10-15).

A further set of interrelationships is thus evident between potassium and metabolic alkalosis at the level of the kidney. These provoke the interesting speculation that sustained metabolic alkalosis leads to a loss of potassium through initial renal correction by excretion of the excess bicarbonate with potassium in an alkaline urine (i.e., simple metabolic alkalosis). If potassium intake is precluded, these urinary losses of potassium (particularly if they are accompanied by additional extrarenal losses of potassium) are met by a loss of muscle potassium, since the later is the principal available store of this ion in the body. Loss of muscle K^+ in turn leads to further alkalosis by the mechanism shown in Figure 10-15. Eventually potassium depletion reaches a point where continuing expenditure of this ion via the kidney is reduced and paradoxical aciduria with its attendant consequences supervenes. Thus metabolic alkalosis begets potassium depletion, and potassium depletion begets metabolic alkalosis.

RELATIONSHIP BETWEEN H^+ SECRETION AND HYPOCHLOREMIA

Until recently it has generally been assumed that the chloride ion per se had no significant or specific physiological effects. Recent evidence, however, demonstrates that chloride exerts an important effect upon the renal tubular mechanisms for the secretion of H^+.

It has been demonstrated in a number of types of experiments that in states of hypochloremia—including those in which chloride is replaced by another anion (such as nitrate)—the quantity of bicarbonate reabsorbed by the kidney is increased. A possible explanation for this effect follows from the mechanism by which sodium and chloride are reabsorbed by the tubular cells. Normally, sodium is returned to the peritubular plasma by a process of active transport in which energy derived from metabolism is the driving force. The transport of this positively charged particle creates an electrical potential difference such that the peritubular plasma is positive to the tubular urine. Chloride movement is passive, being downhill electrically, as the result of the potential difference created by the sodium transport. In a steady state, the net effect of the active movement of sodium and the passive movement of chloride leads to a potential difference of about 20 mv, tubular urine being negative to peritubular plasma. If the concentration of chloride within the tubular urine falls while that of sodium is unchanged, a larger potential difference results. As a consequence, this causes an augmented movement of H^+ from within the cell into the tubular lumen which is now more negative than normal. The augmented secretion of H^+ in turn increases bicarbonate reabsorption.

This phenomenon of augmented H^+ secretion in hypochloremia is of importance in both metabolic alkalosis and respiratory acidosis. In the former case, the chloride concentration is most often reduced because of loss of gastric acid due to vomiting. In the latter case, the chloride concentration falls because of renal excretion as a compensation, the chloride accompanying NH_4^+ as the kidney acidifies the urine and generates new bicarbonate (see Figure 10-28 where in this case X^- represents Cl^-). In either case hypochloremia, once established, provides a potent stimulus for continuing bicarbonate reabsorption, causing a perpetuation of the metabolic alkalosis but providing an additional compensation in respiratory acidosis.

Notes on References

The basic reference for the physical chemistry of the blood buffers is Peters and Van Slyke (1931), Chapters 12 and 18 of which have recently been reprinted separately. Clark (1952) contains a very readable discussion of the elementary aspects of blood buffers, while Edsall and Wyman (1958) discuss the same topic on a more advanced level. O. Siggaard-Andersen (1964) gives an excellent discussion of modern concepts of blood buffers, including ingenious new techniques for measurement of the acid-base status of blood.

A comprehensive treatment of the subject of blood gas transport is given by Roughton (1954). Gray (1950) summarizes his hypothesis concerning the chemical control of respiration. An article by R. A. Mitchell et al. (1963) discusses the role of the medullary chemoreceptors in the control of respiration.

Discussion of intracellular buffers and their role in acid-base metabolism can be found in Welt (1957). The DMO method for measuring the intracellular pH is presented by Waddell and Butler (1959). The role of the kidney in acid-base balance is authoritatively explained by Pitts (1963). An excellent survey of both physiological and physicochemical aspects of acid-base equilibrium is presented by Welt (1959).

Research in the authors' laboratories is supported by Grant HD-00117 from the National Institutes of Health and Grant U-1127 from the Health Research Council of the City of New York. The authors also acknowledge with thanks the following awards: Dr. Winters, Career Scientist award I-309 from the Health Research Council of the City of New York; and Dr. Dell, Fellowship 5-F2-AM-19, 779 from the National Institutes of Health.

Chapter 11

ENERGY EXCHANGE IN STRIATED MUSCLE CELLS

SAUL WINEGRAD

The functions of a single cell in a multicellular organism are of two types: first are those which it carries on for its own survival, assisted by extracellular fluid of controlled composition, and which are similar for all cells; second are those, more highly developed or specialized, which may be important for the survival of the organism as a whole. These divisions are somewhat arbitrary and are far from mutually exclusive, but they serve as a framework for discussing the control of cellular functioning. Almost all cells carry on those *general* activities such as energy production, protein synthesis, and active transport of materials across the cell boundary. Regulation of these processes is in accordance with the activity of the given cell, and the control mechanisms are localized within that cell.

The more *specialized* activities such as contraction, impulse conduction, or the secretion of digestive enzymes are controlled by extrinsic systems and so involve intercellular and interorgan reactions. The general functions of the cell are influenced to varying degrees by the level of activity of the more specialized functions. Extrinsic systems, therefore, exert an indirect influence over these cellular reactions concerned more with survival. For example, in muscle cells the rate of energy production is closely dependent upon the degree of activity of its specialized function, contraction (Carlson, Hardy, and Wilkie, 1963), while its rate of oxidative metabolism is also determined, like that of all the cells, by the secretion of the thyroid gland. Nevertheless, in the muscle cell the distribution

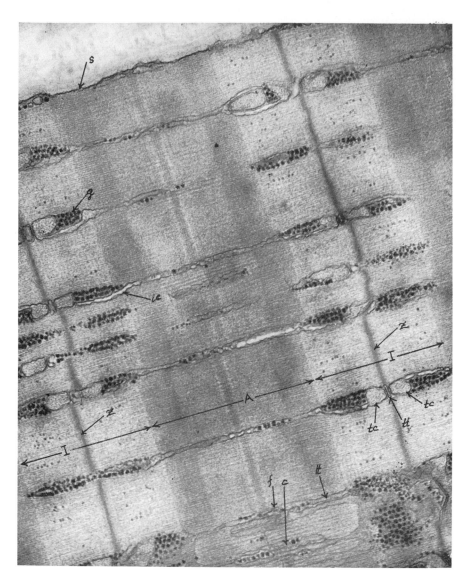

FIGURE 11-1. Longitudinal section of a portion of frog *(Rana pipiens)* sartorius muscle fiber. This muscle was fixed first in 6 per cent glutaraldehyde, then in 1 per cent osmium tetroxide, and embedded in epoxy resin. Dense A bands (A), lighter I bands (I), and dark Z lines (Z) of several myofibrils are seen, with sarcoplasmic spaces between. The sarcolemma (s) appears at the top of the figure. The dark granules (g) are glycogen. Various portions of the sarcoplasmic reticulum are indicated, including terminal cisternae (tc), intermediate cisternae (ic), longitudinal tubules (lt), and fenestrated collars (fc). Also indicated is a transverse tubule (tt) in the center of a triad at the lower right. Longitudinal tubules and fenestrated collars are best seen in the "face view" at the bottom of the figure, where the plane of the section grazes the surface of a myofibril. X 50,000. (Peachey, L. D., *Fed. Proc.* 24, in press, 1965.)

between non-specialized and specialized function is relatively clear. The resting fiber may be said to be egotistical, and the contracting fiber to show community spirit.

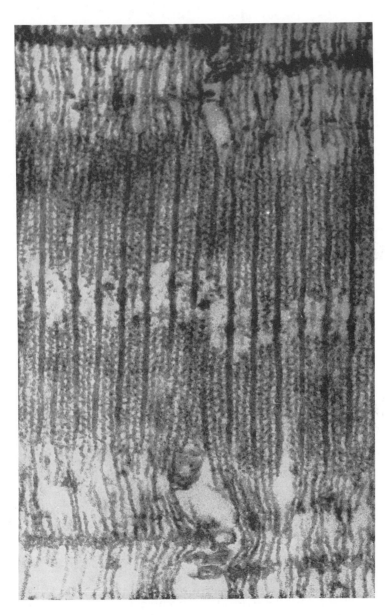

FIGURE 11-2.   A high-magnification electron micrograph of a longitudinal section of vertebrate skeletal muscle showing the overlapping thick and thin filaments. Cross bridges are visible on the thick filaments. (Huxley, H. E., In Brachet, J., and Mirsky, A., *The Cell*. Academic Press, New York, 1960.)

MUSCLE STRUCTURE

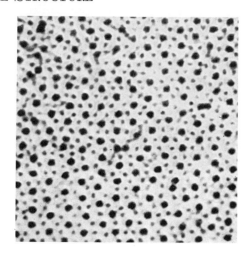

FIGURE 11-3. Electron micrograph of a cross section of vertebrate skeletal muscle. The double hexagonal array of thick and thin filaments that exists in the A band can be seen. Each thick filament is surrounded by six thin filaments, and each thin filament, by three thick filaments. (Huxley, H. E., in Brachet, J., and Mirsky, A., *The Cell.* Academic Press, New York, 1960.)

Whether the muscle cell is relaxed or contracting, its internal activity is closely related to the fiber's highly developed structure. The cell boundary is a membrane believed to consist of two lipid layers covered on each side with a layer of protein. The cell interior contains, in addition to nuclei and mitochondria (which are, respectively, the control centers of protein synthesis and the sites of oxidative metabolism), a large amount of protein organized in such a way as to produce a repeating pattern of alternating light and dark bands called I and A bands (Fig. 11-1). In the center of the I band is a dark line, the Z line, and in the center of the A bands is a light zone called the H zone and a dark line, the M line. Striations are produced by an interdigitation of thick and thin filaments (Fig. 11-2). The thick filaments are composed primarily of the protein, myosin, and the thin filaments of another protein, actin. A third protein, tropomyosin, is present in the Z line and possibly in the thin filaments. The A band is produced by the thick filaments, the I band by the thin filaments, and the darker portions of the A band on either side of the H zone by the overlapping of the thick and thin filaments. The filaments, which are gathered into small bundles called myofibrils, are highly oriented transversely as well as longitudinally (Fig. 11-3). The thick filaments are arranged in a hexagonal array, and each thick filament is surrounded by six thin filaments.

Between the myofibrils are the mitochondria and a longitudinal tubular system called the endoplasmic reticulum, which forms two sacs, the lateral cisternae, at each Z line (Fig. 11-1). These pairs of sacs are separated by a transverse structure, the central tubule, which runs across the entire diameter of the cell and opens into the extracellular space. The central tubule is believed to be an invagination of the surface membrane.

A skeletal muscle cell is about 50 to 150 μ in diameter and may be as much as several centimeters long. These are considerable distances for the process of diffusion. Rapid transmission of information from the surface to the interior and along the cell length requires some process faster than diffusion.

THE RESTING STATE

The resting muscle cell may be considered to be a multicompartment system in a stationary state in which reaction rates and substrate concentrations remain relatively constant. All cellular activity is controlled by the rates of various chemical reactions occurring within these compartments, which include the nucleus, the mitochondria, the endoplasmic reticulum, and the sarcoplasm. Function within each compartment is not independent, since the membranous boundaries are permeable and movement of permeant cellular substances across them occurs continuously. The rate at which various substances cross the membrane depends on the nature of the substance and the nature of the membrane.

It is believed that, in general, substances cross membranes by solution in the substance of the membrane, by passage through a pore, by means of a carrier molecule within the membrane to which the transported molecule is bound, or by a process called pinocytosis, in which a small volume of the solution containing the substance is engulfed by a small part of the membrane and carried from one compartment to the other. The last mechanism transports solution; the others transport specific substances. Therefore, the rate of movement of a substance across the membrane will depend upon factors such as the size of the pores of the membrane. Since the rate of transport can change, the pore must change.

The velocities of all reactions depend upon the concentrations and the physical state of the reactants and the energy requirements for the reactions. Effective control of the chemical reactions can be exerted at any one of these three levels. The enzyme is probably the most characteristic and the most complex of the compounds which participate in the reaction. It is a protein molecule which catalyzes a specific chemical reaction by lowering the energy requirements for the initiation of the reaction.

Enzymes are synthesized within the cytoplasm by a series of reactions which assemble the appropriate amino acids in the proper order dictated ultimately by the genetic material, deoxyribonucleic acid (DNA). The DNA is in the nucleus from the time of formation of the cell, and is a polymer composed of a long series of relatively small molecules called nucleotides. There are four varieties of these molecules. The specificity of a given DNA polymer is determined by the proportion of each of these molecules and their order within the nucleic acid. In the nucleus the DNA governs the synthesis of ribonucleic acid (RNA), a polymer of similar composition in which the building blocks differ from those in DNA only in the nature of a 5-carbon sugar group present in each of the monomers. This RNA, so-called messenger RNA, diffuses into the cytoplasm where it determines the structure and the rate of synthesis of enzymes. An enzyme is assembled on RNA, which acts as a template. Each adjacent three nucleotides within the RNA determine a specific amino acid which will be incorporated into the protein enzyme.

Enzyme concentrations within a cell are not constant, but are profoundly sensitive to the composition of the cell and of its environment. Their concentrations are ultimately regulated by the concentration and the condition of the appropriate RNA moieties. A DNA or RNA molecule may be converted from an active state, in which it can participate in the synthesis of a given protein, to an inactive state, in which it cannot, by a relatively small change in the structure or the composition of the nucleic acid. This process is called repression. The converse, conversion from an inactive state to an active one, can also occur and is called induction. It has been suggested that induction and repression may be caused by the action of a small molecule on the DNA responsible for the synthesis of the RNA, or on a specific repressor or inducer already attached to the DNA. Synthesis of the enzyme could also be affected by interference or by the removal of interference with the activity of the RNA which, either by transporting the genetic information from the nucleus to the cytoplasm or by acting as a template for amino acid assembly, is involved in that synthetic pathway.

Regulation of enzymatic activity by control of the rate of synthesis of the enzyme is a slow process, as it is dependent on the rate of breakdown of the existing enzyme. The activity of the already existing enzyme, however, is not fixed. Small molecules or ions, especially those of the multivalent metals, have a marked effect on the activity of an enzyme without destroying it. For example, the activity of certain enzymes which facilitate the removal of the terminal phosphate from adenosine triphosphate (a substance that is highly important in both the general and specialized functions of muscle) is very sensitive to Ca^{++} and Mg^{++} concentrations (Weber and Herz, 1963). One or more of the end products of an enzymatic reaction may inhibit the enzyme; thus, the accumulation of the end product in certain cases decreases the rate of the reaction. The most reasonable explanation for such phenomena is that the small ion or molecule causes a change in the configuration of the protein enzyme in a part of the molecule where a specific configuration is needed for catalytic activity.

Both the inhibition of an enzymatic reaction by the accumulation of the products of the same reaction and the activation or inhibition of an enzymatic reaction by cations can function to maintain the concentration of the reactants or the products within a given range. In the former case, the mechanism is evident. In the latter, a cation acts as an activator or an inhibitor by its action on the enzyme or the substrate. It is also possible that one of the products of the reaction may bind the active cation and prevent the interaction of the latter with the enzyme or substrate. The rate of the reaction, in this case, varies with the degree to which its ionic activator or inhibitor is bound by one of its products.

The concentration of the substrate, in addition to the concentration and activity of enzymes, is critical in determining the rate of a chemical process. The series of reactions concerned with energy metabolism is an example in which the concentration of the substrate for one reaction controls the rate of several reactions. Muscle cells store energy in the form of high-energy phosphate compounds, having obtained the energy by a combination of the anaerobic and the aerobic metabolism of proteins, fatty acids, certain ketones, and carbohydrates. Anaerobic metabolism involves the stepwise, enzymatic breakdown of glycogen (a process called glycolysis) or of glucose to pyruvate, with the associated reduction of nicotinamideadenine denucleotide (NAD to NADH). NADH can also be produced by the enzymatic breakdown of two-carbon fragments previously derived from protein, lipid, or carbohydrate by other chemical activity. NADH is oxidized back to NAD by enzymatic reactions which reduce pyruvate to lactate.

In the mitochondria, however, in the presence of oxygen, two electrons from each molecule of NADH can be transferred through a series of highly specialized molecules to an oxygen molecule to form hydroxyl ions. The hydroxyl ions combine with hydrogen ions to form water. The transfer of electrons involves the stepwise transformation of the electron from a high-energy state to progressively lower energy states. At each step a large part of the potential energy released is stored as chemical energy by the synthesis of the high-energy compound, adenosine triphosphate (ATP), from the adenosine diphosphate (ADP) and the inorganic phosphate present in the cell. The respective rates of the anaerobic and aerobic series of reactions are controlled in the cell by the concentrations of ADP and NAD and the partial pressure of oxygen. The influence of the partial pressure of oxygen, called the Pasteur effect, refers to the inhibition of glycolysis when a cell shifts from anaerobic to aerobic metabolism. The current explanation for the shift is that the binding of ADP by some of the intermediates of active oxidative phosphorylation is sufficiently strong to keep the cellular concentration of ADP low. This inhibits anaerobic metabolism because the high-energy phosphate bond receptor is not available to the reactions of the glycolytic cycle.

The phase boundaries of the compartments within the cell act as diffusion barriers limiting the movement of impermeant or relatively impermeant substances between the compartments. The exchange of material across these interfaces is determined by the permeability of the interfacial membrane and the electrical potential gradient across them. Since the properties of intracellular membranes have been much more difficult to study than those of the surface membrane, current ideas of the former are generally extrapolations from the knowledge of the latter. It is known that lipid-soluble substances tend to cross the surface membrane of muscle much more easily than more polar molecules. Charged molecules do, however, pass through the membrane; their rate of penetration is a function of the size of the particle plus the water shell that it attracts by virtue of its charge. Hydrated K^+ is a smaller ion than hydrated Na^+ and passes through the membrane more rapidly. In the resting muscle an electrical potential gradient exists across the cell membrane. This potential is probably a potassium diffusion potential arising from the large K^+ concentration gradient across the membrane, and it has a strong influence on the rates of penetration of charged particles. In addition to the forces derived from concentration and electrical potential gradients, there are special transport systems or "pumps" which move molecules against their electrochemical gradient and are called, therefore, active transport mechanisms. Although much less is known about the intracellular membranes, the best estimate is that they influence the passage of particles by the same mechanisms as the surface membrane.

The characteristics of a membrane and of transport across it can be altered considerably by changes occurring on either side of the membrane. An increase in the concentration of potassium in the extracellular fluid will cause a decrease in the transmembrane potential, and consequently will influence the movement of charged particles across the membrane. The permeability to smaller, non-charged molecules may also be altered. The permeability of the resting membrane is very sensitive to the concentration of calcium in the surrounding bath, an increased concentration causing an increase in membrane resistance. Certain drugs, such as the veratrum alkaloids and the local anesthetics, influence the permeability of the resting membrane by enhancing or decreasing the likelihood of a change in the state of the membrane in response to alterations in cellular environment.

CONTRACTION

Departure of the cell from the stationary state is initiated either by neural activity or by changes in the composition of the fluid bathing the cell. The nervous system and the extracellular fluid coordinate the more specialized functions of a cell with the activity of the rest of the organism by means of neural transmitters or humoral substances which alter the cell membrane or the cell interior. The nervous system initiates muscular activity by releasing, at the endings of some nerves, the neural transmitter, acetylcholine, which reduces the postsynaptic membrane potential by increasing the membrane permeability (Katz, 1962). In addition to this control, the metabolism of the muscle cell is continuously under the influence of circulating thyroid hormone, which alters oxidative phosphorylation so that the process is less efficient in storing chemical energy and produces more heat. This change is called uncoupling of oxidation and phosphorylation, and it increases heat formation at the expense of the synthesis of ATP. The actions of acetylcholine and thyroid hormone represent mechanisms whereby one cell qualitatively or quantitatively alters the physical and chemical reactions of another cell.

In the case of the effect of acetylcholine on skeletal muscle, the alteration is qualitative, a transition from a resting cell generating little or no tension to an actively contracting cell performing mechanical work. The first events occur at the myoneural junction, a complex structure in which the nerve supplying the muscle cell and the muscle cell itself are separated by only a few hundred angstroms $(1A = 1 \times 10^{-8}cm)$. Infolding of the muscle cell surface at this point greatly increases the surface area of near contact with the nerve ending. During the resting state the muscle membrane is permeable to potassium and chloride and relatively impermeable to sodium. In addition, it actively pumps sodium out of the cell. The result is the maintenance of a potential difference of 95 mV and concentration gradients of sodium, potassium, and chloride across the resting cell membrane. The sodium and chloride concentrations on the outside and the potassium concentration on the inside are the larger. Stimulation of a motor nerve initiates a release of acetylcholine at the myoneural junction, which diffuses across the gap between the nerve and muscle and causes a change in permeability of the postjunctional muscle membrane.

The released acetylcholine changes the permeability of the muscle membrane only within the junction. Under certain conditions, as in the embryonic junction, in the chemically denervated muscle, and in junctions poisoned with botulinum toxin (which prevents the presynaptic release of acetylcholine), sensitivity is not restricted to only a small part of the muscle membrane, and the entire surface membrane then responds to acetylcholine (Axelsson and Thesleff, 1957). The repeated exposure of the membrane to acetylcholine appears to be necessary for the restricted localization of acetylcholine sensitivity, presumably by initiating some inhibitory process in the non-junctional membrane. In situations in which the entire membrane becomes sensitive to the chemical transmitter, presumably because repeated exposure of the membrane to acetylcholine does not occur, the sensitivity spreads gradually from the myoneural junction to the ends of the muscle fiber.

The change in the permeability of the postjunctional membrane that occurs following nervous stimulation produces a rapid depolarization of that membrane and a net movement across it of ions in accordance with their new electrochemical gradients. The local current flow between the depolarized junctional membrane and the polarized non-junctional membrane partially depolarizes the latter and initiates a permeability change consisting of successive, transient periods of high permeability to sodium and potassium (Hodgkin, 1958). The permeability change of the junctional membrane, therefore, is induced by acetylcholine, whereas that of the non-junctional membrane is brought about by current flow through it. A decrease in potential causes an exponential increase in permeability to sodium and becomes self-generating at a certain level of depolarization. These events produce an action potential which is conducted over the entire muscle surface.

The normal resting muscle cell does not actively generate tension, but the depolarization of the membrane sets off a series of processes which result in a large transient increase in tension, the muscle twitch (Hodgkin and Horowicz, 1960). The electrical disturbance of the action potential is conducted to the cell interior along the transverse system of tubules known as the T-system (Huxley, 1959) (Fig. 11-1). Elements of this system are within 1 to 2 μ of the contractile filaments of the cell, and therefore quickly conduct the electrical disturbance to the proximity of the tension-producing structures. The electrical disturbance of the T-system probably causes some change in the immediately adjacent lateral cisternae of the endoplasmic reticulum, resulting in the release of large stores of calcium contained in the cisternae of the resting cell (Weber et al., 1963; Winegrad, 1965).

The endoplasmic reticulum contains a system for actively transporting calcium from a surrounding medium into the lumina of the reticulum against a large concentration gradient (Hasselbach and Makinose, 1961). This process is associated with a splitting of adenosine triphosphate, which furnishes the energy for the transport. This "pump" is able to maintain over a 500-fold difference in calcium concentration between the reticulum phase and the surrounding fluid. It is for this reason that, in the resting cell, there is a large store of calcium in the endoplasmic reticulum.

The contractile filaments are extremely sensitive to the concentration of ionic calcium. When the concentration of ionic calcium is less than 10^{-7}M, the contractile filaments are in a relaxed state. As the calcium concentration is increased above this level, however, the filaments contract and produce a tension which increases progressively with the calcium concentration until maximum tension is reached at about 10^{-6}M calcium.

In skeletal muscle, where there are two sets of interdigitating, contractile filaments, small cross bridges are present on the thick or myosin filaments. It is generally believed that tension development and shortening occur as a result of the interaction of the actin filament with the cross bridges of the myosin filament. The myosin filaments contain sites, probably located in the cross bridges, which function as enzymes having the ability to split the terminal phosphate from adenosine triphosphate

(ATP) and to release a large amount of chemical energy. This is presumably the energy used in muscle contraction. The rate at which the myosin ATPase can split ATP is greatly enhanced by ionic calcium and by the interaction of myosin with actin. It would appear that the calcium ion controls the level of tension of the contractile filaments by the degree to which it enhances myosin ATPase activity and promotes the myosin-actin interaction. The exact way in which the splitting of ATP is related to the interaction of the filaments and in which chemical energy is converted to mechanical energy is not known (Cain et al., 1962).

The relaxation of muscle is thought to occur by a reaccumulation of released calcium in the endoplasmic reticulum after the electrical activity associated with the action potential is over. The calcium pump of the reticulum appears to be capable of transporting enough calcium with sufficient rapidity to account for the time course of relaxation.

The resting muscle contains, in the form of high-energy phosphate, only enough energy to support about six or seven twitches. Some mechanism by which the ATP is resynthesized as it is used is necessary. One efficient mechanism of coupling energy production to energy utilization would be to have the energy production reactions activated by the same agent, Ca^{++}, which activates the contraction. This would produce equivalent activation of the two processes. Another mechanism would be to have the rate of energy production controlled by one of the end products of the contractile process, such as ADP or inorganic phosphate. Accumulation of the end product would activate the synthesis of high-energy phosphate in accordance with its utilization.

Evidence favors both the hypotheses. ADP is known to stimulate oxidative phosphorylation in intact and isolated muscle mitochondria, while calcium ions activate phosphorylase kinase, which activates the enzyme phosphorylase to increase the rate of glycolysis. These two mechanisms for energy production, aerobic (oxidative phosphorylation) and anaerobic (glycolysis), are not completely independent of each other because pyruvate, which provides much of the substrate for oxidative metabolism, is one of the products of anaerobic glycolysis. When the muscle is performing work its oxidative phosphorylation is stimulated by the increase in intracellular ADP that results from the contractile process. The amount of glycolytic activity will be determined by the circumstances of the contraction, since the stimulus to increased glycolysis is a complex one, involving the combined effect of the actions of NAD, ADP, Ca^{++}, and inorganic phosphate on the activities of several of the enzymes involved. Both anaerobic and aerobic energy production are linked by at least one common activator, ADP, by substrate, and by the Pasteur effect. Both are stimulated by chemical changes associated with contraction, but as the rate of one increases the rate of the other decreases. This coordination promotes the delicate balance between total energy production and energy utilization that should exist in an efficiently functioning muscle cell.

This analysis of current knowledge of muscle contraction suggests some principles about the control of cellular activity. In the resting cell surrounded by an environment of relatively constant composition, reactions occur at steady rates to maintain a relatively constant concentration of the cell constituents. Control tends to be graded and to depend on the concentration, diffusibility, and accessibility of key substances. It is not enough for a substance to be present; it must be able to reach the site where it is active.

The control of a specialized function such as contraction may not be graded, may have a threshold and an all-or-none response, and may involve mechanisms such as electrical conduction which can act more rapidly than diffusion. Under the influence of the chemical transmitter, acetylcholine, and a decrease in the electrical potential, the membrane undergoes an abrupt change in permeability. The current flow initiated at the cell surface is rapidly conducted to the cell interior, and the contractile filaments undergo an abrupt change in state. Part or all of the cell changes from a stationary state to a new, dynamic state. The control mechanism involved is a trigger in which a relatively small change or input of energy provokes a considerable change in the state of the cell and the release of a large amount of energy.

The magnitude of the response of a cell may be relatively independent of the size of the change in the control mechanism, as long as that change exceeds a certain threshold. The control is an on-off switch. Once the contractile machinery is triggered into action, however, the actual energy output of the cell is determined by an external factor, the load on the muscle. One of the fundamental properties of muscle is the hyperbolic relationship between the tension generated by the muscle and the velocity of shortening. At maximum tension the muscle cannot shorten, and at maximum velocity of shortening the muscle cannot develop significant tension. Since work is computed as the product of force and distance, it is evident that the work done by the contracting muscle will be determined by the load, since under physiological conditions, the velocity is generally the dependent variable.

Moreover, the restoration of the muscle to its precontraction or resting state requires a degree of chemical activity that quantitatively parallels that which occurred during contraction. The activator of the contraction, calcium, must be reaccumulated in the endoplasmic reticulum against a concentration gradient. The amount of energy necessary for this depends on the amount of calcium to be transported. The concentration of compounds containing high-energy phosphate was decreased during the contraction and must be restored. The rate of synthesis of new high-energy phosphate will depend on the concentration of ADP. Both of the restorative processes, therefore, have graded controls.

Summary

A muscle cell, like a multicellular organism, is a highly organized structure, and hence a change in the functional state of one cellular system may influence others. This information is communicated by means of changes in concentration, permeability, electric field, or physical state. The coordination of one reaction with another, a phenomenon called coupling, is an essential part of the control of the cell. The coupling may be fast or slow. The cell is sufficiently large so that the distances between parts can be significant. The nature of the mechanism transmitting information will influence the speed with which other cellular systems react. If transmission occurs by diffusion and the change in local concentration of a substance, the response time may be relatively slow. It will be considerably faster if a change in electric field or a current flow is involved. The amount of the cell involved in the response is also influenced by the control mechanism. Local potential changes or localized current flow can restrict the regions of the cell which respond to the initial change, whereas a change in membrane permeability will affect the concentration of substances in most or in all of the cell. Intracellular diffusion barriers can limit the region responding to a given change in concentration.

The size of the response may depend on the magnitude of the initial change; this occurs when oxidative metabolism is increased as a result of an increase in the ADP concentration. The extent of the contractile response to the action potential, however, may be independent of the electrical event. The energy available for contraction and the energy released during contraction are not determined primarily by the action potential, the major function of which is the initiation of the conversion of chemical energy to mechanical energy. The amount of heat and work resulting from the splitting of ATP during the muscle twitch is determined by events independent of the trigger, such as the resting length and the load on the muscle. The initiation of the event is controlled by one system, the trigger, and the size of the event primarily by others.

The tightness of coupling and hence the degree of response of a reaction to a control mechanism varies with the nature of the control mechanism. The ATPase activity of actomyosin is very tightly coupled with the intracellular concentration of calcium because the latter is presumably effective on the active site of the enzyme, either directly or indirectly. The resting oxidative metabolism is loosely coupled with the resting length. The former is increased when the resting length is increased, but it is also influenced by diverse things such as extracellular K^+ concentration, resting potential, and intracellular substrate concentration.

Notes on References

I have noted in the text where specific information from the research literature was required. These references are listed in the bibliography for the whole volume. For the reader who wishes to pursue the matters discussed in this chapter in a more general sense, the following suggestions may be made: A detailed discussion of the nature and the control of energy metabolism in cells is that by Jöbsis (1964). Davson (1964) discusses the phenomena of muscle contraction with regard to both the electrical and mechanical events. An excellent review of the structure and functional implications of structure in muscle is that by H. E. Huxley (1960). Supplement 2 of Volume 15 of the journal, *Circulation Research,* presents a similar discussion, pertinent in particular to cardiac muscle. Several symposia on the control of biochemical reactions (Bonner, D.M., 1961) and the biochemical problems in the study of muscle (Gergely, J., 1964; Huxley, A. F., and Huxley, H. E., 1964) provide an excellent picture of contemporary thinking along these lines. An excellent review of current theory about the mechanism of function of excitable membranes is that by Hodgkin (1958). Finally, I would recommend to the reader an interesting exposition of the thought of one of the most imaginative workers in muscle biochemistry (Szent-Györgyi, A., 1953).

Research in the author's laboratory is supported by Grant NB-04409 from the National Institutes of Health and by a grant from the Foundation for Neuromuscular Diseases, Inc.

Chapter 12

CONTROL OF WATER EXCHANGE
REGULATION OF CONTENT
AND CONCENTRATION OF WATER
IN THE BODY

JAMES A. F. STEVENSON
University of Western Ontario

Life proceeds only in a watery environment. The potential for life may persist in the dry state, as in the seed or the spore, but for the functional processes of life to occur water is necessary. Moreover, the concentration of water within the cell must remain within certain limits for efficient function of the cell's machinery. One of the important ways in which this has been accomplished is by the development of a selective permeability of the membrane that surrounds the living unit, e.g., the cell membrane. Some of this selectivity depends upon the physical characteristics of the membrane, some on active mechanisms in the membrane that use energy to carry substances across the membrane, usually against concentration gradients. This selectivity affords some control of the composition of the aqueous medium of the life machinery within the cell.

With the development of multicellular organisms, the external fluid environment continued to circulate in the microscopic spaces between the cells. As these organisms became more highly developed, this fluid between the cells was contained within the body and thus became separate from the general, external, aqueous environment of the organism. Nevertheless even today in the mammal the fluid surrounding the cells appears to have retained much the same composition and concentration of salts that were present in the sea when the evolving organism captured it. This extracellular fluid Claude Bernard termed "the internal environment."

For an individual cell to be successful in any environment, there must be opportunity for the provision of foodstuffs and oxygen and for the removal of various products excreted by the cell. When the cells live densely packed together, diffusion is not enough, and adequate circulation must also be assured. This, in turn, if the organism as a whole is not to face deficits and excesses, requires that the circulating extracellular fluid must be able to exchange relatively large amounts of many substances with the external environment of the organism. These exchanges are provided by the lungs, the gastrointestinal tract, the kidneys, and so on. Many involve the exchange of water itself between the organism and the external environment. The ability to provide some control of such exchanges is required for the independence of the external environment that, as Claude Bernard (1878) pointed out, the internal environment gives to the organism. We are here concerned with these exchanges of water and their control in that most complex organism, the mammal.

DISTRIBUTION AND MOVEMENT OF WATER

Let us first look at some of the factors that control distribution of water in the body itself, for, as we shall see, the control mechanisms for exchanges with the external environment are built on these. As we have seen, the water of the body is divided into two main compartments: the intracellular and the extracellular. The latter is divided into two subcompartments, the interstitial fluid lying between the cells and outside the vascular system, and the plasma and lymph water moving within the vascular and lymphatic systems. In the adult mammal, about 70 per cent of the lean body weight is water; some two thirds of this is within the cells while the remaining third is the extracellular fluid.

As far as is known, the movement of water itself between the various compartments is relatively free and depends upon the effective osmotic pressure on the two sides of the membrane separating the compartments. The effective osmotic pressure derives from the permeability characteristics of the membrane as well as from the composition of the solutions on each side of it. In the case of the cell membrane, we find that in most mammals it is, in effect, relatively impermeable to sodium—that is, a metabolic mechanism keeps pumping this cation out of the cell so that its concentration inside is maintained at a very low level. On the other hand, potassium appears to permeate this membrane relatively easily and, to maintain electrochemical equilibrium across the membrane, reaches a high concentration within the cell equal to that of sodium in the surrounding extracellular fluid. These cations are, of course, balanced by an equal concentration of anions inside and outside the cell. The anions within the cell are primarily phosphate and protein, those outside primarily chloride and bicarbonate. There are other specifically important cations and anions as well as undissociated compounds on both sides of the membrane, but their osmolar concentration is comparatively low. Thus potassium and particularly sodium appear to be by far the most important factors in the control of the effective osmotic pressure across the cell membrane. One must add, however, that unusually high concentrations of glucose, urea, or other substances outside the cell can change the effective osmotic pressure across the membrane temporarily, until their concentrations have come to equilibrium on the two sides of the membrane. Although this phenomenon is rare under natural conditions, it is important in the understanding of certain experimental results.

As a result of these conditions, the distribution of water between the cell and the internal environment of the body is controlled by the effective osmotic pressure across the cell membrane, which in turn depends primarily upon the concentration of sodium in the extracellular fluid. If the concentration of water in the extracellular fluid increases, the concentration of sodium falls, and there is a net movement of water into the cells; if the concentration of water in the extracellular fluid decreases, the concentration of sodium increases and there is a net movement of water out of the cells. With this net movement of water there will be a swelling or shrinking of the cell and its membrane, and as we shall see later, this may be an important mechanism for the detection of the content and the concentration of water in the cell and thus in the extracellular fluid.

The effective osmotic pressure across the capillary membrane is an important factor in the control of the distribution of water between the vascular and interstitial subcompartments of the extracellular fluid. Here, however, because of the different permeability characteristics of the capillary membrane with its pores, the solute that creates the effective osmotic pressure is the concentration of plasma proteins in the vascular fluid (plasma), a concentration which is much greater than that in the interstitial fluid (7 gm % vs. 1 gm %). In other words, the concentration of water is greater in the interstitial fluid than in the plasma, and thus there is a constant osmotic pressure to cause the movement of water from the interstitial fluid into the vascular fluid in the capillary. On the other hand, the hydrostatic pressures of the blood produced by the pumping action of the heart tends to force fluid out of the capillary. As Starling (1896) showed, at the arterial end of the capillary the hydrostatic or blood pressure pushing fluid outward is greater than the net effective osmotic pressure drawing fluid inward. With the resistance that the capillary provides to the flow of blood, this hydrostatic pressure falls below the inpulling osmotic pressure as the venous end of the capillary is approached. Thus, water tends to flow out of the capillary at the arterial end and into it at the venous end, so that there is a constant circulation of water and the substances dissolved in it between the fluid surrounding the cells and the blood.

Under normal conditions there is a nice balance of this inflow and outflow so that the volumes of plasma and fluid are kept relatively constant. If, however, the concentration of plasma proteins falls below a critical level because of disease or undernutrition, or the hydrostatic pressure of the blood rises above normal because of an increased venous pressure (e.g., from obstruction to venous flow or from failure of the heart; the arteriole protects the capillary from increases in systemic arterial pressure) the balance is upset. Fluid tends to accumulate in the interstitial space (extracellular or pitting edema) until the pressure from distention of the tissues equals the hydrostatic pressure pushing fluid out of the capillary. The fluid that accumulates in this way is isotonic and similar in concentration to the rest of the extracellular fluid. The

TABLE 12-1. AN AVERAGE 24-HOUR WATER EXCHANGE OF A HUMAN ADULT, MODERATELY ACTIVE IN A TEMPERATE ENVIRONMENT
(ML OF WATER)

GAIN FROM ENVIRONMENT		Major Movements of Water Out of and Into Body Proper		LOSS TO ENVIRONMENT
		Out (Secreted)	In (Reabsorbed)	
Gastrointestinal Tract				
Water as fluid	1300	Saliva	1500	
Water in food	1000	Gastric juice	2500	
Water of oxidation	300	Bile	500	
		Pancreatic juice	700	
		Intestinal juice	3000	
Total	2600	+ 8200 −	10,600	→ 200
Respiratory Tract	0			400
Skin	0			500
Kidney	0	Filtered 172,000 −	170,500	→ 1500
TOTAL	2600			2600

body appears to have no way of detecting, correcting, or compensating for this upset in its internal exchange and distribution of water.

From this brief survey of the controls of the exchanges of water between the several compartments of the body water, we observe that water moves freely everywhere under the influence of osmotic or hydrostatic forces. Its movement and distribution within the body is thus a passive result of other conditions. The most important of these conditions are: the concentration of the electrolytes, particularly sodium (across the cell membrane), and the hydrostatic pressure and the concentration of proteins of the capillary fluid (across the capillary membrane). It is important to remember that the electrolytes have no, or only a transient, effect across the capillary membrane because it is relatively permeable to them; and that the plasma, interstitial, and cellular proteins have little effect across the cell membrane because, although it is impermeable to them, their contribution to the effective osmotic pressure is minor compared to that of the electrolytes which operate here.

Exchanges of Water With the External Environment

We may now turn to a consideration of the exchanges of water between the body and the external environment (Table 12-1). Water is taken into the body through the mouth and absorbed from the gastrointestinal tract. It is lost from the body to the external environment through the skin, respiratory tract, kidney, and, to a slight extent, the gastrointestinal tract. Because mammals tend naturally to live in environments where potable water is available, water intake is usually the dependent variable, being adjusted or controlled to balance whatever is lost by the several routes mentioned. (There is one important exception to this generalization—a mechanism which permits the body to reduce the loss of water through the kidney to a minimum when a deficit occurs and environmental water is not readily available for ingestion.) It will, therefore, be a simpler story if we consider the losses of water from the body first and then the means whereby the body can reduce or replace these losses to maintain a constant concentration or content of water in its various compartments.

Loss of water through the respiratory tract and skin is not under any direct control that is related to the content of water in the body. In the case of the respiratory tract, the loss is a consequence of the fact that the membranes of the tract are by nature moist, and that water evaporates from them to the inhaled dry air, to be exhaled and lost to the body. The rate of loss will, therefore, vary directly with the dryness and the temperature of the air inspired and the rate at which the air is moved, i.e., the depth and rate of ventilation. Ventilation varies directly with the production of carbon dioxide and this, in turn, is proportional to the metabolic rate. Thus an increased water loss from respiratory evaporation aids in the loss of heat produced by increased metabolism. Further, some mammals increase heat loss by panting. This shallow respiration greatly increases the evaporation of water, and so the loss of heat, from the upper respiratory passages without upsetting the exchange of respiratory gases. Here then, we see the control of water loss sacrificed to the control of exchanges of carbon dioxide and heat which have even greater priority in the homeothermic mammal.

Nor is loss of water from the skin controlled in relation to the body's need for water. This is not a serious factor for the many species that do not have sweat glands over most of the surface of their bodies, but for man, the horse, and other species that do perspire, this can be very important and a demanding route of water loss. In man there is a constant secretion from the sweat glands which, under most environmental conditions, is immediately evaporated from the skin when, although the skin may feel slightly moist to the touch, there is no subjective sensation of its being wet. This loss and the respiratory loss of water are referred to as the insensible water loss.

Central control mechanisms, in response to an increase in body temperature, cause a marked increase in the secretion of sweat, which enhances the evaporative loss of heat. The rate of secretion often exceeds the rate of evaporation and one becomes aware of droplets of sweat on the skin—sensible water loss. The loss of water by this route may reach dramatic proportions in man (e.g., 2 L/hr). Sweat contains about one half the concentration of sodium present in the extracellular fluid. If equal volumes of water are lost through sweating or through respiration, there will be less change in the relative concentrations of extracellular water and sodium in the case of sweating. This will mean less stimulation of any detector sensitive to a change in such concentration. Nevertheless, the losses in sweat are often of such magnitude that the concentration of sodium in the internal environment is significantly increased. There is no reduction in the loss of water as sweat upon adaptation to a hot environment. This is another example of the priority of the regulation of body temperature over that of body water.

The total heat load, the sum of environmental heat gain and metabolic heat production, are proportional to surface area, and the surface area relative to body mass, including body water, increases as mass decreases. For this reason, the smaller animal faces a greater proportional loss of water by evaporation than the larger animal; this is an important factor within a species (e.g., the infant) as well as between species. Some animals have made adaptations to reduce the evaporative loss. For example, the kangaroo rat, which lives in the hot dry desert where evaporative water loss is enhanced, reduces this loss by (1) a countercurrent cooling of the air leaving the nose, (2) an absence of a sweating mechanism, and (3) spending the hot day in its underground burrow which is relatively cool and humid (Schmidt-Nielsen, 1964). The camel, on the other hand, has adapted to some extent by permitting a much greater circadian cycle in its core temperature than most other mammals exhibit. During the cool night this temperature drops so that the animal starts the hot desert day with a low heat load in its large bulk. It can, therefore, accumulate relatively much more heat before it needs to increase its evaporative loss. Nevertheless, in spite of these ingenious and economical adaptations, the fact remains that respiratory loss of water is not controlled in relation to the supply of water in the body.

Although tremendous amounts of water move into the gastrointestinal tract each day, practically all of this water is reabsorbed lower in the tract and only a very small proportion is lost with the feces (Table 12-1). It should be remembered that, from the point of view of the internal environment, the lumen of the gastrointestinal tract is outside the body proper and that water in the tract is not immediately available for distribution throughout the body compartments. Water moves into the lumen of the tract in response to the effective osmotic gradient between the internal environment and the contents in the lumen. When solid food or hypertonic solutions are ingested, water moves into the upper tract until its contents are isotonic with the fluids of the body. This causes a temporary dehydration of the internal environment and the cells. This may be a signal of some importance in the inhibition of the ingestion of solid food and hypertonic solutions. Water that moves into the tract is later (and lower) reabsorbed through osmotic forces as the foodstuffs and salts are absorbed. If substances which cannot be absorbed are ingested, these create a continuing osmotic pressure within the tract which results in the retention of water and its loss in the stool, e.g., the osmotic cathartics such as magnesium sulfate. Similarly, vomiting and diarrhea, whether of central or of local origin, can result in the loss of tremendous amounts of fluid from the body. This fluid is isotonic with the fluids of the body, but differs in composition from that of the extracellular fluid because there is a selective secretion of ions in different parts of the tract. For example, vomiting usually results in a relatively greater loss of hydrogen than sodium, whereas diarrhea usually results in a relatively greater loss of potassium than sodium ions. These particulars are important in understanding the control of water balance, for unusual losses affect the distribution of the several ions between the cell and the internal environment and thereby influence the compensatory controls invoked in response to the water loss.

The final route of water loss from the body which we shall discuss is through the kidneys. The mammalian kidney excretes water both as a vehicle for other material, and as "free" water when the content or concentration of the water in the body is in excess. The urine is originally formed at the glomerulus, a special capillary net vested with a single layer of glomerular epithelium. This serves as a filter through which the capillary blood pressure forces a fraction of the plasma water flowing by. All of the substances with a molecular weight of less than about 70,000 dissolved in this fraction also filter through the membrane. About 80 per cent of the filtered sodium and chloride is immediately reabsorbed in the proximal convoluted tubule with water following it proportionately on an osmotic basis. The fifth of the glomerular filtrate that remains in the lumen of the tubule now passes into the thin loop of Henle, which dips down into the medulla of the kidney. The hairpin shape of this loop provides for a countercurrent multiplication of the effect of a moderate extrusion of sodium from the lumen to the surrounding interstitial fluid. Thus, a very high osmotic pressure is maintained in the medullary interstitium here. The tubular urine is then returned to the convoluted tubule in an isotonic or slightly hypotonic state.

In the distal convoluted tubule and, perhaps, in the collecting duct, there is a selective reabsorption of sodium, and potassium or hydrogen moves into the lumen. This is under the control of the adrenocortical hormone, aldosterone. The urine now passes down the collecting duct toward the impermeable urinary passages that will conduct it to the external environment. It is in the collecting duct that the final and selective reabsorption of water in relation to the body needs is made. This duct passes through the hypertonic medulla produced by the countercurrent multiplier system of the loop of Henle. The antidiuretic hormone (ADH) of the neurohypophysis causes the reabsorption of water by increasing the permeability of the wall of this duct to water. Water is drawn into the interstitium by the very high osmotic pressure there, and is thereby returned to the body. We here see for the first time a mechanism designed specifically for the control of an exchange of water between the body and the external environment. It will be noticed, however, that this control involves the osmotic movement of water and not, as was previously thought, a direct active movement of water molecules by the tubular cells.

In considering the role of the kidney in the control of body water, it is important to remember that the movement of water in and out of the kidney tubule is always in response to osmotic forces. Thus, if there is a solute in the tubulary urine that cannot be reabsorbed or the quantity of solutes in the urine exceeds the maximum rate of reabsorption that the tubular cells can achieve for those solutes, the substances will remain in the lumen and create an effective osmotic pressure that will retain water and thus increase the urine flow and water loss to the body.

Examples of such an effect are: the osmotic diuresis caused by the intravenous infusion of manitol or inulin, whose molecular size is small enough to pass the glomerular filter but too large to be reabsorbed from the tubule. As a result these substances retain water in the tubule and cause a marked diuresis. Glucose is normally completely reabsorbed in the proximal convoluted tubule, and urea is passively reabsorbed by diffusion along the whole length of the tubule. When the blood levels of glucose, as in diabetes mellitus, or of urea, as when a high protein diet is eaten, are unusually high, the filtered load may exceed the maximal reabsorptive capacity of the tubule. This increases the solute load remaining in the tubule, and thus the obligatory loss of water in the urine results.

This effect will persist even in the face of a maximum secretion of ADH, for this control mechanism can achieve only a limited osmotic concentration of the urine. This limit varies among species; in man it is represented by a specific gravity of the urine of about 1.035 or a concentration of 1.400 milliosmoles per liter. On the other hand, when water is in plentiful supply in the body, little or no ADH is brought to the kidney and the distal convoluted tubule remains relatively impermeable to water. As a result, the water entering the collecting duct flows on through it and is excreted in large volume.

Let us examine what is known about this specific control mechanism for the excretion of water that operates through the action of the antidiuretic hormone on the collecting tubule of the kidney. It was recognized since early times that certain diseases are characterized by a large output of urine and intake of water, and the Greek term diabetes, a syphon, was applied to this group of diseases. In the sixteenth and seventeenth centuries, diabetes insipidus was distinguished from diabetes mellitus by the insipid, rather than sweet, taste of the urine.

It was not until the beginning of this century, however, that Magnus and Schäfer (1901) showed that an extract of the posterior pituitary had an antidiuretic action that would control diabetes insipidus. The demonstration by Ranson and his colleagues that bilateral destruction of the supraoptic nucleus of the hypothalamus would produce diabetes insipidus although no direct damage was done to the posterior pituitary gland led to an intensive experimental analysis of this system (Fisher, Ingram, and Ranson, 1938). It is now known that the neurons of the supraoptic nucleus and the paraventricular nucleus produce the octapeptide hormones, ADH and oxytocin, which then move down the axons of these neurons in the neurohypophysial tract to be released from their endings in the posterior pituitary. From the posterior pituitary the blood carries these hormones to their target organs: in the case of ADH, the renal collecting tubule and perhaps the smooth muscle of the wall of the arterioles; and in the case of oxytocin, the smooth muscle of the uterus and that surrounding the lacteal ducts of the breast.

In any situation in which there is an increase in the concentration of salt relative to that of water in the extracellular fluid, evidence appears of an increased secretion of ADH and oxytocin. Changes in the effective osmotic pressure of the extracellular fluid apparently serve as a signal to some detector in the ADH control system. Verney (1947) demonstrated that the injection of slightly hypertonic saline into the carotid artery of the hydrated dog causes a marked reduction in urine flow, evidence of an increased secretion of ADH. The internal carotid artery is the direct source of the vascular supply to the hypothalamus, and Verney postulated that the dehydration or shrinkage of the cells in this region, as the result of the increase in the tonicity of the surrounding extracellular fluid, stimulates some special cells. This information causes the release of ADH from the neurohypophysial system. It may be that the cells which secrete ADH are also the detectors, but this is not known, nor is the precise nature of the signal to these detectors known. For instance, the signal could be the change in osmotic concentration within the cell or the resulting change in the stretch of the cell membrane as water moves in or out. Further, it is not known whether the receptor mechanism involved is one which stimulates ADH secretion in response to dehydration or inhibits its release in response to hydration. General evidence tends to favor the former possibility.

In addition to responding to variations in the effective osmolar concentration of the internal environment, the ADH-control system appears also to respond to changes in the volume of the extracellular fluid. These changes are monitored as variations in the distention of the vascular system caused by changes in the plasma volume, a compartment of the extracellular fluid. Stimulation of stretch or distention receptors in the walls of the atria and great veins activates vagal afferents, and the information is relayed to the hypothalamus to inhibit ADH secretion. There is also experimental evidence that when a reduction in extracellular fluid volume results in a decreased stimulation of receptors in the carotid sinuses and aortic arch, the secretion of ADH is increased. Hemorrhage causes a much greater increase in the secretion of ADH than does hyperosmolality of the extracellular fluid. Indeed, such hypovolemia will cause an increase in the secretion of ADH even if the effective osmotic pressure of the extracellular fluid is low.

Thus we see that control of the excretion of free water by the kidney is influenced by feedback mechanisms that can appreciate both the concentration and volume of water in the body and particularly the volume of water in the vascular system. In addition, the secretion of aldosterone, the hormone of the adrenal cortex that causes the kidney to retain sodium in the body, appears to be under the control of systems which are sensitive to the concentration of sodium and to the volume of the extracellular fluid; a decrease in either factor results in an increased secretion of aldosterone and retention of sodium; conversely, an increase results in a reduced secretion of this hormone. The resultant changes in the concentration of sodium and thus the effective osmotic pressure of the extracellular fluid will, in turn, affect the retention of water by the mechanisms that we have just described. By these mechanisms, the body is supplied with feedback and control systems that provide a precise adjustment of the kidney's losses of water and sodium. These are the major items in providing constancy of the osmotic environment of the cell and thus of its internal concentration of water.

These controls can, however, go only so far in maintaining homeostasis of the internal environment; the uncontrollable loss of water through the lungs and skin continues, and the kidney itself is always obliged to excrete sufficient water to meet the minimum osmotic requirements of the solutes that it excretes. Therefore, although these controls are of the utmost importance in the immediate control of the water economy of the body, the long-term maintenance of this economy depends upon the intake of water. We shall now examine the evidence for the body's control of this aspect of its water exchange.

WATER INTAKE

The inevitable continuous loss of water from the body creates a deficit. This deficit acts as a stimulus to water intake and the behavior necessary to achieve this intake. This suggests that there is a feedback system in which the water deficit turns on drinking behavior, whereas replenishment of the deficit, satiation, turns it off. Under natural circumstances, water enters the body only through the mouth and gastrointestinal tract, but it may do this as water per se, as the vehicle in various fluid mixtures, and as the water contained in solid food. There is also another source of water to the body—metabolic water produced in the oxidation of foodstuffs. Water obtained from any of these sources serves to restore a deficit; but it must be remembered that many fluids and most foods contain salts and other solutes that require water for excretion, so that none, or only a portion, of the water so ingested may be available to replace a deficit. A water deficit is, in man, associated with the subjective feeling of thirst, and often the term thirst is used to describe the state of an animal in water deficit. Nevertheless, water intake is the operant response to the deficit state, and provides a quantitative operational measurement.

What are the control systems for water intake and what are the signals that operate in the feedback to these systems? Historically, there have been three groups of scientific hypotheses proposed to explain the stimulation of water intake: (1) the peripheral origin group in which sensations of dryness from the mouth, throat, and perhaps other tissues serve as the stimulus (dryness of the mouth is inversely related to salivary flow and this in turn is usually directly correlated with the body water content); (2) the general origin group in which it was postulated that some deeper tissues or perhaps all the cells of the body are stimulated by lack of water; (3) the central origin group in which it was postulated that dehydration specifically stimulates some part of the central nervous system. As we shall see, all three general hypotheses apparently have an element of truth in them. The experimental evidence presently available, however, suggests that the major feedback systems for the control of water intake lie in the central nervous system.

It has been known since the early nineteenth century that thirst, or the intake of water by animals previously deprived of water, can be allayed by the intravenous injection of water and "other fluids." Claude Bernard, among others, showed that an animal with an esophageal or gastric fistula was not satisfied when it drank water unless the fistula was closed so that the water reached the stomach and was absorbed. Thus, it could be deduced that important signals for the inhibition of drinking and, perhaps, for the stimulation of drinking, occurred in the body proper.

At the turn of the century, the investigations of A. Mayer (1900) and Wettendorf (1901) revealed that the osmotic pressure of the blood increased when animals were deprived of water, and they suggested that peripheral nerves carry information of this change to the brain. That such a change in the effective osmotic pressure of the internal environment is an important signal of the water content of the body was substantiated by the observation that, when extracellular salt loss exceeds water loss and produces a hypotonic internal environment, water intake is much less than it would be from a similar reduction in volume without the loss of salt. This suggests that the important signal is the net movement of water into or out of the cells in response to the effective osmotic gradient, a concept which has received further experimental support. Urea distributes equally between the cells and their environment and does not appreciably affect the distribution of water. Injection of this substance was found to be a very much less potent stimulus to water intake than is the injection of salt. This is similar to the rather slight effect that the injection of urea has on the ADH control system. Any increase in water intake following a urea load is probably in response to the water deficit created by the diuresis that urea causes.

It is an old and general observation that hemorrhage, an isosmotic reduction of the blood and total extracellular fluid volume, often results in thirst and the increased intake of water. The experimental decrease of extracellular fluid volume without change in its effective osmotic pressure has also been found to stimulate water intake. This has led to the postulate that stretch receptors with, or similar to, those involved in the control of the cardiovascular system and of the release of ADH, are involved in a feedback system that influences the control of water intake.

In addition to the signals of effective osmotic pressure across the cell membrane and of the volume of the extracellular fluid (or at least of its vascular compartment), there are probably several other important signals, and many minor modulating ones, to the control systems that are responsible for the behavior involved in water intake. Man has recognized for a long time that his water intake increases in a hot environment or when he increases his own heat load by exercise. This is also true of other animals. In both situations there is an increased evaporative loss of water and, in certain species such as man, a loss through hypotonic sweat. The increase in effective osmotic pressure of the extracellular fluid that results could be considered a sufficient signal for water intake, but as we shall see, there now is evidence that the central nervous system may have receptors that influence water intake directly in response to a thermal signal. This suggests that the mammalian body has developed a feedback system that initiates preparatory compensations (i.e., for the increase in water loss that an increase in heat load will require).

In contrast with these general signals of water deficit that lead to water intake, the rather localized stimulus of a dry mouth which results from the reduction in salivary flow that occurs in dehydration is still considered by some to be the important signal to water intake (Gregerson and Cizek, 1961). However, failure of adequate water intake does not result when this signal system is obliterated. Experimental extirpation or congenital dysfunction of the salivary glands or denervation of the mouth and pharynx does not impair water intake, and cocainization of the mouth in man gives only brief relief from thirst. Nevertheless, this dry mouth mechanism may play a role as a reinforcing signal to a central state already alerted by the signals we have discussed, and may lead to the conscious definition of thirst in man.

What do we know about the mechanisms that handle the information provided by the signals to water intake that we have just discussed? During the nineteenth century several investigators suggested various regions of the brain as the site of control systems for water exchange. The demonstration some thirty years ago of the role played by the supraoptic and paraventricular nuclei of the neurohypophysial system in the secretion of the antidiuretic hormone focussed attention on the hypothalamus as an important site of the systems involved in water regulation. When it was later observed that most rats with bilateral lesions in or near the ventromedial nuclei of the hypothalamus (rats which usually show hyperphagia and obesity) failed to release a significant proportion of a large water load as urine, drank much less water in proportion to their food intake than did their controls, and regularly excreted a very concentrated urine, it seemed reasonable to postulate that the lesions had interfered with some control of water intake. This was supported by the observation that these defects in water exchange occurred whether or not these animals were obese. The further discovery that they showed a significant elevation in serum sodium supported the view that a chronic inadequate intake of water with relative dehydration had been produced by the destruction of some receptors or the systems that they served in the control of water intake.

Further exploration revealed that lesions somewhat more lateral in the hypothalamus, which do not affect food intake, do cause impairment of water intake; and it was finally demonstrated that lesions in the lateral hypothalamus of the rat, in the caudal plane of the ventromedial nucleus, produce an animal that no longer drinks water. Such animals fail to show any spontaneous intake of water even though maintained for many months after operation. These animals eat spontaneously and adequately when their hydration is maintained by regular administration of water through a stomach tube. If water is not administered, the animals do not drink spontaneously, their food intake soon falls, and they die from dehydration and starvation (Montemurro and Stevenson, 1957). The effective lesions are very close to those Anand and Brobeck (1951) had found to produce aphagia in the rat. Indeed, some investigators have concluded that the aphagia produced by these lateral lesions is secondary to the adipsia. It is not surprising, however, to find the systems that control food intake and those that control water intake as closely interrelated anatomically as they are functionally.

Several investigators have shown that stimulation in this lateral region will induce the well-hydrated rat to drink more water. It may be concluded, therefore, that a vital part of the system controlling water intake lies in this region of the hypothalamus of the rat. Whether a similar system is identically located in other mammals remains to be determined.

The statement that rats with lesions in the lateral hypothalamus no longer drink water spontaneously must be qualified. It has been shown that if these animals are maintained by the intragastric administration of water until they are well recovered from the operation, they can then gradually be brought to ingest palatable and nutritious fluids, sweetened water and, finally, ordinary water (Epstein and Teitelbaum, 1964). However, even in this final stage they will only drink water when they are eating dry food. For this reason they have been called "prandial drinkers." They do not adjust their intake of water to the demands presented by a hot environment or an osmotic load. One explanation is that they have learned to take water to help in the mastication and swallowing of the dry food. In other words, the fundamental central systems for the control of water intake in relation to body needs have indeed been destroyed. Central representation of the sensation of a dry mouth may still remain, or careful nursing has provided the animal with time to learn a drinking behavior which maintains it as long as its water deficits can be covered by prandial drinking.

In the goat, Andersson and his colleagues have found that stimulation of a more anteromedial site between the descending column of the fornix and the mammallothalamic tract, induces drinking (Andersson,

Gale, and Sundsten, 1964). There is some evidence that ablation in this general region will produce adipsia in the dog. In the pigeon, electrical stimulation even more rostrally (in the medial aspect of the preoptic and anterior hypothalamic regions) has been successful in evoking the drinking of water. Although there presently appear to be species differences in the precise location of this control system, it may be that different ramifications of the same system have been destroyed or stimulated in the different species. Indeed, various parts of a control system could have varying importance in different species.

This hypothalamic system can be activated by several different types of stimulation. In addition to electrical stimulation, the injection of minute amounts of hypertonic saline in the appropriate regions in the rat and the goat has resulted in drinking by the hydrated animal. The question, of course, arises as to whether the hypertonic solution acts as a specific stimulus to an osmoreceptor or whether its effect is due to a non-specific irritation and excitation of nerve cells. Injection of micro-amounts of acetylcholine in the lateral hypothalamus of the rat appears to have a specific effect in activating the water intake system; a similar injection of the adrenalines activates feeding in the fed animal but has little effect on water intake (Grossman, 1964). This observation supports the view that there are discrete, although intimately interrelated, drinking and feeding control systems in the lateral hypothalamus of the rat, the former having cholinergic, and the latter adrenergic synapses at this point.

Thermal stimulation of the hypothalamus has also been shown to affect water intake. Andersson and his colleagues placed thermodes permanently in the anterior and preoptic regions of the hypothalamus of the goat and changed the temperature of these thermodes by running cool or warm water through them (Andersson, Gale, and Sundsten, 1964). Heating this region induced drinking, and cooling inhibited it; converse effects on feeding behavior were obtained. These local effects of temperature in the hypothalamus fit well with the effects of environmental temperature on water intake and food intake.

Evidence of the presence of control systems for water intake has also been found in other parts of the hypothalamus and the central nervous system. Bilateral ablation in the mammillary region frequently results in a permanent increase of water intake, but it is not yet clear whether this is a specific and direct interference with a water-control system or a secondary effect on water intake from interference with systems regulating temperature or food intake. There is also some evidence that the region of the area postrema and dorsal motor nucleus of the vagus in the floor of the fourth ventricle, which has a distinctive histological structure, may contain osmoreceptors or other elements involved in the control of water exchange. Evidence is contradictory concerning the role of the subcommissural organ (which lies beside the pineal gland) in the control of water and electrolyte exchange, and even a preliminary conclusion on this question is hazardous. It has been suggested that the adipsia purported to occur after ablation of the subcommissural organ was, in fact, due to the destruction of the dorsal longitudinal fasciculus which runs nearby, for destruction of this outflow tract of the hypothalamus apparently causes a severe hypodipsia.

In general, stimulation or ablation of the neocortex does not have marked effects on water intake. This, perhaps, is to be expected because this region of the brain is most likely to be the site of a diffuse origin of modulating influences relating the control of water intake to the immediate exigencies of the external environment. In the rat, but not yet in the cat or other species, cholinergic stimulation at many points in the visceral brain, particularly along the limbic circuit described by Papez (1937), evokes drinking. All these regions have connections with the hypothalamus, and it is probable that they play a role in the modulation of drinking behavior. The "system" that has been traced by this cholinergic stimulation is very similar to that in which self-stimulation is reinforced. It has been suggested that, in the rat at least, this system is involved in the thirst drive. Ablation here does not markedly affect water intake. The microinjection of acetylcholine into the amygdala is said to reinforce drinking behavior if this is already in progress, although it will not induce such behavior de novo as does injection into the lateral hypothalamus. Complete section of the frontal pole at the level of the orbital sulcus does not appreciably affect the spontaneous water intake of dogs or their response to the intravenous injection of hypertonic saline. It does, however, prevent amphetamine-induced inhibition of water intake in response to hypertonic saline, and also eliminates other kinds of cortical inhibition of hypothalamic activity.

EVALUATION OF CONCEPTS

The foregoing provides a summary of our present knowledge of the structural substrates of physiological functions in the control of water intake. The reader should consult detailed reviews and the original reports in this field to comprehend fully the complexities and nuances of this subject (Wolf, 1958; Andersson and Larsson, 1961; Stevenson, 1964). We have mentioned several different types of experimental stimulation or excitation in the central nervous system that will evoke water intake. Electrical and chemical excitations imitate events which occur when information passes through the system, and do not represent stimuli that activate the system under natural conditions. On the other hand, osmometric, volumetric and thermal excitation appear to be natural signals to the receptors that feed information into the control system for water intake.

In the case of local osmotic excitation in the hypothalamus it is difficult to be sure that the osmotic event, or the change in cell volume that it causes, is not, like electrical stimulation, an unspecific stimulus. Nevertheless, we do know that minute changes in effective osmotic pressure of the internal environment are associated with a specific and quantitative excitation of this system. Present evidence suggests that the receptors for this signal exist in the hypothalamus and perhaps in other parts of the central nervous system (e.g., area postrema). The existence of peripheral osmoreceptors in the viscera must also be considered (Inchina and Finkinshtein, 1964). There is no evidence, as yet, of volumetric receptors in the hypothalamus; the volume receptors that influence water intake may be the same as those in the vascular system which affect the ADH and cardiovascular control systems when the vascular or extracellular fluid volume changes.

The evidence for specific thermoreceptors in the rostral hypothalamus is impressive. It is not yet clear whether there are specific receptors for cold and for heat here as there are subcutaneously at the periphery. One receptor whose range of response extended on both sides of the homeostatic level (i.e., normal core temperature) could probably provide sufficient information, as do the vascular baroreceptors, for a control system. There is some physiological evidence of a major association pathway between the known rostral thermoreceptor region of the hypothalamus and the more posterolateral focus of the food intake control system in the rat. It is possible that, in those species in which stimulation and ablation in the anteromedial hypothalamus have abolished or severely interfered with water intake, such a vital association system has been destroyed.

There is other evidence that the hypothalamic systems can integrate information from the local thermoreceptors and from the peripheral thermoreceptors as shown by the experiments in which the information from these two sources is made contradictory (q.v., Hammel, Chapter 4). We have also seen some evidence that there are elements in other parts of the nervous system and in the periphery which provide information reflecting other demands upon the organism. By reinforcing (or inhibiting) the activity of the water intake control system, these influence the priority, or degree of motivation, for water intake in the immediate situation. Sensations from a dry mouth are probably a specific example of this, while information from the exteroceptor systems and from the motivational systems also has its effect.

The foregoing discussion has dealt with the factors and systems involved in the activation, or turning on, of water intake behavior. The turning off of this behavior, i.e., when the animal stops drinking, would result from (1) a reduction in the activity of the receptors that turn on the system, as the concentration and content of water in the body return to their neutral level, (2) positive inhibitory signals, or (3) a combination of these two types of information. An important positive inhibitor of water intake, as of food intake, is gastric distention. Experiments in the dog have shown that distention of the stomach with water will inhibit further drinking even before there is time for the water to be absorbed into the body. This signal, arising in stretch receptors in the stomach wall, appears to be carried in the vagus and along pathways which pass through the ventromedial region of the hypothalamus on their way to the lateral and, perhaps, anterior focus of the drinking system. That the inhibitory effect is due to gastric distention is confirmed by the observation that similar distention with a balloon has the same effect. The importance of this as a safety signal to prevent embarrassment of the gastrointestinal tract is obvious; this is probably particularly important in the infant in whom regurgitation and aspiration, when the stomach is overfilled, are tangible dangers.

This signal may also become a learned, or conditioned, cue which permits the metering of water intake before the water has been absorbed into the body. Some species are able to replace a water deficit rather accurately in such a short space of time that significant absorption could not have occurred. Thus, this turning off of drinking cannot be due to significant changes in the signals that activate the turning-on systems that we know about. It is also probable that information comes from taste and smell receptors that indicate that water is being drunk, and perhaps from muscle receptors that indicate the amount of swallowing. On the basis of experience, this information could be titrated centrally against the intensity of the turning-on signals. With this learned metering, an accurate replenishment of a water deficit could be made quickly, as soon as the animal encounters and drinks water. The animal is then soon free to leave the source of the water to deal with other demands.

Most species replace the major part of the water deficit rather quickly when water becomes available, but in some, such as man, immediate drinking ceases before the replenishment is complete. This would suggest that the distention signal has inhibited the response to the turning-on signals before the final necessary volume has been achieved. Drinking now becomes more gradual and intermittent; man usually waits until he takes food to complete the last part of his rehydration. One apparent benefit of this type of response in man is that large water deficits are often associated with sweating, which includes some loss of salt. To replace the total volume of water lost without the intake of any salt would result in the development of a hypotonic extracellular fluid and embarrassment of all the cells, particularly the muscle cells. Waiting until food is taken to complete rehydration means that the salt deficit is likely to be replaced before the last part of the water deficit is restored. In this way, the danger of facing a hypotonic internal environment is avoided.

There are of course many more modulating influences on the final product of any behavior system. There are the other demands in the immediate environment; needs of greater priority, e.g., oxygen or escape. Information from the systems responsible for these must play upon the system that controls water intake at some point between the sensors that activate it and the final motor behavior that is evoked. The modulating influence of taste and smell is a special case for the water intake system. Excessive intake of water may also occur. Ordinarily, there is a fairly constant ratio between water intake and food intake under otherwise constant conditions; the ratio appears to be largely determined by the osmotic load that assimilation of the food presents to the body and kidney. When the only food available is a dilute fluid, the demand for food intake forces the animal to take an excess of water coincidentally. In this case the animal is essentially eating the water as a vehicle of its caloric and other requirements.

Adolph (1964) has suggested the following equations for the formulation of the integration of multiple factors in a control system, such as that for water intake:

$$\text{Water intake} = f(A, B, C, \text{etc.})$$

or

$$\text{Water intake} = f_1(A) + f_2(B) + f_3(C) + \text{and so forth.}$$

He defines f as a coefficient, operative whenever the corresponding factor (or stimulus) is not zero. For example, if A represents water deficit, B food intake, and C salt excess, the three factors sometimes operate with the same coefficients (positive or negative), whether alone or together. For many factors, within limits, the coefficient is linear. For other factors it may be a matter of being on or off. The coefficients of known factors may be evaluated empirically, in terms of multiple correlations among the factors influencing water intake. In the equation, the activity ceases whenever none of the factors differs significantly from zero or when positive factors (activating signals) are equally antagonized by negative factors (inhibitions, satiations).

SUMMARY

Figure 12-1 is a schematic summary of the important fundamental factors that have been demonstrated to activate or modulate the control systems for the regulation of body water content. It will be remembered that the distribution of water between the cell and the internal environment, as well as between the vascular and interstitial compartments of that environment, is determined by the algebraic sum of the effective osmotic gradient and hydrostatic pressure across the membranes separating these compartments. The body faces uncontrollable losses of water from the lungs and skin as the maintenance of the respiratory membranes

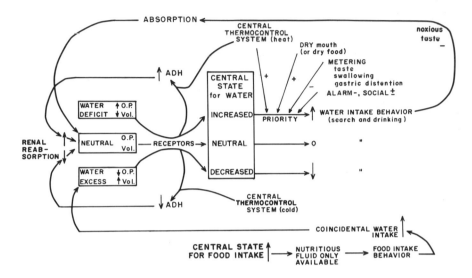

FIGURE 12-1. A schematic summary of the known feedback systems and modulators in the mammal that are involved in the facultative control of water exchange with the external environment. At the left are listed the signals that particularly activate the receptors which send information to the central systems that control these exchanges. The behavior evoked by the signals is modulated by various factors that enhance ($+$) or inhibit ($-$) it. The response evoked, if successful, restores the internal environment to its normal or neutral condition, and the signal diminishes. Water deficit develops continuously because of the inevitable losses described in the text. Water excess is much less common; one cause is shown in the diagram. It should be realized that there are situations in which the osmotic pressure (O.P.) and volume signals may be opposed. $\blacktriangle$ = increase, $\blacktriangledown$ = decrease.

and body temperature take precedence over the maintenance of body water. The kidneys also demand a certain irreducible volume of water for the excretion of the solute load presented to them. Water available beyond this can be excreted or retained by the kidney under the influence of the antidiuretic hormone, whose rate of release from the neurohypophysial system is governed by a control system that responds to information concerning, particularly, the effective osmotic pressure and volume of the internal environment.

Water intake is controlled by the central nervous system, and is usually activated by a deficit in body water content. The receptors that provide information for this control system include, on the positive side, those sensitive to changes in the effective osmotic pressure, volume of the internal environment or its vascular component, and probably those sensitive to peripheral and central temperature, and to oral and pharyngeal hydration, and so on. On the negative side, there are those receptors sensitive to gastric distention and other learned gastrointestinal cues for metering intake. In addition to these, the control system for water intake behavior is influenced by information reflecting the priority of other demands on the animal's behavior.

Regulation of the water economy of the body has a high priority, but in the table of precedence it stands below several other, more acutely vital regulations.

Experimental work cited from the author's laboratory and, in part, the preparation of this chapter were supported by a grant from the Medical Research Council of Canada. The author wishes to express his appreciation of the bibliographic, editorial, and secretarial assistance given by Miss B. Box and Mrs. D. Cox.

Chapter 13

CONTROL OF FOOD INTAKE

CHARLES L. HAMILTON

Control of food intake in mammals has been variously attributed to sensations originating in the stomach, including the so-called "hunger pangs" of stomach contractions; to conditions of glucose utilization; to levels of circulating amino acids; to conditions of body temperature regulation; and to levels of body fat depots. While there is no serious question that all these mechanisms are involved in a very intimate way, there is considerable divergence of opinion regarding the supremacy of any one of them as well as the manner in which they drive food intake under the varied circumstances of life. Before we can try to harmonize these opinions, it is necessary first to clarify the relationships among these several reactions. The body of valid observations is large enough so that, for this purpose, one can assemble a preliminary master scheme or organization of physiological mechanisms for food intake. Many pieces of information, however, still appear to be missing, and any such attempt must be, to an appreciable degree, speculative. Nevertheless, it is our intent, in the following paragraphs, to attempt such a speculative association of experimental observations on the hypothesis that food intake is an animal response that is a component of several, not necessarily compatible, systems of physiological regulation.

THE BRAIN AND FOOD INTAKE

All authors agree that the brain is the prime controller of feeding. The search for localization of these functions in the central nervous system has had a long and controversial history, which includes many attempts to learn what occurs when a particular structure is either destroyed, damaged, or stimulated by electrical or chemical means. To summarize these experiments, one can begin at the highest level of the nervous system, the cerebrum, and say that human infants born without a cerebral cortex still show sucking behavior and control of food intake, but the role of the cortex in adult feeding behavior has not been adequately analyzed. Similarly, the influence of the limbic system, phylogenetically the most ancient part of the cerebrum, on the control of food intake is uncertain and contradictory. For example, studies primarily on monkeys and cats have shown that bilateral lesions of the frontal orbital cortex are followed by a decrease in food intake. Lesions involving structures of the temporal lobes are said by some authors to be followed by hyperphagia, and by other authors, by hypophagia in the same species (the cat). Comparable lesions in rats apparently lead to no impairment of control of food intake. Consequently, at the present time, the limbic system cannot be specifically included in the scheme of regulatory mechanisms related to feeding.

Lesions of the hypothalamus produce more obvious changes in food intake than lesions of any other part of the brain. Ablation of its medial portion, in all mammals studied so far, leads to hyperphagia and obesity. Lesions in its lateral region lead to aphagia and, in some cases, to death from starvation. In other cases, and specifically when the diet includes foods preferred by the subject, this aphagia is temporary, and the animal recovers many of its feeding responses. Electrical stimulation of the lateral hypothalamic area precipitates feeding in the sated animal, but stimulation of the ventromedial area is accompanied by cessation of feeding. Chemical stimulation of the lateral hypothalamic area is also followed by eating when the chemical stimulus is an adrenergic drug, but by drinking when the stimulus is a cholinergic drug (Grossman, 1960).

The early effects of medial lesions are quite dramatic; some animals exhibit a three- to fourfold increase in food intake. Since there is also an induced hypoactivity after such lesions, the net and final result is an extremely obese animal. One of the enigmas of the effects of such lesions has been that even though the animals are hyperphagic on an ad libitum feeding schedule, when obstacles are placed between them and food they do not expend the effort that a normal rat does to obtain the food. These hyperphagic rats also seem extremely sensitive to the palatability of diets. For example, addition to the diet of minute amounts of quinine sulfate depresses food intake in hyperphagic rats more than in the normal rat. Thus the rat with ventromedial lesions, hyperphagic on an ad libitum feeding schedule, does not exhibit the hyperphagia when the food is not readily available or when it is relatively unpalatable. These results have led to the conclusions that the ventromedial region of the hypothalamus is related to conditions of satiety and that the initiation of food intake is governed by the lateral hypothalamus. The sum of observations is that particular areas of the brain can be shown to be involved in feeding behavior, although just how they function is not clear.

Food Intake and Sensations from the Stomach

A relation between gastric contractions and feelings of hunger and satiety is an old idea and one for which the first experimental evidence was collected by Cannon and Washburn (1912) and by Carlson and his associates (1916). They laid the experimental foundations for a theory that the state of gastric filling might be related to feeding. Yet the subsequent discovery that patients whose stomachs have been removed or denervated also experience sensations of hunger and eat a normal amount of food makes difficult the conclusion that contractions play the major role in feeding responses.

Gastric distention and the hypothalamus, however, are demonstrably related. Anand and his associates (1951) have shown that when a cat's stomach is inflated by a balloon, there is increased electrical activity in the ventromedial region of the hypothalamus. They found no change, however, in the electrical activity in ventromedial or lateral areas during spontaneous "hunger contractions." Anand also found that, following injection of glucagon, when utilization of glucose was increased, electrical activity of the ventromedial region was also increased and there was a concurrent inhibition of gastric contractions. Mayer (1953) has interpreted such data to indicate that it is the ventromedial area that inhibits the contractions. He has shown, using hyperphagic rats with ventromedial lesions, that glucagon administration is not associated with any marked inhibition of stomach contractions. His interpretation is compatible with other data which demonstrate that hyperphagic rats generally eat no more meals per day than controls but consume more food per day. Perhaps they eat more at each meal because, after injury to the ventromedial hypothalamus, their hunger contractions are more intense or persist longer than normal.

Proper evaluation of the role of stomach distention in the control of food intake requires further experiment. For example, although after denervation or ablation of the stomach the animal still controls caloric intake,* we do not know how such an animal responds to other stresses placed on the feeding system. At present the role of stomach distention is placed within the regulatory systems as shown in Figures 13-1, 13-2, and 13-3. Neural impulses arising because of gut filling are afferent to the CNS and thus serve to inform the controlling system of the degree of filling. It is likely that these signals prevent overloading of the gut and therefore play the role of a safety device. Moreover, the degree of influence of gastric filling seems to be determined by the other priorities of the regulatory system. For example, in a cold environment, in which the

---

* *Ed. Note:* Control of caloric intake means that the net caloric intake remains constant when inert material is added to the diet to increase its bulk. Rats quickly adapt to inert dilutions as high as 50 per cent, and maintain their normal caloric intake.

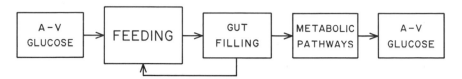

FIGURE 13-1.   Diagram showing the general relationships of feeding to regulation of arterial and venous (A-V) blood glucose levels. The directed arrow signifies "Has an effect upon."

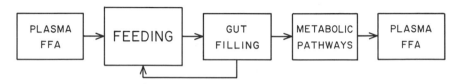

FIGURE 13-2.   Diagram indicating control flow for the relationship of feeding to regulation of plasma free fatty acid concentration (FFA). The directed arrow signifies "Has an effect upon."

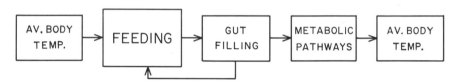

FIGURE 13-3.   Diagram indicating control flow for the relationship of feeding to regulation of average body temperature. The directed arrow signifies "Has an effect upon."

rat can double its bulk intake overnight, the gut-filling feedback is either overridden, its sensitivity is altered, or gastric motility is enhanced. Cold stress may set the stretch receptors of the stomach at levels permitting greater filling before compelling signals are fed back to the feeding mechanism; the opposite may occur in heat stressed animals. These hypotheses are amenable to experiment and represent the types of study needed to add detail to our diagram. It appears certain, however, that stomach distention plays some role in all the systems in which control of food intake is a component of some specific regulation.

CONTROL OF FOOD INTAKE IN REGULATION OF BLOOD GLUCOSE

Food intake is one of the controlling factors in the regulation of blood glucose concentration. Relationships between food intake and blood glucose concentrations have been elaborated by Mayer and his associates into the glucostatic theory of food intake (1953). In consideration of this hypothesis, the several factors influencing carbohydrate metabolism must be taken into account. Obviously, blood sugar levels are the result of ingestion, absorption, storage, synthesis, utilization, and excretion. These processes are under the influence of many hormones.

Most available data on hormonal effects are concerned with the utilization of glucose, possibly because its study has been stimulated by interest in diabetes mellitus, a disease characterized by underproduction or inactivation of the "hypoglycemic" hormone, insulin. Insulin lowers blood sugar by hastening the deposition of glycogen in the liver and muscle and increasing the rate of formation of fat. A primary action of insulin appears to be to change the permeability of the cell membrane to glucose.

Another hormone, glucagon, produces hyperglycemia, apparently by stimulating hepatic glucogenolysis. Other physiological effects of the hormone are an increased metabolic rate and decreased gastric motility. Both the cortex and the medulla of the adrenal glands produce hormones with significant effects upon the regulation of blood sugar. Epinephrine given intravenously causes a prompt rise in blood sugar, primarily as the result of breakdown of liver glycogen to glucose. In this respect, the physiological role of epinephrine and insulin are antagonistic. The glucocorticoids of the adrenal cortex also produce hyperglycemia; their action is antagonistic to insulin in that they increase gluconeogenesis, increase mobilization of fats from tissue, and increase liver fat. These hormones include cortisone, 17-hydroxycorticosterone, and corticosterone. Another corticoid, oxycorticosterone, tends to lower blood sugar levels because it inhibits renal reabsorption of glucose.

Involvement of the adenohypophysis in the regulation of blood glucose was elaborated originally, by Houssay, as follows: Hypophysectomy increases the sensitivity of a normal animal to insulin and decreases the severity of hyperglycemia in a depancreatized animal. Injections of adenohypophysial extract increase the severity of hyperglycemia in hypophysectomized-depancreatized animals, whereas in normal animals, they decrease sensitivity to insulin, and upon repeated injection, can even induce a diabetic condition. Apparently these effects of the gland are related to two other hormones, ACTH and growth hormone (GH), the former affecting glucose utilization indirectly through adrenocortical hormones, and the latter elevating blood sugar levels through inhibition of peripheral utilization of glucose.

The fact that hyperthyroidism intensifies diabetes mellitus in man indicates some relationship between that hormone and blood glucose levels. This effect appears to be due primarily to an increased utilization of glucose followed by an increase in gluconeogenesis.

These brief considerations of the effects of hormones on the regulation of blood glucose only serve to indicate the complexity of the system.

Since the ultimate source of glucose is ingested food, however, one must expect these very complex hormonal arrangements to interact at one or more points with the mechanisms insuring a continuing external supply. Mayer's glucostatic hypothesis of the control of food intake has this interaction as a foundation and provides an explanation of short-term (daily) control of feeding. (Besides this, one may suppose that long-term control over months or years may also operate in the organism, in a fashion related more to body weight than to blood glucose levels.) Blood sugar seems to be a reasonable basis for a short-term signaling system. Among other advantages is the fact that, since the body's stores of fat and protein are relatively large and more constant than its stores of carbohydrate, the short-term changes of the latter might be the more responsive device for initiating states of hunger and satiety.

However, since animals with the hyperglycemia of diabetes mellitus continue to eat food, it is apparent that absolute magnitudes of blood glucose concentrations alone cannot be the sole basis for variations in food intake. This fact has led Mayer to propose that the availability of glucose to the cells is the critical signal, and that this can be measured as the difference between arterial and venous blood glucose levels. When the difference between arterial and venous blood glucose is found to be small, little glucose utilization is implied, and food-seeking or feeding behavior is likely to begin. It has been shown in man that in peripheral blood (forearm) this difference increases after a normal meal (high in carbohydrate), and that in many cases the increasing difference is correlated with subjective feelings of satiety. Nevertheless, feelings of satiety are not invariable, and satiety also follows ingestion of a carbohydrate-free meal when the A-V differences remain unchanged from preingestion levels. Yet the A-V difference hypothesis is consistent with the large appetite of patients with diabetes, since in this disease the A-V differences are small.

Experimenters have been relatively unsuccessful in inhibiting food intake by infusion of hypertonic glucose solution to evoke hyperglycemia. Glucagon, which also elevates blood sugar, does inhibit food intake; but recent evidence points to the fact that glucagon also increases metabolic rate and has other effects. On the other hand, injections of insulin, which produce hypoglycemia, are almost invariably followed by increased food intake. At present, therefore, while it seems certain that elevated levels of blood glucose are not necessarily related to satiety, there is every indication that when blood glucose is lowered, feeding is initiated.

Mayer's hypothesis suggests that cells of the medial hypothalamus are sensitive to their own rate of utilization of glucose and so act as receptors in the control system. To serve this function, these gluco-receptor cells must require insulin to utilize glucose, but at this moment this point lacks experimental verification. There is no evidence that any neurons require insulin for glucose utilization. Nevertheless, one cannot discard completely the possibility that cells in the medial hypothalamus do have the necessary properties to serve as an error detector in the controlling system informed by feedback via the blood stream. Mayer also presents as evidence in support of the glucose receptor hypothesis the data obtained following the injection of gold thioglucose into mice. Hyperphagia and obesity are observed, and histological examination of the brain shows that most, but not all, the damage from the heavy metal occurs in the ventromedial region of the hypothalamus. It is presumed that the susceptibility of these cells to gold thioglucose injury is evidence for their sensitivity to glucose levels or utilization.* In spite of this destruction of the hypothetical glucoreceptors, mice treated with gold thioglucose have no obvious disorder of carbohydrate metabolism. They show a normal regulation of blood glucose levels and a normal insulin sensitivity. In the rat, with comparable lesions of the ventromedial area made by electrolysis, no consistent abnormalities in carbohydrate metabolism have been observed either, and the evidence for increased insulin sensitivity is contradictory. Thus, although the ventromedial hypothalamus may aid in regulation of blood glucose under normal conditions, it does so by controlling intake of food (i.e., carbohydrate) from the external environment rather than by control of the internal glucose economy.

In another effort to discover glucoreceptor mechanisms in the hypothalamus, Anand and his colleagues (1961) have used electrophysiological recordings from electrodes implanted in the lateral and medial aspects of the tuber cinereum of monkeys. They observed that upon the intravenous infusion of hypertonic glucose, the medial areas became more active electrically and the lateral areas less so. Conversely, when hypoglycemia was induced by intravenous injection of insulin, the medial areas became relatively inactive and the lateral areas occasionally showed increased electrical activity. Injections of amino acid mixtures and fat emulsions elicited no change in electrical activity of these two areas of the hypothalamus. Whereas the intravenous injection of glucagon had no immediate effect on the electrical activity, at a later time, when the hyperglycemic effect of the glucagon was evident, there was increased electrical activity of the ventromedial area. In addition, gastric hunger contractions were inhibited during the period of hyperglycemia.

Taken together, all this work leads to the conclusion that food intake is intimately related to the regulation of blood glucose. Of perhaps greatest significance is the fact that food intake is initiated by low levels of blood sugar. Since cells of the CNS are sensitive to such conditions, there can be little doubt that the brain is involved in this control system. In our diagram for blood sugar regulation (Fig. 13-1), we can identify some of the factors; it is plain that both onset and cessation of

---

* In the rat, gold thioglucose produces such widespread damage in the brain that the animal does not survive to show hyperphagia.

feeding are necessary parts of the overall process and that, to some degree, feeding behavior must originate from the operation of this particular system. So far as this part of the regulation is concerned, we assume that the detector is a neural mechanism that is sensitive to glucose or that is affected by neural impulses from peripheral glucoreceptors. The hypothalamus initiates efferent nerve impulses to the motor systems involved in the search for and ingestion of food. The process of eating may trigger the conditions necessary to evoke the responses to gut-filling previously described. Since fasting animals regulate their blood glucose, controlling signals other than the ones involved in feeding must be responsible for the management of the internal economy. Overall, the system operates to initiate food intake when a fall in glucose levels is anticipated. This situation is reflected in small A-V glucose differences.

The study of feeding behavior in a diabetic animal with high levels of blood glucose furnishes pertinent information about some other facets of the glucose feeding mechanism. Despite the assumption that cells of the ventromedial nucleus are responsive to insulin, analysis of the available data suggests that lack of glucose at a cellular level in regions of the body other than the brain may initiate feeding. If so, this may explain the initiation of feeding by the diabetic animal.

While all of these experiments cast some light on the role of food intake in the regulation of blood sugar, it is apparent that certain other areas should also be investigated. Temporal relationships between conditions of glucose utilization and initiation of food intake in an ad libitum feeding situation must be studied. The course of the blood sugar curve in relation to satiety is a question that must be answered. Finally, it would be desirable to pursue Mayer's hypothesis that A-V glucose differences are a sufficient index of cellular glucose utilization, and that they may serve as the critical variable relating feeding to glucose regulation.

CONTROL OF FOOD AND BODY FAT CONTENT

While short-term control of food intake may be glucostatic, the long-term control is probably lipostatic. The relative constancy of body weight in adult animals shows that the body preserves a cumulative balance between ingested food and energy loss. Kennedy (1953) has proposed that this involves a regulation of body fat content and is accomplished by a circulating metabolite that is directly related to quantity of body fat. The concentration of this metabolite is sensed by the brain so as to control food intake and accomplish the adjustments needed for maintenance of body weight over long periods of time. In this hypothesis, the body fat has logically the same position as does glycogen in the glucostatic hypothesis—that is to say, both of them are forms of stored energy. The value of such fat stores is illustrated by the long survival of the hypothalamic, hyperphagic, obese rat during starvation. Such a rat may live more than three times as long as a non-obese control. Another relevant example is the hyperphagia and obesity observed in birds prior to migration. In long-range migrators, the fat content increases to levels of 75 per cent of total dry body weight. During flight, lipids are used predominantly for energy (Odum and Farner, in Tepperman and Brobeck, 1960). A comparable storage of fat occurs in mammals that hibernate, but in other mammals the control system seems not to function so as to accumulate fat in anticipation of great need.

Experimental evidence for the adjustment of feeding to the state of fat stores is of several types. Rats made obese by tube feeding show lower than normal ad libitum food intake after the intubation is discontinued. Their food intake rises to control levels only when their body weight falls again to the normal range. Rats and monkeys that develop hyperphagia after hypothalamic lesions show two stages in the development of obesity. The first (dynamic phase) is that of highest food intake and maximum rate of weight gain. After a varying period, body weight becomes stabilized (static phase) and there is usually a significant reduction of food intake. Some investigators use these data to argue that in the hypothalamic, hyperphagic rat or monkey the body weight is still regulated, but at a higher level. By various artifices, however, even normal animals may be caused to gain or lose weight inappropriately. For example, both intact and hypothalamic, obese rats show an increased caloric intake and body weight when placed on a high-fat diet. Then when the usual high-carbohydrate diet is readministered after a regimen of high fat, both show a loss of weight to previous levels following a reduction in caloric intake. The body weight of the rat can also be increased significantly by immobilizing the animal. For some reason, when physical activity levels of rats are thus restricted, there is a partial "decoupling" of the adjustment between food intake and body weight. This weight gain with inactivity may be a quite general phenomenon, and can be observed in primates, including man. Perhaps, as Mayer has pointed out, evolutionary selection for millions of years has abetted those members of the species which were adapted for fat storage when food was limited and intervals between meals were necessarily long. In the past, the very act of seeking food required much more energy expenditure than it does now in many species for whom food is readily available. Perhaps the ability to store fat is no longer a critical, adaptive mechanism.

The tissue in which fat is stored is metabolically active, and fat is continually being mobilized for use as an important energy source. It is released from fatty tissue as free fatty acid (FFA) bound to albumin. In vivo, norepinephrine and to a lesser degree epinephrine, growth hormone, and ACTH are releasing factors. Insulin tends to block the release of FFA. In addition to these hormones, a substance known as FMS—fat

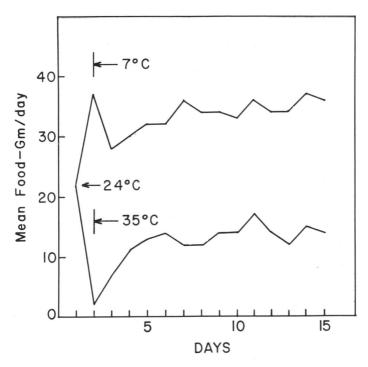

FIGURE 13-4.   Alterations in food intake of rats exposed to warm (35°C) and cold (7°C) environmental temperatures after adaptation to a neutral temperature of 24°C for a control period.

mobilizing substance—has been extracted from the urine of rats and of man while they were actively mobilizing fat (as in the fasting state) (Chalmers et al., 1958). When injected into rats and mice, this substance causes them to increase fat mobilization and lose weight (Stevenson et al., 1964). In these rats at least part of the weight loss is due to reduced food intake. While the relation of FMS to FFA is not clear, one can begin to see the elements of a hypothesis in which increased plasma FFA, which shows a rise in the postabsorptive state, might initiate feeding; but available data on this issue are contradictory. It appears that the formation and release of FFA are under hormonal and autonomic nervous system control and, therefore, are undoubtedly part of a complex control system similar to those regulating blood glucose and body temperature (Fig. 13-2). Thus, body fat appears to be a controlled variable, and food intake is an important mechanism participating in the regulation of body fat exchanges.

CONTROL OF FOOD INTAKE IN THE REGULATION OF BODY TEMPERATURES

Brobeck has proposed that food intake is controlled as if it were an integral part of the systems regulating body temperature (1960). This has been referred to as the thermostatic theory. Certain indirect evidence in its support may be summarized as follows: (1) Homeotherms increase food intake in the cold and decrease it in the heat. A rat, eating approximately 126 kcal a day at a cage temperature of 7°C, reduces its caloric intake to 18 kcal or less when placed in a hot environment (35°C). Fig. 13-4 shows this adjustment of feeding in warm and cold exposure. Deep body temperature in these environments may differ no more than 0.5°C from the average value in a neutral environment. (2) If an animal is force-fed by intubation at a high environmental temperature, it develops hyperthermia; and if enough calories are given by such gavage, the rat may die as a result of the high fever. (3) If the ambient temperature is high enough, animals fed ad libitum will die of starvation in preference to eating, which would compromise body temperature regulation.

There is evidence also that certain neurons of the anterior hypothalamus show sensitivity to direct heating and cooling. These cells lie dorsal to the optic chiasm and ventral to the anterior commissure. Heating them by means of implanted thermodes is followed by activation of heat loss mechanisms, and, conversely, cooling is followed by shivering and activation of heat retention mechanisms. In goats such heating or cooling is also followed by a change in food intake such that feeding increases following cooling and decreases following heating of the hypothalamus (Andersson et al., 1964). Goats and rats with ablations of the anterior hypothalamic area fail to reduce food intake when heat stressed and develop high fevers. Rats with such lesions not only eat more than controls during exposure to heat, in spite of their higher fevers, but when allowed to secure food by pressing a lever in a hot environment, they work for more food than do normal rats at the same temperature (Hamilton and Brobeck, 1964). The specific nature of the deficits produced by these rostral lesions is revealed by contrast with animals having lesions in the ventromedial hypothalamic area. The latter animals continue to alter food intake in the heat and cold in a manner comparable to controls. It appears, then, that there are within the anterior hypothalamic area cells that are sensitive to heat and cold stress, that in some manner control heat exchange, and that also have connections to neural pathways for feeding. The anterior hypothalamus has an action upon the controllers of the peripheral vasomotor system and the respiratory system, and also probably upon the controller of feeding, including the lateral portion of the tuberal hypothalamus.

The heat content of the body is the difference between heat production and heat loss. Food intake is intimately involved in the former and indirectly with the latter; for example, the familiar feeling of warmth after a meal reflects the postprandial increase in metabolic rate (heat production) and also peripheral vasodilation (heat loss). It was once believed that food intake is stimulated by increased metabolic rate. This idea followed from the fact that both metabolic rate and food intake increase in the cold. However, if this idea were extended to include the postprandial increase in metabolism, it would require that each meal provide a stimulus for additional eating. The postprandial increase in metabolic rate would thus serve as a positive feedback that would render the system ineffective and perhaps self-destructive. Similar paradoxes appear if one selects deep body temperature as the variable which determines feeding behavior. Deep body temperature is elevated a bit in the cold, is "normal" at ambient temperatures near thermal neutrality (for the rat, approximately 31°C), and then is elevated again at high ambient temperatures. In the cold, therefore, a high food intake accompanies high body temperature; during mild heat stress, a low food intake is associated with low body temperature; but at high heat stress, the food intake is low despite elevated body temperature.

There is little doubt that a rat responds to changes in cutaneous or peripheral temperature. In a cold environment, rats learn to depress a lever activating a heat lamp. It has been shown that the change immediately before the rat presses the bar to secure external heat is a drop in skin temperature (Weiss and Laties, 1961). There is in this circumstance, no systematic relationship between bar pressing and the deep body temperature. These data show the important role of peripheral receptors in this control system, and suggest that the lower skin temperature is the stimulus producing increased food intake in environmental cold stress when the core temperature is actually elevated. If we are to predict food intake from conditions of body temperature, skin temperature must be considered, as it is in Table 13-1 showing the basic relationships between body temperature, certain thermal stresses, and food intake as we know them at present. It is clear that the most consistent relationship is between skin temperature and food intake, and that temperature regulation and food intake are linked most strongly through mechanisms of peripheral sensation. These include the temperature receptors of the skin and nasopharynx.

The same relation between feeding and skin temperature is seen also during artificial heating and cooling of the anterior hypothalamus of goats. The heating of the hypothalamus that depresses food intake also evokes a peripheral vasodilatation, whereas the cooling that increases food intake also produces a peripheral vasoconstriction. One must add that experiments have not been conducted to learn whether artificial heating of the brain will inhibit food intake in a cold environment. The

TABLE 13-1. RELATIONSHIPS BETWEEN DEEP BODY TEMPERATURE, SKIN
TEMPERATURE AND FOOD INTAKE DURING CERTAIN ENVIRONMENTAL
CONDITIONS

| | TEMPERATURE | | |
CONDITION	CORE	SKIN	FOOD INTAKE
Air temperature			
7° C	↑	↓	↑
32° C	↓	↑	↓
38° C	↑	↑	↓
Exercise	↑	↑	↓
Heated hypothalamus	↓	↑	↓
Cooled hypothalamus	↑	↓	↑

Note: Arrows point in the direction of changes from "normal" conditions taken
at 26°C air temperature.

experiments which use ablation of the anterior hypothalamic area and
study the consequent effect on feeding in the rat may indicate that the
central receptors were either eliminated or that their responsiveness
was altered so that the animal was no longer adequately informed of
conditions of peripheral temperature.

One can rationalize the mechanisms for regulating both temperature
and processes of food intake by using contemporary schemes of regulation of body temperature. The temperature which is regulated is conceived
to be an average body temperature, taking into account skin, subdermal
tissue, and deep core temperatures. Core temperature, alone, will not
explain the regulation, any more than it can be used in predicting food
intake. The similarity between these two conditions leads us to believe
that the relation of food intake to body temperature likewise depends
upon some derived value such as mean temperature. Table 13-1 gives an
example of the type of conditions that demand such a hypothesis.

An outline for the regulation of average body temperature is shown
in Figure 13-3. The average is usually determined by taking one third of
the skin and two thirds of the rectal temperature. Under ordinary circumstances the flow of body heat is from the core to the surface to the
environment. The rate of this transfer must be estimated by the detectors'
yielding information about environmental, peripheral, and core temperatures. Within limits, this gradient can be controlled and the average
body temperature maintained. The two primary controls involved are
rate of heat loss and rate of heat production. Food intake is part of the
latter component, for without food intake, ultimately there can be no
heat production. But what of the starving animal, does it not regulate
body temperature? It does, up to a point; in a rat the first indication of
impending death by starvation is a precipitous drop in core temperature.

Until this time the animal calls on energy reserves and consumes its own body fat. Thus, food intake and energy depots serve not only in the regulation of blood glucose but also in the regulation of body temperature.

Within the context of the priorities and compromises of the body, temperature regulation stands second only to regulation of P_{O_2} and P_{CO_2}. As mammals sacrifice water to produce evaporative heat loss in a hot environment even when water is not readily available, so food intake is decreased during severe heat stress. The organism does not continue to eat food and obligate the body to dissipate the undesirable heat which is produced. This leads us to believe that the temperature regulating mechanism may directly affect food intake, regardless of the body stores of fats and glucose. Yet in normal conditions of temperature regulation (non-heat or cold stress), the control of food intake does depend upon information about the other two variables, glucose and fat. When such is the case, the conditions of temperature regulation are related more to cessation than to initiation of feeding.

Other Factors in Control of Food Intake

We have outlined three regulatory systems associated with control of food intake. Are other systems also involved? One thinks immediately of the body supply of protein. Unfortunately, very little information is available, and certain experiments indicate rather clearly that animals are relatively insensitive to changes in the protein content of diets. For example, at ambient temperature of 22 to 25°C, when the proportion of protein in the diet is lowered from 22 per cent to 5 per cent rats do not increase total food intake to cover the deficit; death ensues in approximately 45 days. When such studies are conducted in a cold room at 8°C, food intake increases enough so that total protein intake is adequate, and the animals survive (Andik et al., 1963). Similarly, rats forced to exercise also increase food intake enough to cover the protein deficit and survive even at ambient temperatures of 22 to 25°C. It appears, therefore, that rats fed low-protein diets will not eat more to obtain more protein unless other experimental contingencies are created that tend to increase the caloric intake.

Another type of observation is that rats decrease total food intake when protein in the diet is high (40 to 50 per cent). This phenomenon has been attributed to the heat production associated with the metabolism of protein, since of all three foodstuffs, protein has the highest specific dynamic effect.* The contrast is plain since rats actually increase caloric

---

* *Ed. Note:* Specific dynamic effect (SDE) or action (SDA) is the increase in metabolic rate observed during assimilation of food. The excess heat is produced in biochemical reactions of the intermediary metabolism.

intake when given either high-fat or high-carbohydrate food (especially fat), both of which have a low specific dynamic effect. Brobeck has used these data to support his argument that some aspects of food intake are controlled in relation to temperature regulation. For example, a rising body temperature and an increased rate of heat loss associated with the specific dynamic effect may indeed be related to the satiety signaling system. If it is, then we might expect to find that rats on a high-fat or high-carbohydrate diet take a longer time to reach satiety and therefore eat larger meals.

Food and water intake are closely associated phenomena. It is well known that when an animal is deprived of water, food intake is significantly reduced, and depending upon the type of food, may be completely eliminated. Conversely, when an animal is fasting its consumption of water is much less. This relationship may signify the presence of some oral factor concerned with ease of ingestion. Perhaps it is difficult to eat a completely dry diet. That such is the case is suggested also by the observation that if the food presented to a rat has a high fat content and a greasy consistency, the number of calories ingested increases and water consumption decreases. Wetting the mouth may be less urgent when the diet is greasy.

Upon ingestion of food, the flow of water into the G.I. tract from the tissues causes a relative dehydration and a demand for water. We are all familiar with the phenomenon of postprandial thirst. Whether this reveals a general relationship between cellular dehydration and food intake is not clear. Insulin injections which increase food intake of rats also increase consumption of water. Yet, insulin causes the movement of water from the extracellular to the intracellular spaces and therefore creates a cellular hydration. These conflicting data indicate a need for further study. In summary, inasmuch as dehydration reduces food intake, we may say that food intake is in part controlled by the availability of water.

The Problem from the Point of View of Control Systems

We have described in some detail the several control systems in which feeding participates. Each of them seems to be directed to some bodily goal other than feeding itself, such as the prevention of overdistention of the gut, the maintenance of suitable glucose concentrations for normal cell function, or the regulation of body temperature. Furthermore, each of the systems involves materials which enter the animal in the form of food; that is to say, although they may be distinct in objective, the systems all require replenishment through the process of eating. Finally, the systems function at their own time and rhythm; two may be at one time in concert and at another in opposition. From the point of view of control theory, these data present a different kind of physiological problem from the phenomena detailed in other chapters of this book. In conventional terms, feeding is an activity related to several different regulations which accomplish relatively unrelated objectives. Each of these regulations—of glucose, fat, or body temperature—can call upon the animal to eat, so that any one of them may appear to be a specific and dominant stimulus to feeding. The control system view, by contrast, is that feeding is one point of junction between many multi-input, multi-output feedback loops. The problem then becomes one of deciding how the interactions and priorities within the total system may be expressed in this one variable. Needless to say, the problem is vastly more difficult than many of the simpler physiological regulations in this book. The evidence required for describing the overall system, however, is physiological in nature, and may be utilized also in explaining the relationships between ill-defined but recognizable subsystems. With a broader vision, one might imagine that the interactions and priorities shown by the subsystems joined through feeding behavior are, in reality, the output of a much larger total system that regulates all phases of the organism's energy exchange. Can a case be made for the idea that there is such a regulation of energy balance.

Regulation of Energy Balance

As an unstable system expending energy to maintain its existence, an animal maintains an energy balance. The point at which balance occurs appears to be not completely fortuitous but a matter of regulation. Since energy enters the animal body as chemical foodstuffs and leaves in many forms (i.e., as heat, motion, chemicals, and growth), it is a point of interest to inquire in what manner appropriate sensing of the state of energy balance is accomplished. It is obvious that one can differentiate grossly the difference in the effort of lifting a 5 kilogram and a 25 kilogram weight, but such judgments are not utilized in the maintenance of energy balance. Nor is it plausible to believe that a rat in a treadmill perceives that it ran two hours today but only one hour the day before, and therefore that it should increase its food intake. The alternative to these is some indirect monitoring of energy expenditure.

Consideration of energy balance involves four general factors: energy source, energy storage, energy expenditure, and rate of exchange. In the animal body at constant temperatures these are expressible as changes in body mass, given symbolically by the expression:

$$\frac{dQ}{dt} = I(t) - O(t)$$ 13.1

When mass of the body and of food is expressed in terms of their energy equivalent, then $\frac{dQ}{dt}$ represents the rate of buildup or breakdown of available body stores of energy, while $I(t)$ is the rate of ingestion and $O(t)$ the rate of energy output. During fasting $I(t)$ is zero, and the body loses weight; conversely, when $I(t)$ exceeds $O(t)$, a gain in body weight is observed. Likewise, with $I(t)$ constant, but with a greater work output,

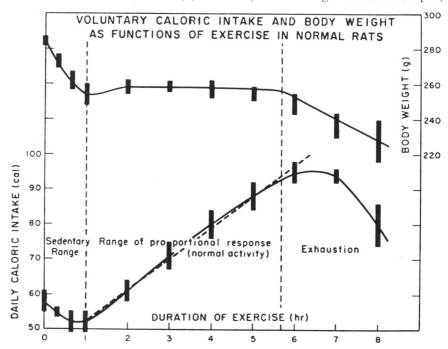

FIGURE 13-5. Relationship between forced exercise, food intake, and body weight of the rat. (Mayer, J., et al, *Am. J. Physiol.* 177:544, 1954.)

$O(t)$ is increased, and a fall of body weight is observed. (Unfortunately, in weight reduction schemes where exercise is used for this purpose, food intake usually increases if the diet is not carefully controlled.) Mayer (1954) has presented data of the type needed in studying the relationships between energy expenditure and food intake (Fig. 13-5). The graphs are a plot of the number of hours of exercise taken vs. the average food intake per day in an ad libitum situation. These data show a positive linear relationship between food intake and exercise over the middle range of energy expenditure, with curvilinear relationships at both extremes. Mayer explains the drop in food consumption at the high levels of exercise as an exhaustion phenomenon; the rat apparently cannot eat enough to keep pace with increased energy expenditures and so loses weight. Under the sedentary, or low exercise portion of the graph, body weight rises, and food intake shows a slight increase as activity falls to zero. Thus, when $O(t)$ is reduced, the animal does not adjust $I(t)$ so as to maintain balance. In the sedentary range, food intake increased no more than 3 or 4 kcal per day, but with the decrease in activity this excess yielded a significant rise in body weight (Fig. 13-5). I noted earlier in this chapter that one can make rats fat by immobilizing them; this was attributed to a decoupling between food intake and activity. Mayer has proposed that the decoupling is a consequence of evolutionary selection in which survival is enhanced by the ability to store energy in intervals of idleness.

In other experimental situations permitting spontaneous running in an activity wheel, rats show an increased food intake and body weight when entrance to the activity wheel is blocked. Conversely, if a rat is at first maintained in the cage without access to the wheel, it reduces its food intake and body weight when permitted to run. Brobeck has postulated that the interaction between running and feeding is related, in turn, to regulation of body temperature. Where the means to be spontaneously active are available, as in the wheel, the rat can run to maintain body temperature. But where the opportunity for free exercise is not available, the animal eats a bit extra to keep warm. Both adjustments result in paradoxical changes in weight. This is perhaps a case in which temperature regulation may have priority over regulation of body weight. These data indicate that there is no one optimal level of body weight.

Whereas most of the preceding statements are valid both with regard to material balance, as stated in the continuity expression, and with regard to general medical experience relating food and body weight, we should like to submit that this is one of the instances in which the continuity expression is too general to be useful. There is a vast difference between questions related to the long-term averages (days, weeks, or years) of energy intake, output, and balance, and those concerning the physiological reactions that determine the onset and cessation of a particular meal. In the long run, the mechanisms that accomplish the latter also achieve the former; but in the process of study, we should recognize that the terms, hunger, satiety, and eating are distinctly different from terms such as energy balance and stored fat. Since we have been rather indefinite about the use of the term feeding, using it primarily in the general sense of average behavior, we should now point out some of the specific problems which become apparent when details of the process are investigated. It appears that in this context we have exceedingly little reliable information about the eating of meals (particular episodes of eating) even under conditions in which this is obviously the pattern by which food is taken.

If feeding is controlled by a system of the type which is error-activated, it must include a feedback loop for transmitting information about output to the control system. Perhaps the search for such error signals is the most neglected aspect of current research endeavors on food intake. One simple clue for initiating such study may be of the following sort: Let us use for the analysis a laboratory rat with food available ad libitum, so that we avoid the more rigid habits of man with his general adherence to three meals a day with coffee breaks. Records show that the rat eats meals throughout a 24-hour period, and that most of the meals are eaten during the hours of darkness, so that approximately 75 per cent of the total intake is consumed during this time. This departure from a random distribution of eating times suggests that the initiation of food intake is an orderly process. It remains for investigators to determine the ordering variables.

From these considerations, it appears that regulation of energy balance is the ultimate concern of the systems controlling food intake. Since energy flow apparently cannot be directly monitored in terms of physical or chemical units, loss or gain in the system is indirectly sensed by components of the systems regulating blood glucose, temperature, fat, and perhaps other substances including body protein stores. There is no conceivable way that the organism may expend energy, for whatever purpose, without the results' being reflected in the body's storehouse of energy. This relationship is inherent in the continuity expression. Any change of

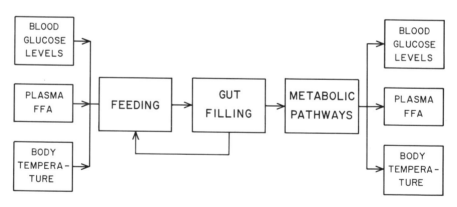

FIGURE 13-6. Composite diagram showing the pathways of control flow common to the four principal systems involving feeding. If one imagines that each box at one edge of the diagram overlaps the corresponding box on the other edge, the representation will lie upon a cylindrical surface; antecedent and consequent effects therefore form a closed-loop system. Any sequence of paths is allowed; hence, the blood glucose levels, for example, may be expected to affect body temperature. The diagram is intended to focus attention upon the importance of strong coupling between the several loops, and upon the central role of feeding in all component control schemes.

O (t) will be reflected in the bodily stores of energy, and these changes in turn are fed back to the food intake control system. No negative feedback is evident from the energy outflow, not only because such a signal and receptors to detect it are unknown, but also because such a feedback is not necessary for the system to function.

Summary

Control of food intake has been attributed to many reactions on the basis of a variety of experiments. Further, because these control factors are in turn subject to complex special control systems, there has been a tendency to conclude that control of food intake may best be explained on the basis of a single, integrated bodily regulation—hence the glucostatic and lipostatic hypotheses. However, most investigators today would agree that, while such hypotheses are correct in general, the major problem is to establish how the many subsystems, which employ feeding as part of their internal workings, function in concord or in discord, as the case may be. A major problem in the physiology of feeding behavior is the need for a quantitative measure for priority and opposition among the many known stimuli to feeding. All the systems which influence feeding may be coupled together to form a mammoth control system, which may be called the controller of energy balance of the body (Fig. 13-6). This synthesis, however, may be an intellectual contrivance rather than a useful representation of bodily organization, since energy flow does not appear to be directly monitored, and since paradoxes of energy balance may readily be produced in the laboratory. In the future we must make a greater distinction between the long-term characteristics of feeding as measured in body weight and grams of food per day, and the short-term reactions such as the initiation and cessation of single meals and the actual rate of eating.

Selected References

The proceedings of the second conference on Brain and Behavior contain further information on the work presented in this chapter. The sections by Pribram and by Brobeck come to grips with the problems of model construction in this field. There is also an excellent review of the work of Chernigovsky and his associates at the Pavlov Institute, Leningrad (Brazier, 1962).

For those interested in the subtleties of behavior following hypothalamic lesions in the rat, Teitelbaum's paper (1961) is an excellent source.

A well written and interesting overview of factors related to energy balance is included in J. Tepperman's *Metabolic and Endocrine Physiology* (1962). The *Brook Lodge Symposium on Energy Balance* also covers many topics of importance to this subject. While none of the authors made attempts to integrate the variables of energy balance, the reference is a source of much useful information on feeding (Tepperman and Brobeck, 1960).

The biochemistry of fats is a rapidly expanding field. The reader can get a good foundation on recent work in Rodahl and Issekutz (1964). Finally, for a most delightful and provocative book on animal energetics, see Kleiber's *The Fire of Life* (1961).

Portions of the experimental work reported in this chapter were supported by Grant MH-10179 from the National Institutes of Health.

Chapter 14

NERVOUS CONTROL
OF GASTRIC SECRETION

FRANK P. BROOKS AND RICHARD A. DAVIS

Study of the nervous control of gastric secretion has been an area of active physiological research since the remarkable achievements of Pavlov's laboratory (Babkin, 1949). Simultaneously, the study of humoral control also developed; Bayliss and Starling first demonstrated the hormonal control of pancreatic secretion, and then Edkins prepared an extract from the pyloric antrum which stimulated gastric secretion (Gregory, 1962). Throughout the next 60 years many investigators tried to isolate this substance and to separate it from histamine. In 1964, Gregory and Tracy succeeded in preparing two peptides or "gastrins" from hog pyloric mucosa, and with the assistance of others, reported the amino acid sequence of the peptides (Anderson et al., 1964).

Uvnas suggested in 1942 that some of the stimulatory action of the vagi on gastric secretion might be due to the release of gastrin, in the same way that the release of acetylcholine from the vagi inhibits the rate and force of contraction of the heart. Conclusive evidence against this hypothesis seemed to come from observations on dogs which had vagally denervated (Heidenhain) gastric pouches. The pouches failed to secrete in response to insulin hypoglycemia, which presumably released gastrin from the pyloric antrum in the main stomach. Subsequent experiments, however, removed this objection. When gastric content from a Heidenhain pouch was collected by an irrigation technique capable of detecting small amounts of hydrogen ion, secretion in response to teasing of the dog by the sight of food or during insulin hypoglycemia could be demonstrated. Furthermore, when the contents of the pyloric antrum was maintained artificially at a pH near 7, nervous stimulation acting upon the main stomach produced considerable amounts of acid secretion from the vagally denervated pouch. Removal of the pyloric antrum abolished the acid and pepsin secretion in response to display of food. These results support the concept that the release of a humoral substance from the pyloric antrum is an important mediator of the stimulation of gastric secretion by the vagi (Grossman, 1963). Nervous and humoral control mechanisms are thus brought together. One should note, however, that the release of a humoral substance from the antrum does not account for the entire stimulating effect of vagal excitation. Vagal stimulation produced by insulin hypoglycemia will increase acid secretion in dogs after removal of the antrum and intestine. Presumably this is due to a direct action of the nerves on the secreting cells.

Similarly, sympathetic stimulation influences gastric secretion, but under most circumstances the action is inhibitory. Norepinephrine and epinephrine inhibit the acid secretion stimulated by histamine in unanesthetized dogs. We do not know whether these inhibitions are direct effects on the parietal cells or secondary to a decrease in mucosal blood flow.

The qualitative evidence for neurohumoral control of gastric secretion, therefore, is conclusive. However, the accomplishments of modern physiology depend in large part upon the ability to express control in quantitative terms. In some systems this has advanced to a considerable degree, as shown by the chapters in this monograph on respiration and temperature regulation. An essential factor for this success is the identification of a regulated variable, e.g., Pco_2 or hypothalamic temperature. In the case of the stomach, the following variables might be considered as candidates for regulation: hydrogen ion concentration, osmolarity or physical state of gastric content, and tension within the wall of the stomach. Moreover, the control of these variables must function under three different conditions: namely, following feeding, when secretions play an important role in the digestion of food; during periods of rest, when the stomach as well as other organs are inactive; and during conditions such as physical exercise, when some parts of the body are at maximum activity while the stomach is at rest. Perhaps it is not surprising that we have found no single variable for gastric secretion which can be shown to be regulated under all of these conditions. We do know that the humoral or neurohumoral portion of these control mechanisms must operate during feeding; since the transplanted fundic pouch, deprived of its sympathetic as well as vagal nerve supply, is still able to secrete in response to a meal.

Because an autonomy of the stomach is shown by its ability to secrete in vitro, some of these control mechanisms may operate entirely within the wall of the stomach. However, the central nervous system can influence the secretion of such components as hydrochloric acid, pepsinogens, mucus, and intrinsic factor. This may represent an overpowering of the local control mechanisms. In the dog, central nervous system stimulation seems selectively to exert a greater effect on the secretion of pepsinogen than on the others.

We shall attempt to define the control of gastric secretion by mechanisms within the stomach itself, then in terms of the autonomic nervous system outside the stomach, and finally in reference to the brain. The goal in our laboratory has been to understand the control of gastric secretion in the intact animal. This has led us to the use of unanesthetized animals, since secretory responses to stimuli may be completely different under anesthesia. How the samples are collected is of some importance. For example, intubation of the stomach through the nose or mouth is well known to influence the volume and composition of gastric content. We have found that in dogs the gastric pH is significantly lower when secretion is obtained by intubation rather than through a gastric fistula. In man, fasting subjects with duodenal ulcers are said to secrete progressively less acid after repeated intubations, while ulcer-free persons may have a pH above 3.5 after intubation, yet secrete acid readily in response to a meal. For these reasons, a chronic gastric fistula offers many advantages for the study of gastric secretion.

There are, however, other complications. Gastric content is normally contaminated with variable amounts of swallowed saliva and regurgitated duodenal content. Even when extragastric contaminations are excluded, the secretion of the fundic or acid-secreting glands is diluted by that from the pyloric glands. With a gastric fistula, a variable amount of gastric content may escape through the pylorus. Moreover, analysis of stomach content from a fistula is difficult after feeding. For these reasons many of the important studies on control of gastric secretion have been done by using pouches made from the fundic portion of the stomach. Pavlov reported that the acid secretion from a pouch faithfully mirrors that from the main stomach. However, the number of actual experiments he reported is small, and his analytical techniques were primitive by modern standards. It is generally admitted that the vagal innervation of his pouches was incomplete. The effects of diverting the flow of chyme from gastric mucosa are unknown. These considerations give rise to some reservations when one tries to transfer information about the control of secretion in pouches to that in the normal stomach.

The desire to study secretion in pouches led investigators to use dogs. As shown by Pavlov, however, dogs in the basal state do not secrete acid. This means that until the purification of canine gastrins is achieved, we unfortunately have no true physiological stimulus to the stomach except a meal. Consequently, histamine and insulin hypoglycemia are used widely as stimulants. In vagally innervated pouches, we have compared the acid and pepsin secretory response to a meat meal with maximal and 25 per cent of maximal doses of histamine, and with 0.15 and 1.5 u/kg of insulin (Long and Brooks, 1965). Pepsin secretion was greatest after insulin; yet, the smaller dose of histamine gave the closest approximation to the secretion after the meal. Therefore, studies on control of secretion in the dog must take the stimulus into consideration.

Although investigators tend to rely upon the results of experiments in dogs, the transfer of these conclusions to other species is uncertain. It is difficult to obtain data on gastric secretion from vagally innervated pouches in cats, and only two laboratories have reported studies on gastric pouches in monkeys. We are not aware of a report upon an innervated pouch in a rat. We have learned of a case report in which a pouch, constructed in a man in the course of additional gastric surgery, in time showed a gradual loss of secretory response to stimulation.

A final limitation on the study of the nervous control of gastric secretion is the lack of a structural basis. We do not know the anatomic details of the relationship between nerve endings and the secretory cells of the stomach. The best evidence for action of the autonomic nervous system on secreting cells comes by analogy from the sublingual glands, in which stimulation of either parasympathetic or sympathetic nerves makes the inside of the cell more negative. Receptors which give rise to action potentials in single vagal afferents when the stomach is stimulated by pressure, changes in pH, or osmolarity have not been identified. The cellular source of gastrin is unknown. Vagal stimulation does cause the disappearance of the zymogen granules from the chief cells, presumably indicating the secretion of pepsinogen. Moreover, insulin hypoglycemia is followed by changes in the microvilli of the intracellular canaliculi of parietal cells as seen by electron microscopy.

Intramural Control of Gastric Secretion

A tentative model of an intramural control system might include a receptor responsive to changes in gastric content or tension in the gastric wall and linked via the myenteric plexus to the secreting cells. Neurohumors could be released both at the receptor and at the parietal cell. The role of the pyloric antrum requires that provision be made for the release of gastrin by the action of stimuli on the mucosa of the antrum as well as in response to reflexes originating in the body of the stomach.

The evidence for receptors in the stomach is indirect (Grossman, 1963). Distention of the stomach causes afferent impulses in the vagi; and stimulation of the central cut end of the vagus causes the secretion of acid and pepsin in a cat, providing the other vagus is intact. The topical application of acetylcholine to the pyloric mucosa, or distention of a pyloric pouch, causes acid secretion from a vagally denervated pouch. Distention of the vagally innervated fundic portion of the stomach after removal of the pyloric antrum causes acid secretion accompanied by a high pepsin output.

Attempts have been made to determine whether a neural receptor lies between the pyloric mucosa and the gastrin releasing cell. The topical application of local anesthetics will block the acid secretion stimulated by acetylcholine. However, when acid secretion is stimulated by acetylcholine, the inhibition which follows perfusion of the antrum with acid solutions is not lost after the topical application of cocaine or lignocaine. Therefore, it is uncertain whether a cholinergic link lies between the mucosa and the gastrin releasing cell.

Gastrin release in the dog is pH sensitive. If the pH of an antral perfusate is less than 1.5 to 3.5, acid secretion in response to a variety of stimuli is inhibited. On the basis of cross-circulation experiments, this type of inhibition has been attributed to the release of an inhibitory humor or "chalone." The fact that acid perfusion of the antrum does not inhibit the secretion induced by gastrin makes it unlikely that release of a chalone is important under normal circumstances.

Acetylcholine is the neurohumor released from the endings of parasympathetic nerves at the postganglionic level in the stomach. Drugs which interfere with this muscarinic action of acetylcholine, by occupying receptor sites on the effector cells, decrease the acid secretion in response to many stimuli. On the other hand, small amounts of acetylcholine will potentiate the stimulating action of gastrin and histamine. The decrease in acid secretion after vagotomy can be partially restored by acetylcholine. These results indicate an intimate relationship between nervous (cholinergic) and humoral (gastrin) influences on the secreting cells.

Extramural Autonomic Control

Vagovagal reflexes, in which the vagi contain both the afferent and efferent fibers of the reflex, constitute the major pathway for central nervous system control of gastric secretion. As already noted, this mechanism includes the release of gastrin. The vagi appear to exert also a "tonic" action on the responsiveness of the secreting cells to stimulation.

Bilateral vagotomy abolishes the secretory response to sham feeding and to insulin. The acid secreted in response to gastrin is reduced, but the maximal histamine response may be normal, providing that the dose is increased to reach a steady, maximal acid output. Vagally denervated pouches, in response to feeding, secrete smaller amounts of acid after a longer latent period than do innervated pouches. Vagal denervation may sensitize a pouch to stimulation by parasympathomimetic drugs.

Denervating only the antrum by "antroneurolysis," described as "submucosal dissection of the entire antrum," blocks the secretion of a Heidenhain pouch in response to insulin hypoglycemia. If the denervation of the antrum is done by transecting the stomach between the antrum and

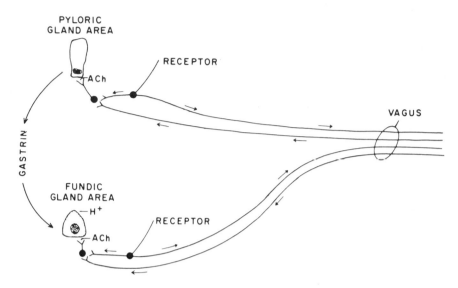

FIGURE 14-1   Diagrammatic representation of the intramural and vagal innervation of gastric mucosa. (Grossman, M.I., *Physiologist* 6:349, 1963.)

body, and by cutting away the major and minor omental connections, gastric acid secretion in response to sham feeding is reduced. The effects of denervation on the inhibition of acid secretion are less clear. For example, acidification of the antrum may inhibit acid secretion after insulin hypoglycemia, whether the nerves to the antrum are intact or not.

The acid and pepsin secretion in a dog during the first hour after a meal seems to be due to a gastrin-like mechanism, but the secretion occurring later is not abolished by antral denervation and is only partially inhibited by antral acidification.

Figure 14-1 is a diagrammatic representation of the pathways of local and vagal nervous control in the stomach as described by Grossman (1963). Note that acetylcholine is released at both the parietal cells of the fundus and the gastrin releasing cells in the pylorus, and that this release may be brought about by local intramural reflexes or by gastric reflexes involving the vagus.

Inhibition of gastric secretion by duodenal mechanisms, i.e., by acid or fat in the duodenum, appears to be primarily via humoral mechanisms. However, the presence of intact vagi makes the pouches more sensitive to inhibition, as in the case of gastrin.

The effects of sympathectomy depend upon the level of denervation. Preganglionic sympathectomy, including denervation of the adrenal medulla, causes an increase in acid secretion after insulin. Coeliac ganglionectomy has no significant effect on the response to most stimuli of either a vagally innervated or a denervated pouch.

CENTRAL NERVOUS CONTROL

We regard the central nervous system as exerting, through the vagi and sympathetics, a variable influence throughout the 24-hour cycle, on an essentially autoregulatory mechanism within the stomach itself. This principle can be illustrated as follows: In fasting unrestrained spider monkeys, we have found a diurnal rhythm in which the pH of gastric content remains between 1 and 2 during the day, and then rises to 6 to 7 during the night. Whether this is the result of conditioned responses to feeding or a basic circadian rhythm remains to be determined. When the monkey is restrained in a chair, secretion is reduced to a steady state throughout the 24 hours.

The central control mechanisms include a reflex pathway in the medulla, sensitive to inhibition from higher levels, and sharply circumscribed areas in the hypothalamus which link gastric secretion to food intake. Receptor mechanisms responsive to hypoglycemia have also been described in the hypothalamus, but their physiological significance is not clear. Upon stimulation of the amygdala, an inhibition of gastric acid secretion occurs coincident with an increase in plasma corticoids.

METHODS

Study of central nervous system control of gastric secretion requires a combination of neurophysiological and gastrointestinal techniques. The usual methods for stimulating gastric secretion via the central nervous system are sham feeding and insulin hypoglycemia. More recently, the neurophysiological methods of producing discrete lesions in the nervous system, or of implanting electrodes chronically in the structure to be studied, and of measuring the effect of the lesion or the stimulation on gastric secretion, are coming into use.

MEDULLA OBLONGATA

Lim and his associates (Langlois et al., 1952) stimulated the seventh and ninth cranial nerves of dogs under chloralose anesthesia, and in the respective experiments, found a small gastric secretory response in 2 of 7 and 9 of 14 dogs. After midbrain or pontine sections which left about 2 mm uncut ventrally in 5 dogs, stimulation of the fifth (buccinator), seventh, ninth, and tenth nerves, as well as insulin hypoglycemia, produced large responses, which usually outlasted the stimulus by 15 to 45 minutes. In 3 dogs with the diencephalon and internal capsules sectioned at the level of or above the hypothalamus, stimulation of the ninth and tenth nerves and insulin caused minimal responses. These results are consistent with the conclusion that gustatory reflex stimulation of gastric secretion acts through the medulla and is susceptible to inhibitory influences from tegmental or hypothalamic areas.

INSULIN HYPOGLYCEMIA

Stimulation of gastric acid and pepsin secretion by insulin hypoglycemia acts through the central nervous system as shown by La Barre and Cespeded (1931), who perfused the isolated head of a dog with hypoglycemic blood from a donor and found acid secretion in the stomach of the recipient. The secretion is abolished by complete vagotomy. The ability of a pouch to secrete acid in response to insulin hypoglycemia is highly correlated with preservation of vagal innervation. This stimulation is due to the hypoglycemia and not to the insulin per se, since it can be blocked by preventing the hypoglycemia with an infusion of glucose. The site of the receptor mechanism for the hypoglycemia is less certain. Early experiments with ablation of the central nervous system in dogs, starting with decortication and proceeding caudally, showed some acid secretory response until the plane of ablation had reached down to the level of the vagal nuclei. However, in our laboratory, rats with chronic gastric fistulas failed to increase acid output in response to insulin when the ventromedial region of the hypothalamus was stimulated electrically. We have also noted the absence of gastric acid secretion after insulin in dogs with lesions in the anterior hypothalamus.

The acid secretory response to insulin can be divided into phases in time. In man, an initial depression of basal acid secretion is the first effect. In dogs, we have also seen the inhibition if the animal was secreting acid at the time the insulin was given. This inhibition is followed by stimulation. There is considerable variability, however, between dogs and even in the same dog in the magnitude of the acid secretory response. Table 14-1 shows the variation among 7 pure-bred beagle dogs, all purchased from the same kennel.

TABLE 14-1.   MEAN MAXIMUM ACID SECRETORY RESPONSE TO INSULIN IN DOGS WITH CHRONIC GASTRIC FISTULAS

DOG	VOLUME ML/30 MIN	ACID CONCENTRATION TITRATED TO pH 7 MEQ/L	ACID OUTPUT MEQ/30 MIN	CHLORIDE CONCENTRATION MEQ/L	BLOOD SUGAR 30 MIN AFTER INSULIN MG/100 ML
13 (6)*	9 ± 3 S.E.	78 ± 4	0.60 ± 0.15	143 ± 5	31 ± 7
18 (3)	23 ± 7	108 ± 14	2.3 ± 0.5	153 ± 13	27 ± 5
19 (8)	20 ± 3	85 ± 12	2.1 ± 0.5	135 ± 5	29 ± 7
20 (11)	16 ± 4	50 ± 7	0.86 ± 0.04	129 ± 4	21 ± 3
21 (6)	25 ± 3	105 ± 10	2.6 ± 0.4	151 ± 7	48 ± 7
22 (9)	22 ± 3	76 ± 11	1.3 ± 0.1	134 ± 5	23 ± 3
23 (7)	22 ± 4	103 ± 11	2.15 ± 0.3	153 ± 7	36 ± 5

*Number of experiments

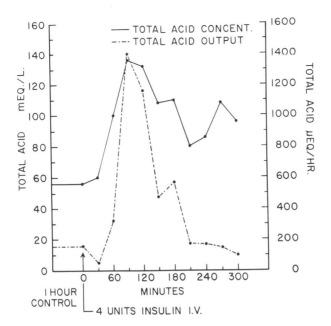

FIGURE 14-2. Comparison of acid concentration and acid output as indicators of the duration of stimulation of gastric secretion by insulin hypoglycemia in the spider monkey. (Smith, G.P., et al., *Am. J. Physiol.* 199:889, 1960.)

The stimulation of secretion has been divided into an early (1 to 3 hours) and a late (3 to 5 hours) phase (Bachrach, 1963). In monkeys (*Macaca mulatta*) the early phase is abolished by vagotomy and the late phase by adrenalectomy. We attempted to confirm this observation in spider monkeys with chronic gastric fistulas. We found the presence of a late phase to be extremely variable. Moreover, acid concentration and acid output as criteria for a late phase gave conflicting results (Fig. 14-2). Acid concentration was well above control levels at 5 hours after insulin, but acid output had returned to control levels at a little over 3 hours. Other authors have reported similarly variable results in pig-tailed macaques.

Using beagles with chronic gastric fistulas, we found late phases (defined as more than 0.5 mEq of HCl/L) in 3 animals on repeated testing, using 1.5 u of insulin per kilogram. In 4 additional animals, no late phases of secretion were observed although we tested repeatedly. One dog showed a single late phase in 11 experiments, and two other dogs showed 4 in 6 and 3 in 7, respectively. We also found a late phase in 2 of 4 dogs with vagally innervated gastric pouches given the same dose of insulin (Long and Brooks, 1965).

In an attempt to relate adrenocortical secretion to the late phase, we measured plasma 17-hydroxycorticosteroids in beagles with gastric fistulas

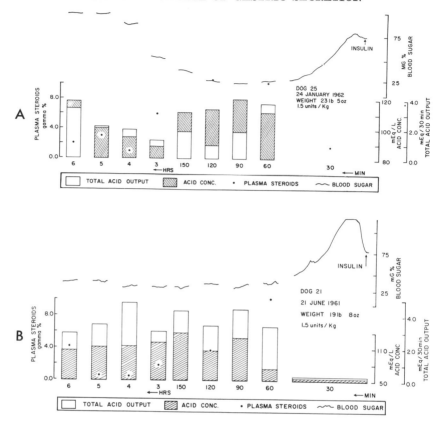

FIGURE 14-3a, b. Relationships between acid secretion, blood reducing substances, and plasma corticosteroids after insulin hypoglycemia in beagles with gastric fistulas. (Davis, R.A., and Brooks, F.P. *Am. J. Physiol.* 204:143, 1963.)

during gastric stimulation with 1.5 u/kg of insulin (Davis and Brooks, 1963). Figure 14-3a, b shows two such experiments in which the blood reducing substances were monitored continuously with a Technicon Autoanalyzer. In one experiment, a late phase occurred at a time when the blood reducing substances and the plasma steroids had reached control concentrations, but in the second experiment, it occurred while the blood reducing substances were still at low levels and the plasma steroid levels were still elevated. We thus found no constant relationship between a late phase and plasma steroids or blood reducing substance.

Although doses of insulin in the range of 1 to 1.5 u/kg had been used for many years, we were unable to find studies over a full range of insulin dosage. We therefore gave insulin in doses from 0.1 to 1.5 u/kg to dogs with chronic gastric fistulas, and to others with vagally innervated pouches (Davis et al., 1965). There was no consistent gastric acid response to insulin in doses below 0.1 u/kg. Considering only the acid secretion that occurred during the period from 30 to 60 minutes after insulin, we can say that in those experiments in which the blood reducing substances reached a level of 40 mg/100 ml (potassium ferricyanide method), an increase in acid output occurred in 18 of 21. If the blood reducing substances did not fall to this concentration, increased acid output occurred in only 1 of 12 experiments. When the rate of fall of the blood reducing substances was calculated as the regression of the concentration of blood sugar on time, there was no correlation between the rate of fall and the acid output ($r = -0.357$). We concluded that the initial phase of stimulation of acid secretion after insulin hypoglycemia depends upon a sensing mechanism in the brain which responds in an off-on manner at a threshold of about 40 mg/100 ml of blood reducing substances.

The gastric secretory response after the 30 to 60 minute period seemed related to the dose of insulin. Doses ranging from 0.1 to 1.5 u/kg produced similar early increases in gastric acid secretion, but the 1 to 1.5 u/kg doses elicited a greater output of acid lasting over a longer period of time than did the smaller amounts of insulin. Plasma steroids attained higher concentrations and remained elevated longer with larger doses, and the duration of hypoglycemia lengthened. The mechanism of the later portions of this response is unknown.

Sham Feeding

An alternative method for studying gastric secretion, introduced by Pavlov, is the technique of sham feeding, in which food eaten by the animal escapes through an esophagostomy. Thus Pavlov demonstrated that the gastric secretory response to food is abolished by vagotomy (Babkin, 1949). Other conclusions suggested by this technique include the following: The acid secretory response to sham feeding can be assigned to conditioned and unconditioned reflex mechanisms. Palatable food is a more potent stimulus than inert material. Removal of the cerebral cortex abolishes the acid secretion in response to tantalizing of the dog with food but not to sham feeding. Removal of the pyloric antrum reduces the acid secretory response to sham feeding, and this can be restored by a small dose of gastrin. Although both sham feeding and insulin hypoglycemia excite gastric secretion via the vagi, considerably less pepsin is secreted in response to sham feeding.

Hypothalamus

To study the influence of the hypothalamus upon gastric secretion, we used a stereotaxic instrument to place lesions in the hypothalamus of 40 rats with chronic gastric fistulas (Ridley and Brooks, 1965). Ten rats developed hyperphagia, obesity, and hypersecretion of acid and pepsin in the fasting state. The rate of secretion was about twice that of the controls, of sham-operated animals, or of animals with lesions that did not become hyperphagic. At sacrifice, the ventromedial nuclei of the hypothalamus were found to be destroyed bilaterally in only those rats with hyperphagia, obesity, and hypersecretion. In other experiments, the hypersecretion was noted to begin on the day the lesions were made. Rats which hypersecreted when food was restricted in amount became obese when food was available in unlimited supply.

The technique of electrical stimulation of the hypothalamus has also been utilized in studying gastric secretion. Using cats, dogs, and monkeys, comparisons have been made of stimulation of the "anterior" and of the "posterior" hypothalamus. These experiments suggest that anterior hypothalamic stimulation enhances acid secretion in cats and monkeys but not in dogs. Posterior hypothalamic stimulation has been reported to increase acid secretion in monkeys, and acid and pepsinogen secretion in cats, but to inhibit acid secretion and gastric blood flow in cats and dogs. We have correlated feeding behavior with gastric acid–pepsin secretion following stimulation of the hypothalamus in rats with implanted electrodes (Misher and Brooks, 1964 and 1965). Rats with electrodes in the ventromedial area where stimulation stopped feeding even though the animals were "hungry," showed also a significant decrease in the output of acid and pepsin when stimulated in the fasting state. If feeding was not inhibited, no decrease in secretion occurred. When the electrode was placed in the lateral hypothalamus and feeding was tested in sated animals, those rats which fed during electrical stimulation subsequently showed a significant increase in acid and pepsin output with electrical stimulation. Once again, if the animals did not feed on stimulation, there was no change in gastric secretion.

These results demonstrate for the first time that there are sharply localized areas of the hypothalamus concerned with both food intake and gastric secretion. The changes are appropriate for the behavior of the animal. Since, when the vagi were cut, the increase in acid secretion in response to lateral stimulation was blocked, it is likely that the stimulatory effects from the hypothalamus reach the stomach via the vagi. The hypothalamus may represent an important point in the control of gastric secretion through the autonomic nervous system.

Limbic System

The limbic portion of the brain includes the hippocampus, fornix, amygdala, septum, olfactory tubercle, cingulate gyrus, and stria terminalis. It is closely identified with behavior oriented toward the alimentary tract and reproduction. Ablations within the limbic system in dogs have given variable results on gastric secretion. Moreover, stimulation by means of implanted electrodes has been reported to cause a stimulation of gastric secretion in cats with innervated pouches and no change in secretion from vagally denervated pouches. The monkeys, dogs, and cats used in these experiments had only a negligible level of basal acid secretion. In contrast, Smith and McHugh (1964) found an inhibition of secretion in response to electrical stimulation of the amygdala in unanesthetized macaques having a basal secretion of acid.

Summary

Nervous control of gastric secretion can be studied in terms of mechanisms located within the stomach itself and also in terms of the modification of the secretion by the central nervous system. Based upon the response to insulin hypoglycemia, there is good evidence, in rats and dogs, that a sensing mechanism in the anterior hypothalamus is connected to the vagal nuclei. Efferent pathways to the secreting mucosa also originate in the hypothalamus. The close association of regions concerned with the stimulation and inhibition of both food intake and gastric acid and pepsin secretion suggests that this portion of the brain functions to integrate the several visceral responses to nutritional needs.

Research in Dr. Brooks's laboratory has been supported by the following grants: Grants RG-5007 and AM-03596 from the National Institutes of Health; and Grants from the Smith Kline & French Laboratories. Dr. Brooks also acknowledges with thanks the granting of the Research Career Development Award AM-K3-2983 by the National Institutes of Health.

Chapter 15

CONTROL AND REGULATION
OF THE CARDIOVASCULAR SYSTEM

LYSLE H. PETERSON

In 1865 Claude Bernard published his famous *L'Introduction à l'Etude de la Médecine Expérimentale,* which contains the concept that the well-being of the body's cells depends upon the maintenance of an appropriate environment. This environment is actually smaller than the cell itself, and yet it must sustain large variations in the metabolic functions of the cell. Consequently, the environment of each cell must be continuously exchanged, and at a rate proportional to cellular demands. This extensive and complex exchange depends predominantly upon the cardiovascular system.

The physiology of this system concerns the manner in which blood composition is maintained and how it is circulated through the tissues with respect to metabolic requirements. To carry out its functions the cardiovascular system relies upon the nervous, endocrine, renal, respiratory, and gastrointestinal systems. Indeed, the functions of the cardiovascular system interact with almost every other function in the body. This means that the properties and behavior of the cardiovascular system cannot be understood unless one knows how its actions are related to other bodily functions. Consequently, this chapter will emphasize the physiological integration essential for the normal control and regulation of the cardiovascular system.

For many years physicians have attempted to explain the manifestations of each disease as some alteration or malfunction of normal physiological mechanisms. Unfortunately, for most cardiovascular diseases this is not possible. This does not mean, however, that these diseases are not abnormalities of biological functions. Rather, the problem is that we do

308

not comprehend the integrated biological processes relating to cardio-vascular function. For example, there are physiological mechanisms that affect the cardiovascular system in disease, but which normally do not play a primary or predominant role in its control and regulation. Under certain circumstances these secondary functions may become primary effectors of cardiovascular behavior, and abnormal functioning develops.

Most of this chapter is concerned with the properties and behavior of the cardiovascular system and with those mechanisms which normally, from day to day, predominate in its control and regulation. Also, since the normal exchanges of the body's cellular environment depend upon the lymphatic system, the latter part of the chapter includes examples of abnormal lymphatic functioning. These illustrate the important points that manifestations of disease may involve functions that are not con-spicuous under more normal conditions, and that it is possible to gain a considerable insight regarding normal physiology through the analysis of such abnormal functions. Ultimately, of course, the concepts of physi-ology are vital to the effective practice of medicine.

It is necessary to emphasize that the behavior and properties of the cardiovascular system are determined by many functions outside the system. A major challenge to the physiologist and the physician alike is the development of new tools and methods for analysis of multifactor systems like this one, whose many functions are related in complex ways. Some discussion of these new approaches will be given later in this chapter. The major topics of the chapter are considered in the following order: Exchange at the capillary level (including lymphatics), circula-tion in different organs with regard to organ function (local and reflex control), cardiovascular factors affecting blood flow (hemodynamics), and regulation and control of the circulation.

Exchange at the Capillary Level

Exchange of substances between the capillary lumen and interstitial "spaces" is predominantly a function of two processes—diffusion and filtration. (A third process, pinocytosis, may account for limited ex-changes of large or specific molecules.) The capillary wall permits the ready passage of small molecules, ions, and to a lesser extent, plasma proteins. While the exact nature of the mural pathways for the exchange of non-lipid soluble substances is not known, they are commonly referred to as capillary pores (Pappenheimer, 1953). Their effective diameter permits the exchange of a volume equal to the entire vascular water volume and its small solutes (particle sizes 30 to 35 Å) in one passage through the capillaries, i.e., in one circuit or about one minute. The number of larger pores, which permit the passage of protein, averages only about 1 per cent of the number of smaller sized pores, and the relative pore-size distribution varies from one vascular bed to another. For example, the capillaries of the liver, and to a lesser extent of the small intestine, permit a greater exchange of plasma proteins than those in skin and muscle. Nevertheless, an amount equal to the entire quantity of plasma protein may be exchanged in 24 to 36 hours.

Lipid soluble substances, e.g., gases, are exchanged throughout the entire capillary wall surface. It is estimated that the total capillary surface exceeds 10,000 square meters, as implied by Krogh's descriptive "4 miles long and a yard wide" (Krogh, 1929). Diffusion accounts for a major proportion of transcapillary exchange. Since no mechanical forces are involved, it normally does not result in net volume shifts between vascular and interstitial compartments. Diffusion of gases and small molecules is so rapid that the interstitial concentrations differ only slightly from plasma ultrafiltrate. Limiting factors are capillary blood flow and tissue utilization. A special consideration, however, relates to diffusion of large molecules and their return to the blood via the lymphatic circulation. Normally, the back diffusion of such large molecules, e.g., plasma proteins, is less than outward diffusion because there are concentration differences (the ratio of concentration of interstitial to plasma protein ranges from 0.02 to 0.80). The interstitial osmotic pressure tends to be maintained at a low level by removal of protein in the lymph. Thus, an inadequate circulation of lymph, or any other situation which changes the interstitial-plasma osmotic balance, tends to result in volume shifts.

The second mechanism of exchange does involve net differences of mechanical forces acting upon the water of these fluids. As the blood enters the capillary lumen, it is under a certain head of pressure (P_{hc}), which decreases along the capillary length because of fluid friction or viscosity. An opposing pressure exists in the interstitial fluid (P_{hi}). The difference in pressure tends to cause movement (filtration) across the capillary wall and through those passages. The resistance to this movement relates to the properties of the movable substances and the properties of the pores. There is also a second set of forces affecting filtration, viz., the osmotic pressure of fluids in the capillary lumen ($P_{\pi c}$) and in the interstitial spaces ($P_{\pi i}$). The net filtration, then, is a function of four pressures (P_{hc}, P_{hi}, $P_{\pi c}$, $P_{\pi i}$). Furthermore, these tend to vary along the length of the capillary and its surrounding tissue. At equilibrium (i.e., under conditions in which there is no net volume shift across the capillary walls) there is a gradient of these forces along the arterial (proximal) half of the capillary which favors a net filtration outward ($P_{hc} - P_{\pi c} > P_{hi} - P_{\pi i}$.) Conversely, along the venous (distal) half of the capillary length the net forces favor back-filtration. This mechanism is usually referred to as Starling's principle of capillary exchange (1894). Later work by Landis (1927), Pappenheimer (1953), Mayerson et al. (1960), and Renkin (1964), as well as many others, served to elucidate this mechanism and its relationship to diffusion, and to define the properties of the capillary wall, lymphatic circulation, and the fluid.

The magnitudes of some general properties of the capillary wall may be enumerated as follows: area is in excess of 10,000 square meters; passages for water, its solutes, and suspensoids occupy about 0.5 per cent of that area; the cytoplasm, which occupies more than 90 per cent, permits diffusion of gases and other small lipid-soluble materials; and "pore" diameters are distributed in a population that peaks sharply in the range of 30 to 50 Å and falls off rapidly outside this range, but includes a relatively small proportion up to about 250 Å. The distribution of pore

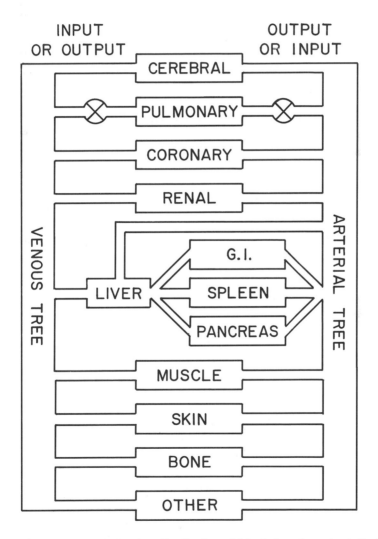

FIGURE 15-1. Model showing distribution of blood flow throughout the body.

sizes varies considerably among capillary beds, as judged by the passage of tagged molecules of graded dimensions. Thus, the liver capillaries tend to have a high distribution of large pores; intestine, somewhat larger; and skin and muscle, still larger. The glomerular capillaries actually pass protein; hence, the renal lymphatics are probably important in its return to the circulation. The "blood-brain barrier," by contrast, represents a small-pore limitation to exchange.

In addition to these variations in capillary wall properties, fluid pressures within and surrounding the capillaries vary considerably from one vascular bed to another and from time to time. Thus, as will be noted in the section on regional circulation, the circulation in various organs differs greatly in those factors which affect capillary blood flow and pressure. Furthermore, other factors such as posture, respiration, exercise, alimentation, clothing habits, and even the weightless conditions of space travel affect the exchange of materials across the capillary walls.

CIRCULATION IN DIFFERENT ORGANS WITH REGARD TO ORGAN FUNCTION

REGIONAL BLOOD FLOW

The variety of functions performed by the various organs of the body suggests that the properties and behavior of their associated circulatory mechanisms also vary significantly. Indeed, the biological and circulatory functions of organs must be considered together since they are mutually related. Figure 15-1 is a schematic diagram of how they may be considered. The diagram indicates that, with regard to their blood supply, the organs (except for the liver) are arranged in parallel. The liver lies in series with the circulation from the gastrointestinal tract (from esophagus to anus), the spleen, and the pancreas; this array of organs is supplied by the visceral circulation. The diagram is drawn so as to suggest that these circulatory organ-units may be considered as "black boxes", i.e., the diagram shows the input and output of blood rather than the internal details of the organ. The following discussion will be concerned with (1) blood flow through the organ, (2) the arteriovenous oxygen content difference, (3) tissue utilization of oxygen and metabolic rate, (4) resistance to blood flow, and (5) major physiological (and, to some extent, pharmacological) factors which influence these variables. When appropriate, these variables will be normalized to unit weight of the organ.

The flow of blood through a vascular bed is determined, of course, by the perfusing pressure (arteriovenous pressure difference) and the resistance to flow, provided by the geometry of the vasculature itself and the effective viscous properties of the blood. In this discussion, the resistance R to flow will include all factors which determine the pressure-flow ratio, $\triangle P/Q$. The quantity expressed as millimeters of mercury per milliliter of flow per second is often described as the peripheral resistance unit (PRU). Analysis of the factors concerned in R is complex indeed. Furthermore, they are not completely understood, as will become apparent in more detail later in this review. It is adequate at this point to consider that R is predominantly influenced by the degree of vascular "tone" as it determines the effective radii of the vessels for any given circumstances of intra- and extravascular pressures.

The parallel arrangement shown in Figure 15-1 implies that the total vascular resistance in the body, and the changes in total resistance from a change in any one organ, are considerably less than if the organs were in series. An approximation of this system may be deduced by the analogy of electrical or simple fluid networks.

In evaluating flow and metabolic properties, the Fick principle is important. For any one of the boxes in Figure 15-1, consider the arterial inflow as input and venous outflow as output. Thus, there are simple but convenient relationships, as follows:

1. By continuity, when the volume of the organ is constant, $Q_{in} = Q_{out} = Q$.

2. Concentration of any metabolite entering (C_i) minus the concentration leaving (C_o) multiplied by rate of flow equals the amount utilized by the organ (U). If the concentration out exceeds the concentration in, the organ has produced this amount. It is apparent that these are all steady state considerations. Thus if $Q = \dfrac{V}{t}$ (volume flow with respect to time), $C = \dfrac{M}{V}$ (amount M of material with respect to the volume in which it is contained), and $\dfrac{V}{t} \times \dfrac{M}{V} = \dfrac{M}{t} = U$ (amount of material utilized or produced with respect to time),

$$(Q_i \times C_i) - (Q_o \times C_o) = Q(C_i - C_o) = U \qquad 15.1$$

or

$$Q = \frac{U}{(C_i - C_o)} \qquad 15.2$$

or

$$C_i - C_o = \frac{U}{Q} \qquad 15.3$$

By knowing any three of the factors, the fourth can be computed. These relationships have found wide use in renal and nutritional physiology, as well as in cardiovascular physiology. Indeed, most of the methods currently used to assess organ blood flow in intact, unanesthetized man utilize the Fick principle.

Flow of blood to a given organ serves two distinct functions, viz., to supply its metabolic requirements, and also to participate in special functions of that organ. For example, renal blood flow supplies the metabolic needs of kidney tissue and also maintains the filtration necessary for urine formation. Skin blood flow supplies the metabolism of skin, but also sustains the body's heat transfers through the integument. The flow of blood for these two separate purposes is controlled by different mechanisms within the body. This control may be local, systemic, or both.

Local control is generally determined by related metabolic factors such as Pco_2, Po_2, and temperature, whereas centralized or systemic control is exerted by nervous or endocrine functions. At times the local and systemic controls may compete. For example, locally increased Pco_2 usually results in vasodilation, yet breathing 5 to 7 per cent CO_2 will raise the arterial blood pressure because of net reflex vasoconstriction, since an increase in Pco_2 causes an increased activity of the sympathetic nervous system mediated by the brain and spinal cord. Exposure to cold may cause such intense vasoconstriction that the skin's metabolic needs cannot be met; thus, the reflex control associated with temperature regulation may predominate over local metabolic effects.

CIRCULATION OF BLOOD IN MAJOR ORGANS

Cerebrum. Cerebral blood flow has been a subject of interest since antiquity, because of its supposed relation to intelligence, sleep, psychosis, and other aspects of mental behavior. Since World War II, when Kety and Schmidt (1948) first described an application of the Fick principle to the measurement of cerebral blood flow (CBF) in intact, conscious man, a great deal has been learned. Currently, in addition to N_2O, the radioisotopic, inert gases, krypton and xenon, are used in methods by which measurements can be made from outside the body. In this way the flow to the two sides of the brain can be measured separately. However, even these are averages of flows through many different parts of the brain with widely varying vascularity. For example, the differences between white and gray (cortex) matter are significant, with the former estimated to have a flow of about 20 ml/min and the latter (gray) of 78 ml/min. About 60 per cent of the brain mass is gray and 40 per cent is white matter. Thus, there must be great differences of flow from region to region in the brain. As noted in Table 15-1, the brain, which comprises 2 per cent of the body weight, sustains a blood flow of 12.5 per cent of the total flow. The average flow per 100 gm (CBF) is remarkably constant under most circumstances. Apparently, there is little effective autonomic nervous system control over the resistance to cerebral blood flow. Moreover, fluctuations of 10 to 20 mm Hg in perfusing pressure do not affect the flow appreciably.

One of the most effective dilators of the cerebral vasculature is CO_2; by breathing 7 per cent CO_2 the cerebral blood flow can be tripled, while hyperventilation may reduce it by more than 50 per cent. Changes in O_2 content, on the other hand, have considerably less effect. Epinephrine in small doses acts as a vasodilator; norepinephrine as a vasoconstrictor.

TABLE 15-1. TYPICAL DATA ON BLOOD FLOW IN RESTING MAN*

ORGAN	TOTAL BLOOD FLOW (ML/MIN)	ORGAN MASS (KG)	BLOOD FLOW (ML/MIN/ 100 GM)	ARTERIO- VENOUS DIFFERENCE (VOL. %)	TOTAL O₂ TISSUE UTILIZA- TION (PER MIN)	TISSUE O₂ UTILIZA- TION (PER MIN/100 GM OF TISSUE)	TOTAL TISSUE METABOLIC RATE (WATTS)	TISSUE METABOLIC RATE (WATTS PER 100 GM OF TISSUE)	TOTAL VASCULAR RESISTANCE	TISSUE RESISTANCE (PER 100 GM OF TISSUE)
Brain	750	1.4	54	6.6	50	3.6	18	1.25	7.2	100
Heart	240	0.3	80	16	38	13	13	4.5	22.5	675
Viscera	1500	2.6	58	3	45	1.7	15.8	0.61	3.6	930
Kidney	1300	0.3	433	1.1	14	3.7	5	1.63	4.15	1250
Muscle	1000	31	4	5	50	0.2	18	0.06	5.4	1350
Skin	500	3.6	4	5	25	0.7	8.75	0.24	10.8	1350
Remainder	710	30.8	2.3	(4.3)	(30)	0.1	10.5	0.035	7.6	2350
Total	6000	70	10	4.1	250	0.35	88	0.123	0.9	540

* The following equivalents were used in computing these figures: 1 L of O_2 ≡ 5 kcal of heat, 1 kcal min ≡ 70 watts of power per minute, 1 kcal min ≡ 420 kg M/min where there is 100 per cent conversion of work to heat.

Severe "alcoholemia" acts as a vasodilator but also decreases cerebral metabolism. No significant gross changes in blood flow have been found in psychoses except in severely deteriorated patients. Also, no great differences (a slight increase) are found in sleep. The metabolic needs* and blood supply are such that a few minutes of inadequate circulation irreversibly destroys brain function. It is fortunate that usual fluctuations in blood pressure do not embarrass cerebral blood flow, that there is little effective autonomic nervous system influence on cerebral vasculature, and that the ambient air concentrations of O_2 and CO_2 are essentially constant.

Kidney. The renal circulation is under the control of the sympathetic nervous system, and its blood flow may be partially shunted to other areas of the body. During heavy exercise, for example, the renal blood flow is reduced by 75 to 80 per cent. During dehydration, in the upright (standing) position, in agitated emotional states, and in certain other stressful situations, renal blood flow is also substantially reduced by way of the splanchnic nerves. Furthermore, the renal circulation mainly fulfills the secretory and excretory functions of the kidney rather than serving only the metabolic needs of the kidney. Thus, if the kidneys were to remove most of the O_2 from the blood they receive, to supply their metabolism they would require only about 50 ml of blood per minute; yet the normal resting blood flow is approximately 1300 ml per minute. Thus, renal blood flow, which at rest is almost 25 per cent of the total cardiac output, serves mainly to maintain the chemical properties of the blood, and only about 4 per cent of the renal flow is necessary for kidney metabolism.

The values in Table 15-1 express only the average, total blood flow per unit time to the kidney under resting conditions. The detailed structure of the renal vascular and lymphatic systems is quite complex and the patterns of flow are poorly understood. As with the cerebral vasculature, the renal blood flow tends to remain grossly unaffected by moderate alterations in perfusing pressure. The details of this mechanism are elusive. It has been suggested that it is due to "autoregulation," a mechanical, servo-control lying within the vessel walls themselves; or to some extrinsic control of resistance; or to a redistribution of the plasma and erythrocyte content of the blood (plasma skimming); or to arteriovenous shunt mechanisms. The renal lymphatic circulation's importance is being increasingly emphasized; it appears to be normally about equal to the rate of urine formation, but may increase by an order of magnitude. The renal lymph has a high protein content (about one half the plasma concentration) and accounts for a significant exchange. It also has been suggested that the lymphatics may play an important role in the "countercurrent" mechanism of the kidney.

---

* The metabolism of the brain has an RQ of 1.0, and is equivalent to a glucose utilization of approximately 76 mg/100gm/min. The brain is an efficient "computer," having a power consumption of only 18 watts.

Muscle. Skeletal muscle, which constitutes almost 50 per cent of the body weight, is also served by a circulation that is highly variable. It tends to be minimal during rest and maximal during heavy exercise. Control of the vessels is both central (reflex) and local. For example, the muscle blood flow of a runner will almost double while he is anticipating the starting gun before he begins to run; this is due, apparently, to reflex vasodilator (cholinergic) nerve activity. During very heavy exercise the flow may increase tenfold because of a decrease in sympathetic vasoconstrictor activity, and an increase in local vasodilator stimulation presumably resulting from heat, metabolic products, and the mechanical "massaging" of the vessels. Thus, the tenfold range of muscle blood flow can be accounted for by reflex vasodilation (twofold), reduced vasoconstriction (twofold), and local vasodilator factors (sixfold). Fainting is usually due to a sudden massive reflex vasodilation in skeletal muscle, and thus to a reduction of blood pressure and a marked decrease in cerebral blood flow. Again, Table 15-1 contains a summary of normal resting values. It is evident that most exercise or work does not simultaneously involve all muscles. If total maximal muscular vasodilation occurred, the heart would be unable to sustain the potential flow. It appears that a large fraction of skeletal muscle vasculature, however, does dilate in the fainting syndrome, in which the initial increase in flow may be three to four times the resting value.

Heart. The coronary circulation is of interest because inadequacy or failure of myocardial blood supply accounts for almost one million deaths per year in the United States, and for billions of dollars in loss to the nation's economy through the premature infirmity of skilled workers and executives. The capillary supply to heart muscle is unusually profuse; it is estimated to be more than 5000 capillaries per square millimeter in cross section, or about one per muscle fiber (Krogh, 1929). Because the metabolism of the heart is high in relation to its blood supply, the oxygen supply is somewhat marginal. The oxygen extraction is normally high, and the arteriovenous difference is the largest of any organ of the body. Under resting conditions the difference is about 16 to 17 volumes per cent, and during exercise may become 18 to 19 volumes per cent. Thus, since extraction is almost maximal, an increased oxygen demand can be met only by an increased blood supply, i.e., by coronary vasodilation. (One should note that blood flow is actually reduced during systole, when the tension surrounding the vessels increases markedly and thus tends to reduce their radii.)

Most information about coronary flow and myocardial metabolism in intact unanesthetized man has been obtained from application of the Fick principle. Again, as with the measurement of cerebral blood flow, N_2O or some other inert gas is breathed and the arteriovenous difference obtained (integrated between onset of rise in blood and equilibrium). The utilization coefficient of the gas is obtained from measured solubility constants in the tissue. The coronary sinus is catheterized for output samples, collected together with arterial or input samples. This method provides only gross data averaged over about 10 minutes, and does not provide information about flow in any local region of the heart or at different times in the cardiac cycle. Table 15-1 contains average normal resting values for coronary blood flow, metabolism, and vascular resistance in man.

Recently Gregg (1964) and his colleagues have acquired interesting data by the ingenious implantation of miniature electromagnetic flowmeters into the coronary arteries and aorta of dogs trained to do heavy exercise and to respond to excitatory stimuli. They have studied these animals for long periods of time under various conditions of activity, exercise, and excitement. Briefly, they have found, contrary to previous beliefs, that the coronary blood flow is not small and fixed during systole. Indeed, at rest the total coronary flow during systole is almost one third of that during diastole; since systole is briefer than diastole during rest, the flow rate per unit time is about 65 per cent of that during diastole.

These investigators also made observations regarding an old controversy as to the effect of heart rate on coronary flow. These workers found that during exercise an increase of "stroke" coronary flow (flow per heart beat) accounts for about 50 per cent of the total increase and the increase in pulse rate for the other 50 per cent. In excitement, marked vasodilation also occurs, such that about 70 per cent of the increased coronary flow is the result of increased stroke coronary flow. Thus, with both excitement and exercise, the coronary blood flow increases during both systole and diastole. Since tachycardia limits the duration of each cycle, there is evident a marked coronary vessel vasodilation. It is usually assumed that the predominant vasodilator stimulus is hypoxia, since any increased metabolic demand produces a relative hypoxia which tends to dilate coronary vessels. Stimulation of the cardiac sympathetic nerves and addition of epinephrine and norepinephrine to coronary blood are among the factors resulting in increased coronary vasodilation. It is usually presumed that this vasodilation is not directly due to the vasodilator action of sympathetic catecholamines, but is indirectly due to the resultant elevated oxygen utilization following increased cardiac performance. One of the interesting and still unsettled questions, however, concerns the heart performance function(s) which determine oxygen utilization.

Skin. Cutaneous blood flow, like renal blood flow, is controlled almost entirely reflexly, even though marked heating of the skin may cause a local vasodilation. At rest and in a comfortably cool enviroment, skin blood flow mainly serves metabolic needs. Under heat stress, either internal (excess heat produced within the body, as by exercise) or external (a hot environment), skin blood flow may increase three to five times. This increase, again, is predominantly reflex, although part of it may be due to altered tissue metabolism. Thus, skin blood flow serves the metabolic needs of skin and also serves an important role in temperature regulation by carrying heat to the surface of the body.

Table 15-1 includes the characteristics of the normal resting skin circulation in a neutral thermal environment. Like skeletal muscle blood flow, skin blood flow has been evaluated in intact unanesthetized man principally by use of volume plethysmography. The finger and toe are usually used for this purpose, whereas with appropriate correction factors, the forearm and thigh are used for measuring flow through muscle. Other indirect methods, such as venous gas saturation and "calorimeter" needles, are also utilized so as to differentiate between muscle and skin blood flow in the limbs.

It is evident that the reflex control of skin and of muscle vasculature is different. For example, placing a foot or hand in warm water normally causes a reflex vasodilation in skin in other limbs but not in muscle. It is also evident that the reflex is mediated via the sympathetic supply to the skin, since nerve blocking abolishes the effect. Furthermore, when the legs of a supine person are raised, there occurs a reflex increase in blood flow to the muscles of the legs but not to the skin. This reflex is thought to arise via receptors in the thorax. During exercise, the blood flow in skin overlying working muscles does not necessarily increase until thermal receptors are stimulated.

Blood flow in skin may increase fourfold as a result of reflex vasodilation, but it increases only about twofold if the sympathetic nerve supply has been blocked. This suggests that an "active" vasodilatation is brought about by the sympathetic nerves. Fox and Hilton (1958) have provided considerable evidence to suggest that the effector substance of this sympathetic vasodilator mechanism is bradykinin. They propose that sympathetic nervous activity stimulates sweat glands to elaborate a bradykinin enzyme, which by its proteolytic action then produces the potent vasodilator substance, bradykinin, from interstitial proteins.

The vascular patterns are quite complex with regard to temperature regulation. For example, arterial blood flow going to a limb may in certain instances return via superficial channels, so that body heat is brought close to the skin surface to exchange with the surrounding environment. Under other conditions it may return by deep veins, thus conserving body heat. Indeed, because the deep veins lie close to the arteries, a "countercurrent" exchange of heat between arterial and returning venous blood reduces the heat distribution to the distal part of the limb. In this arrangement the temperature of arterial blood progressively decreases distally.

Another interesting difference between skin and muscle vasomotor patterns is with regard to the action of low concentrations of epinephrine or norepinephrine. Both are contrictors to skin vessels, whereas in skeletal vessels only norepinephrine is a constrictor while epinephrine is a dilator. The physiological significance of these differences, however, is not clear.

Viscera. Blood flow to the spleen, pancreas, and gastrointestinal tract from esophagus to colon is also very complex. At rest it represents about 25 per cent of the cardiac output. Most of the flow to these organs empties into the portal vein and then flows through the liver. This vein carries more than 75 per cent of the total liver blood flow. The hepatic arterial flow constitutes the remaining quarter; it supplies the metabolic needs of the liver, since its obstruction results in hepatic failure.

The values for the visceral circulation in Table 15-1 are derived from many measurements on resting unanesthetized man and utilize, again, the Fick principle. In this case Bromsulphalein is infused continuously until it reaches an equilibrium concentration in arterial blood. In low concentrations (about 1 to 5 mg/100 ml) the liver removes essentially all of the dye circulating through it in one passage. Given the infusion rate (which is equivalent to utilization) and the arterial concentration (hepatic venous concentration is zero) the total hepatic flow can be measured. As previously noted, this flow includes all blood channeled into the portal vein plus that entering from the hepatic artery.

In addition to supplying the metabolic needs and biological functions of all these organs and tissues, the blood in the portal vein and the liver has a significant volume, which may be shifted into or out of the general circulation. The portal vein wall is contractile, and the hepatic vasculature is highly distensible. This portal-hepatic blood volume of man is estimated to be about 1 liter under normal circumstances. In conditions such as reflex compensation for hemorrhage, the volume may decrease to 500 ml (autotransfusion); yet with an abnormal increase in central venous pressure (congestion) the stored volume may exceed 3 liters.

Like the kidney, the visceral vasculature is innervated by the splanchnic sympathetic supply and undergoes constriction in exercise, in the upright posture, and during emotional excitement. In these conditions visceral blood flow may decrease by 50 per cent or more. Fainting was once thought to be associated with acute visceral vasodilation; this has been shown to be incorrect, in that the visceral blood flow actually decreases as the blood pressure falls in fainting.

The hemodynamics of the visceral circulation are complex, since the circulation along the gastrointestinal tract returning to the portal vein extends almost from esophagus to anus. It is in parallel with the circulation to the spleen, pancreas, and hepatic arterial vessels, and in series with the remainder of the liver vasculature. As noted earlier, the capillary permeability of the visceral circulation varies widely. Indeed, the function of each of the organs and structures served by the visceral circulation is significantly different and variable.

"Remainder." Data in this line in Table 15-1 refer to all remaining vascular beds. In bulk weight, however, it is predominantly bone, weighing about 30 kg. It is estimated that total flow to the skeleton averages about 1 ml/100 gm/min, although it is highly variable from one bone area

to another. Thus, about 300 of the remaining 700 ml/min flows through bones. Glands and other organs of the body account for the last 400 ml/min of flow.

Résumé. These examples of several types and combinations of control of the blood flow through the body's organs serve to illustrate several important physiological principles, as follows:

1. In certain vascular beds the circulation is essentially invariant. The metabolism tends to be constant and there is little effective reflex control (e.g., cerebrum).

2. In other organs the circulation is highly variable and predominantly under reflex control (e.g., kidney, skin).

3. Still other vascular beds are controlled by both the local metabolism and by reflex control (e.g., skeletal muscles).

4. The blood flow to other organs is highly variable and predominantly controlled by the organ's metabolic needs (e.g., heart).

In spite of this complex array of local and central or reflex effects on the blood vessels of each organ and tissue, the body's needs, as a whole, are remarkably well served by the circulation. The following section briefly describes the organization of this overall control and regulation.

Cardiovascular Factors Affecting Blood Flow

Motion of blood through the vascular system depends entirely upon force from the beating heart. Its range of performance is remarkable indeed, ranging from a total blood flow (cardiac output) of about 5 liters per minute in a normal resting adult, to about 25 liters per minute during heavy exercise or work. The two major variables are, of course, the volume pumped with each heart beat (stroke volume) and the number of strokes per unit of time (heart rate). The heart's conversion of chemical to mechanical energy is quite efficient. The mechanical energy associated with external work (cardiac output with regard to arterial load or pressure) is about 20 per cent of the total amount of energy derived from oxidative metabolism (chemical energy consumed). Heart muscle, in contrast to skeletal muscle, can sustain only an insignificant oxygen debt, the equivalent of perhaps 10 to 20 beats. Consequently, it must be continuously supplied with oxygen by coronary blood flow. Although the right and left ventricles normally move the same amount of blood, the work performed by the left ventricle exceeds that of the right by about eightfold, since aortic pressure exceeds pulmonary arterial pressure during systole by about that amount.

Cardiac performance in intact man and animals has been the subject of extensive investigation. In recent years this research has employed electromagnetic flowmeters implanted semipermanently around the root of the aorta, together with indwelling catheters to record pressure. Thus, instantaneous intracardiac, intravenous, and intra-arterial pressures and cardiac output have been measured using unanesthetized, active animals, principally dogs.

A Nobel prize was given in 1956 to Drs. Forsmann, Cournand, and Richards for their application of the Fick principle to the evaluation of cardiac function in human subjects, and for the rapid expansion of information about cardiovascular and pulmonary function which developed therefrom. During the intervening years, our understanding has continued to improve; yet many basic and important questions remain unanswered even today. Productive basic research often creates more questions than it answers.

Different workers have proposed various indices to characterize the mechanical performance and its relation to the metabolic functions of the heart. A favorite basis for such a comparison is Starling's "law of the heart." Starling (1918) suggested that the energy of contraction (i.e., mechanical work performed) is related to myocardial fiber length. Because the dynamic geometry and muscular construction of the heart are so complex, however, it is difficult to obtain direct relationships between work and fiber length; in place of the latter, Starling measured the volume of the heart. Moreover, it is not easy to determine the factors that directly influence fiber length. The various external factors which usually influence heart function are (1) venous pressure, (2) arterial pressure, (3) nervous control and, (4) humoral influence. During abnormal states, many other hormonal, chemical, and mechanical factors, as well as abnormal structural relations, also influence heart performance. One must inquire whether these several factors exert their action through changes in fiber length. It is difficult to assess directly the role of venous or filling pressure on myocardial length, since the distensibility or elasticity of the heart wall is variable and alters the pressure-length relationship. Likewise, the effect of arterial pressure (load) is difficult to analyze, even though it is well known that a normal heart possesses an amazing facility for maintaining its usual stroke volume if arterial pressure is elevated. It does this by developing an equivalent increase in mechanical work. The mechanism for this is in doubt; in some cases, at least, it may result from an initial "stretching" of the myocardial fibers. In general, one can say that an increased cardiac output is not correlated with an increased end diastolic volume. Indeed, end diastolic volume tends to decrease with increased heart rate and activity of the sympathetic nervous system, both of which usually accompany exercise and excitement. The most important point seems to be that the heart of Starling's preparation, isolated from nervous, chemical, and humoral influences, gives results which do not obtain in the intact animal. It is difficult, therefore, to assess the role of Starling's "law" in day-to-day situations in intact man.

Linden (1963) is probably correct in saying that the inconsistencies in relating external cardiac work and oxygen consumption disappear if only the steady state is considered. This is because in going from one level of work to another, various cardiac functions have different time courses and, thus, yield different gross input-output patterns. He points out that in steady-state exercise the oxygen consumption is linearly related to work and heart rate, over the full range of exercise capability between rest and a maximum of O_2 consumption of 3 to 3.5 liters/min.

Detailed relationships of mechanical force produced by the heart, appearing as pressure, and of blood flow are complex and only general principles are understood. The first approximation usually presented to relate the pressure and flow in blood vessels is that known as Poiseuille's equation

$$Q = \frac{\triangle P r^4 \pi}{L \mu 8} \qquad 15.4$$

where Q is flow, $\triangle P$ the pressure gradient along a length L of straight rigid tube with radius r. The liquid is assumed to have an ideal viscosity μ, with a linear and homogeneous relationship between velocity and viscosity, and is assumed to flow in a streamlined manner having no accelerations or turbulence. From Equation 15.4, it is noted that the flow is linear and is directly related to the pressure difference and also to the fourth power of the radius; it is inversely related to the viscosity of the liquid and the length of the tube. Thus, flow is markedly altered by relatively slight changes in radius; e.g., a twofold increase in vessel radius produces a sixteenfold change in flow, if all other factors remain constant. This relationship emphasizes the importance of blood vessel radius and vasomotor activity as a mechanism for controlling blood flow. Thus, it appears that in vasomotion nature has "selected" the most effective factor by which to control and regulate the flow of blood to tissues, namely, by altering vessel radius rather than by changing blood pressure or blood viscosity.

These relationships also help to explain the pressure drop along the capillary noted previously. As a viscous liquid flows along a tube, a pressure drop occurs as mechanical energy is converted to heat in overcoming the fluid friction. Beyond these generalizations, however, Poiseuille's equation is of little value in relating pressure and flow in the vascular system since the vessels are neither straight nor rigid. Also, blood is a very complex rheological liquid, and blood flow is seldom streamlined.

During the past hundred years, many models of hemodynamic parameters and variables* have been described. Later models are more complex than earlier ones and tend to simulate more reliably the actual

---

* Characteristics of a system: e.g., in an electronic device the parameters are capacitance, resistance, and inductance, while the variables are voltage and current.

properties and behavior of the cardiovascular system. As yet, however, a complete model of the hemodynamic (mechanical) characteristics of the cardiovascular system is far from a reality. The following is a brief summary of the present status of the situation, recognizing that the conditions described by Poiseuille's equation are not characteristic of the cardiovascular system.

Poiseuille's equation assumes a steady flow. Blood flow is pulsatile, however, and the velocity profiles are complex and vary from one site to another. In the aortic arch, during systole at least, flow velocity oscillates from 0 to 150 cm/sec, and is turbulent with eddies and back-flow. Most equations which have been developed to describe blood flow are inadequate because they presume only slightly pulsatile flows which are axially symmetrical. Furthermore, the ratio of velocity of axial flow to pulse wave velocity varies along the system, in such a way that velocity of flow decreases in magnitude as the velocity of wave propagation increases.

Poiseuille's equation assumes that blood possesses a single, unique coefficient of viscosity. It is well known, however, that the effective viscosity varies with rate of flow and is influenced by the geometry of the vessel. Blood is certainly non-Newtonian and its properties depend upon hematocrit and plasma protein content. The hematocrit of blood in small vessels is lower than that in larger vessels. Erythrocytes have been described to have shear-stress properties like airplane wings, since they tend to be directed or "lifted" in flowing fluids. This means that in axial flow, erythrocytes tend to orient themselves toward the center of the stream, leaving a plasma "sleeve" toward the outer radius of the vessel.

One additional, important limitation to use of the Poiseuille equation is the fact that the geometry of the vessels and their arrangement as a vascular "tree" are complex. Their parameters are distributed rather than "lumped." And the vessels certainly are not rigid.

Much work has been done on the properties and behavior of blood vessels. Because many of their biological functions are related to vascular radius, any variable having an effect upon radius can be of major importance. The relationships between pulsatile pressure and blood flow are very sensitive functions of the instantaneous radius of the vessel wall. Impedance to flow is a power function of radius (where the exponent may be between —3 and —5) and also of the rate of change of radius. Consequently, the radius and changes in radius of the vessels of the body largely determine the distribution of blood flow. Blood volume within a given vessel is a function of the square of the radius. Finally, the activity of sensory receptors, which lie within the vessel walls and which take part in the regulation of the cardiovascular system, depends upon the radius and its rate of change.

Thus, because of these general, radius-dependent functions of blood vessels, and because any model presuming pulsatile flow must take vessel radius into account, we shall consider now these properties of vessel walls which determine the instantaneous and time derivatives of vessel radius. Obviously, in the absence of internal pressure, each vessel has a radius dependent upon its anatomy. What this radius actually is as blood flows through it is determined by the stresses applied to the wall and by the stiffness of the wall itself.

In considering, first, the applied stresses, it is evident that the intra- and extravascular pressures are radial whereas the effective stress tending to stretch the wall is tangential. For a cyclindrical vessel under equilibrium conditions, the relationship is

$$T = \frac{a^2 \, P_i - b^2 \, P_o}{b^2 - a^2} + \frac{(P_i - P_o)a^2 b^2}{r^2 \, (b^2 - a^2)} \qquad 15.5$$

in which T is tangential stress (force per unit area) at a point in the wall where the radius is r; a is the inner radius and b is the outer radius of the cylinder; δ is wall thickness and equals b—a; P_i is inner radial pressure; and P_o is outside pressure (Peterson, 1962). If P_o is equal to zero, the equation becomes

$$T = \frac{a^2 P_i}{b^2 - a^2} \left(1 + \frac{b^2}{r^2}\right) \qquad 15.6$$

When the vessel wall is thin, such that b — a is small compared with r, this relationship can be stated to a first order approximation. Thus:

$$T = \frac{rP_i}{\delta} \qquad 15.7$$

Equation 15.7 applies only to a thin-walled vessel whose radius to wall thickness ratio is large, and where tangential stress is assumed to be uniform throughout the wall. When the ratio is greater than 10, the errors introduced by these assumptions are small; but below a ratio of 10 the error increases rapidly, since the radial distribution of tangential stress is minimal at the outer radius and increases exponentially toward the inner radius (Eq. 15.5). It should be noted that the use of the equation

$$T = P_i r \qquad 15.8$$

is inappropriate when applied to blood vessels, since among other limitations it ignores the wall thickness. Furthermore, the practice of considering organs such as thick-walled vessels and the heart as a series of cylinders of increasing radius is also incorrect, since it suggests a distribution of stresses which grow in the wrong direction and in the wrong manner.

By application of methods soon to be discussed, ratio $\frac{r}{\delta}$ has been found to vary from approximately 16 to 3 in the large arterial vessels. From the root of the aorta to second-order branches from the aorta, the ratios tend to decrease peripherally, since there is both a tapering and an increase in wall thickness along the aorta and major branches.

It became apparent through several lines of reasoning that the relationship between arterial stress P and strain ϵ could be satisfactorily described by a linear, first-order, differential equation, as follows:

$$P = E_p + R_p \frac{d\epsilon}{dt} \qquad 15.9$$

The coefficient E_p represents the proportionality between pressure and strain, and is analogous to a distensibility coefficient. It is a function of both the elasticity of the wall and the geometry of the vessel. The coefficient R_p represents the proportionality between pressure and rate of strain and is, therefore, a distensibility-rate coefficient. It is a function of both the viscosity of the wall and its geometry.

In order to compute the elasticity and viscosity of the wall itself it is necessary to account for the geometry. Since we are dealing with vessels whose $\frac{r}{\delta}$ approximates that of a thin-walled cylinder, we can use Equation 15.7, i.e., $T = \frac{rP_i}{\delta}$. Thus:

$$T = \frac{r}{\delta} (E_p\epsilon + R_p \frac{d\epsilon}{dt}) = E_T\epsilon + R_T \frac{d\epsilon}{dt} \qquad 15.10$$

It should be noted that the absence of a second-order term implies that the mass effect of the vessel wall has been omitted. This is justified because the mass of a nominal artery wall, expressed in the units of the equation, has a numerical value of 3×10^{-4}. This value is negligible compared to the coefficients for ϵ (of the order of 10^3) and R (of the order of 10^2), and for the wall accelerations occurring under physiological circumstances.

Strains encountered in arteries are small; although in normal arteries the pulse pressure may vary from 30 to 100 per cent of the mean pressure, the resulting strains vary only from 1 to 3 per cent. Even the ascending aorta normally undergoes variation in strain of less than 5 per cent during the cardiac cycle. The carotoid sinus is one of the most distensible parts of the arterial system, with a relatively larger radius and thinner wall; and yet it also undergoes a strain of less than 5 per cent during each cycle.

The $\frac{r}{\delta}$ ratio varies from 5.3 to 9.4 in the carotoid artery, from 4.5 to 6.0 in the thoracic artery, from 5.7 to 14.0 in the abdominal aorta, and from 3.4 to 8.8 in the femoral artery. It must be stated that current methods used for estimating wall thickness leave much to be desired. It is interesting that as the walls of the arteries become stiffer with advancing age, there is a parallel tendency for $\frac{r}{\delta}$ to increase. Since the pressure-radius or volume-distensibility relationship E_p is equal to the product of the wall modulus of elasticity multiplied by the reciprocal of $\frac{r}{\delta}$,

$$E_p = E_T \frac{\delta}{r} \quad \text{or} \quad \frac{r}{\delta} E_p = E_T \qquad\qquad 15.11$$

these changes in $\frac{r}{\delta}$ tend to compensate for the increased stiffness of aging.

Usually the dissipative factor associated with hemodynamics is considered to be the blood viscosity. Nominally, blood viscosity may be considered to be approximately 5×10^{-5} gm sec cm^{-2}, i.e., five times that of water.* The equivalent viscosity of the artery wall is nominally 500 gm sec cm^{-2}, i.e., of the order of ten million times that of blood. Because an appreciable amount of energy is dissipated in stretching the vessel walls with each heart beat, this viscous effect of the wall should affect the arterial pulse wave velocity. The vessel wall is known to be stiffer for higher harmonics of strain, and therefore, arterial pressure, than for lower harmonics. Since the arterial pulse is composed of several harmonics, the higher harmonics are propagated more rapidly than lower harmonics. This dispersion tends to cause the arterial pulse wave to alter its shape as it traverses the arterial tree (Peterson, 1964).

Structures within the vessel wall are capable of altering in such ways as to cause rather large changes in the moduli which determine vascular radii. The application of vasodilators and vasoconstrictors to the vessel wall has been shown to cause, respectively, a reduction of wall elasticity E_T to 50 per cent of the "normal" values, or to cause a tenfold increase. However, as previously noted, the vessel strain remains small. It is also interesting to note that over these ranges of strain, with constriction and dilation, the vessel stiffness remains essentially constant over any given cardiac cycle. It is evident, of course, that the small strain permits us to use the concept of an elastic modulus, and also is responsible for the manifestation of constancy of moduli over the cardiac cycle. When vasoactive substances are applied, whether variations of moduli will occur during any given cardiac cycle depends upon the rapidity of the effect of the substance.

These considerations apply mainly to larger vessels, i.e., from the aorta to its third- or fourth-order branches. In more peripheral branches extending to the capillaries, the vessels tend to become *relatively* thick-

---

* *Ed. Note:* Here gm refers to grams force. Viscosity is also expressed in dyne sec cm^{-2} = poise. Water at 20° has a viscosity of 0.01 poise.

walled, and the ratio of radius to wall thickness becomes smaller. This is due primarily to a reduction in radius. Therefore, the vessels become stiffer, in that a given distending pressure tends to cause less strain. The resistance to flow within the vessels increases, the mean pressure falls more abruptly, and the oscillation of pressure with each heart beat tends to become reduced toward the periphery of the arterial system. So far, little information is available regarding the wall material itself in small arteries; what evidence there is suggests that the moduli of elasticity and viscosity are grossly similar to those of larger arteries. In the dilated or "resting" state, the collagen content of the vessel walls chiefly determines wall elasticity and viscosity, but during vasoconstriction the vascular smooth muscle properties predominate.

Variations in arterial blood pressure during the cardiac cycle are just as complex as are its determinants. In the root of the aorta the pressure normally oscillates between approximately 80 to 120 mm Hg, with a mean pressure of about 90 mm Hg. As the wave is propagated along a vessel, the pulse pressure increases progressively, so that in the femoral artery it may have *increased* by 100 per cent. At the same time the mean and diastolic pressures fall slightly (by less than 5 per cent). Thus, the pressure pulse undergoes a marked distortion in its transmission, both because of numerous small pressure reflections, and because of a dispersion of the pressure components, since the higher frequency harmonics of the pulse travel at higher velocities than lower ones (Peterson, 1964).

Along the venous vasculature, the situation is also complex. The pressures are lower (ranging from 20 to 10 mm Hg to essentially zero), and the veins may not be cylindrical in cross section. External pressures due to the action of skeletal muscle and to intra-abdominal and intrathoracic forces associated with respiration have relatively greater effects on the walls and blood flow in the veins than in the arteries. The importance of these external forces is demonstrated by the fact that for a man in a quiet resting upright position, the heart alone may not provide enough force to overcome gravity and maintain an adequate circulation of blood. Skeletal muscular action thus helps to circulate the blood during ordinary standing posture. When effective gravity is increased further, the circulation tends to fall even more readily; to prevent this in aircraft and spacecraft, "anti-g suits" are used to keep blood from pooling in the lower parts of the body. The opposite condition, weightlessness, accompanies space adventures, but its long-term effects are not well understood. Apparently, the cardiovascular system tends to become "deconditioned" so that an astronaut has difficulty in withstanding even one "g" upon returning to earth.

For all these reasons, then, the dynamics of the blood and blood vessels are dissimilar to the conditions implied in Poiseuille's equation. It remains to be seen, however, how these differences may be used to modify that model or to construct one more nearly like the cardiovascular system.

REGULATION AND CONTROL OF THE CIRCULATION

It is evident that the control of the resistance to blood flow within a tissue or organ is a function of both local and systemic factors, the latter being either neutral or humoral reflexes. It is also evident that many of the local factors in one organ are independent of those in another. If all vascular regions were to dilate simultaneously to the extent that they do individually during periods when the several organs are active, the circulation would surely fail. This total "potential" flow would exceed 40 liters per minute, whereas the known maximum cardiac output is approximately 25 liters per minute. If the reflex patterns were likewise independent, there would be an even greater likelihood of cardiovascular inadequacy. Clearly, there must normally be an integrated control of cardiovascular functions which permits the total organism to function adequately under a variety of stressful conditions as well as during normal day-to-day activities.

The subject of regulation and control of biological functions is one of the most exciting areas of modern physiology. It is unlikely that a complete understanding of cardiovascular function in health and disease can be achieved until the multiplicity of interactions of the cardiovascular system with the nervous, endocrine, renal, respiratory, gastrointestinal, and locomotor functions are understood. For example, we cannot adequately explain the interactions that occur in exercise, prolonged weightlessness, prolonged subjection to one or more "g's" in the upright position, prolonged bed rest, thermal stress, and most, if not all, diseases of the cardiovascular system.

By definition, it is implicit that if a function is regulated, the regulating system has certain parameters. These include, first, appropriately located sensors that are uniquely responsive to the function being regulated and second, a mechanism which "compares" the real value of the function as indicated by the sensor with what the value should be for appropriate behavior of the function, and which sends information to the mechanisms controlling the real value of the function. (This is often referred to as the error signal, because it represents the difference between what its value really is and what it ought to be. The physiological analogy to mechanical, thermal, or electrical error-generating devices is subtle and difficult to conceive. We realize that the body temperature is regulated, but the biological analogs of the wall thermostat and the desire to be comfortable are not obvious.) The third characteristic implied in a regulating system is a set of mechanisms which control the function by increasing or decreasing its value. Fourth is the coupling of the sensing, setting, and controlling mechanisms in closed communication, as a result of which the system operates as a "closed loop."

With this brief statement of the characteristics of regulation, it is instructive to consider the extent to which they can be applied to cardiovascular functions. Classically, it is considered that arterial blood pressure is regulated. The system is thought to include pressure receptors (usually exemplified by the carotid sinus mechanisms) which transduce arterial pressure within the vascular system into a proportionate frequency of nerve impulses transmitted to the central nervous system. Within the central nervous system a transposition occurs such that if the impulse traffic from sensory nerves indicates an excessively high pressure, the efferent traffic over the sympathetic nervous system is reduced. Conversely, if the afferent traffic signals an abnormally low pressure, there results an increased sympathetic activity. Thus, the carotid sinus receptor–central nervous system mechanisms represent a "pressorstat" and the message sent to alter the sympathetic outflow from the central nervous system is an "error signal." Likewise, the cardiac output and the blood vessels' resistance to flow represent the control variables for blood pressure, since the product of flow and resistance is proportional to pressure. A decreased sympathetic activity results in a slower heart rate with decreased stroke volume and a net decrease in cardiac output. Similarly, the innervated vascular smooth muscle tends to dilate as sympathetic nervous activity decreases, thus causing a decreased resistance. The overall effect is a fall in blood pressure. The converse set of circumstances would result in a rise of blood pressure.

The foregoing concept is supported by ample evidence that, other factors notwithstanding, the output of the sympathetic nervous system is inversely related to the input activity of Hering's nerves from the carotid sinuses. The quantitative relationships, however, have not been worked out. It is also established that the sympathetic efferent activity affects heart rate and stroke volume and vascular smooth muscle in the manner we have already indicated. Furthermore, although the relationships among these factors (viz., pulse rate, stroke volume, and vessel wall properties) are very complex, there is little reason to doubt that a net increase in cardiac output along with vasoconstriction will cause a net increase in arterial pressure.

In addition to this supporting evidence, however, one must consider certain limitations of this proposed mechanism. One of these is the question of how marked alterations in cardiovascular functions can occur as they do, with little change in arterial pressure, under natural conditions. Another is, what causes the abnormal condition of hypertension?

As indicated in the previous section of this chapter, many functions of the cardiovascular system are related to the radii of blood vessels. An additional example may be added now. The receptors which lie in the walls of blood vessels are not pressure receptors or blood volume receptors, but are strain receptors, and the electrical activity they generate relates to the extent to which they are stretched and to their rate of stretch. Since they are stretched by the vessel wall in which they reside, the degree of stretch is determined by the strain and rate of strain of the vessel wall, together with the manner in which they are located in and coupled to the wall structures. Whether they are considered pressure, volume, or stretch receptors would be only academic if the stress-strain properties of the receptor-containing vessel walls were constant and linear. Such is not the case, however, since there are several factors that are likely to alter the mechanical properties or stiffness of these vessels. There is evidence that the walls of the carotid sinus can, and do, undergo changes in stiffness, both reflexly and by alterations in the chemical and water content of the wall. Also, in disease the mechanical properties of the tissues (collagen, elastin, smooth muscle, and so forth) of the wall may undergo changes. Thus, the relationships between blood pressure and vessel volume at the site of the receptors may be altered or "reset." For example, it has been shown that the electrical activity in Hering's nerve, with respect to arterial blood pressure, is quite different in experimental hypertension from what it is in normal animals (McCubbin et al., 1956; Peterson, 1962). In hypertension, increases in the sodium and water content cause the arterial walls to stiffen. There is also evidence that the carotid sinus wall may undergo alterations in stiffness due to vasoconstriction and dilation associated with variations in sympathetic nervous activity (Peterson, 1962; Jones et al., 1964). For these reasons, it is evident that the regulation and control of the cardiovascular system, which include the role of receptors such as those in the carotid sinus, can be altered so that its "set point" appears to be changed.

Within the vascular tree and even within the heart wall there are many other areas containing sensory receptors which generate "coded" signals transmitted to the central nervous system, where this "information" is integrated, and where, in turn, sympathetic nervous activity is generated and transmitted along the vasomotor, cardiac, and other pathways. As indicated, the effective vasomotor innervation varies significantly from one vascular bed to another. Furthermore, skeletal, respiratory, and other motor and sensory pathways are, to some extent, coupled to those of the cardiovascular system, and the endocrine system is also involved in this control and regulation. For example, the kidneys affect cardiovascular functions in several ways, one of which is endocrine in nature. Under certain conditions the juxtaglomerular cells of the kidney produce and release an enzyme, renin, which acts upon a plasma protein substrate to produce a very potent vasoconstrictor, angiotensin. This substance not only causes generalized vasoconstriction, but also may cause the adrenal cortex to release increased amounts of aldosterone, which, in turn, affects the total body water and electrolyte content and concentration. Aldosterone release may cause these alterations in the vascular water and electrolyte content both by direct action and by indirect effects following renal tubular excretion of sodium. Another vascular mechanism affecting the kidney is a reflex system arising in the thoracic vessels and carried via the vagi to the central nervous system, and thence efferently along the sympathetic system to cause significant changes within the kidney and in urine formation.

It is most challenging to try to integrate all these factors capable of controlling the functions of the cardiovascular system. Some of them are reflexly coupled through numerous "closed loops" involving complex nervous and endocrine pathways. Others are "open ended"; i.e., they tend to be, at least in a direct sense, independent of each other and relatively unaltered by cardiovascular behavior. The problems of "putting the body back together again" are truly formidable, because of the complexity of the models that will include all of the variables and parameters. Since the pathways and effects are so complicated and since there are so many, it will be necessary to solve the equations describing such models by means of a computer, because no human brain or group of brains can contemplate their interactions. Several teams of investigators are already attempting this, using computer oriented models to approach an understanding of regulation of cardiovascular functions.

CONCLUSIONS

Late in 1964, several hundred of the world's foremost scientists and physicians, concerned with cardiovascular function in health and disease, met in Washington, D.C., to review the progress in these fields since a previous conference held fifteen years earlier. Their task was also to identify significant areas of current ignorance and problems which "should" be investigated. It was apparent that in the intervening period a great deal of progress had been made, yet significantly more current problems and areas of ignorance were identified. (As we have already noted, research produces more questions than it answers.) The remaining problems range from molecular interactions at the cellular level to the entire system of organs and their regulation and control. As stated earlier, there are few cardiovascular functions which are not also functions of most, if not all, other bodily systems. Furthermore, the environment of the body and its genetic heritage are also significant factors in determining cardiovascular properties and behavior. Their study has been only barely started.

The brief account of cardiovascular functions which has been developed here does not include many of these important aspects of cardiovascular physiology. Little has been said about the microstructure of the cardiovascular system or its "microbehavior." Thus, the prime mover (cardiac muscle) and the prime controller of flow and pressure (vascular smooth muscle) have each been given very little attention. The excitation-contraction mechanisms have not been discussed, although they determine the heart rate and synchrony of contraction. Little has been said of the coupling of the sympathetic nervous system with the cardiovascular effector sites, nor have exciting discoveries regarding the chemistry of the heart and vessel walls been discussed. Determinants of blood volume and of the composition of blood have been neglected. Indeed, this review has only scratched the surface of subject matter related to the cardiovascular system.

It is obvious that cardiovascular diseases are prevalent because they kill more persons than the next three most common causes of death combined. Their socioeconomic importance is illustrated by the statement that "The 365,000 Americans between the ages of 25 and 65 who died of these diseases in 1962 would have earned wages totaling more than 1.5 billion dollars and paid close to 200 million dollars in federal income taxes had they lived *one* more healthy working year" (President's Commission on Heart Disease, Cancer and Stroke). President Johnson has convened a commission to recommend Federal programs to combat them. If the recommendations of the commission are executed, the effect upon research and upon scientific, clinical, and educational institutions will be enormous. It has been suggested that the cost for only five years will amount to $3,000,000,000.

Research in the author's laboratory is supported by the following grants: Grant HE-07762 from the National Institutes of Health and Grant N ONR-551 (18) from the Office of Naval Research.

References

Adolph, E. F. 1943. Physiological regulations. Jaques Cattell Press, Lancaster, Pa.
Adolph, E. F. 1961. Early concepts of physiological regulations. Physiol. Rev. 41:737–770.
Adolph, E. F. 1964. Regulation of body water content through water ingestion. pp. 5–17. In M. J. Wayner [ed.] Thirst. First Florida State University Symposium, Tallahassee. 1963. Pergamon Press, Ltd., Oxford.
American Standards Association. 1963. (ASA C 85.1). American standard terminology for automatic control. Am. Soc. Mech. Eng., United Eng. Cntr., New York.
Anand, B. 1961. Nervous regulation of food intake. Physiol. Rev. 41:677–708.
Anand, B. K. 1963. The internal environment and alimentary behavior. Vol. II. pp. 43–116. In Mary A. B. Brazier [ed.] Brain and behavior, proceedings of the conference. AIBS, Washington.
Anand, B. K., and J. R. Brobeck. 1951a. Hypothalamic control of food intake in rats and cats. Yale J. Biol. Med. 24:123–140.
Anand, B. K., and J. R. Brobeck. 1951b. Localization of a "feeding center" in the hypothalamus of the rat. Proc. Soc. Exp. Biol. Med. 77:323–324.
Andersen, H. T., H. T. Hammel, and J. D. Hardy. 1961. Modifications of the febrile response to pyrogen by hypothalamic heating and cooling in the unanesthetized dog. Acta Physiol. Scand. 53:247–254.
Anderson, J. C., M. A. Barton, R. A. Gregory, P. M. Hardy, G. W. Kenner, J. K. MacLeod, J. Preston, and R. C. Sheppard. 1964. The antral hormone gastrin—synthesis of gastrin. Nature 204:933–934.
Anderson, L. L., A. M. Bowerman, and R. M. Melampy. 1963. Neuro-utero-ovarian relationships. pp. 345–373. In A. V. Nalbandov [ed.] Advances in neuroendocrinology. University of Illinois Press, Urbana.
Andersson, B., C. C. Gale, and B. Hökfelt. 1964. Studies of the interaction between neural and hormonal mechanisms in the regulation of body temperature. pp. 42–61. In E. Bajusz and G. S. Jasmin [eds.] Major problems in neuroendocrinology. The Williams and Wilkins Co., Baltimore.
Andersson, B., C. C. Gale, and J. W. Sundsten. 1963. The relationship between body temperature and food and water intake. pp. 361–375. In Y. Zotterman [ed.] Olfaction and taste. Pergamon Press, Ltd., Oxford.
Andersson, B., C. C. Gale, and J. W. Sundsten. 1964. Preoptic influences on water intake. pp. 361–379. In M. J. Wayner [ed.] Thirst. First Florida State University Symposium, Tallahassee. 1963. Pergamon Press, Ltd., Oxford.
Andersson, B., and S. Larsson. 1961. Physiological and pharmacological aspects of the control of hunger and thirst. Pharmacol. Rev. 13:1–16.
Andik, I., Sz. Donhoffer, M. Farkas, and P. Schmidt. 1963. Ambient temperature and survival on a protein-deficient diet. Brit. J. Nutr. 17: 257–261.

Assenmacher, I., and J. Benoit. 1958. Quelques aspects du contrôle hypothalamique de la fonction gonadotrope de la préhypophyse. pp. 401–427. In S. B. Curri, L. Martini, and W. Kovac [eds.] Symposium internationale sul diencephalo, Milano. Springer-Verlag, Vienna.

Axelsson, J., and S. Thesleff. 1957. A study of supersensitivity in denervated mammalian skeletal muscle. J. Physiol. 147:178–193.

Babkin, B. P. 1949. Pavlov, a biography. The University of Chicago Press, Chicago.

Bachrach, W. H. 1963. On the question of a pituitary-adrenal component in the gastric secretory response to insulin hypoglycemia. Gastroenterology. 44:178–189.

Bazett, H. C. 1949. The regulation of body temperatures. pp. 109–192. In L. H. Newburgh [ed.] Physiology of heat regulation and science of clothing. W. B. Saunders Co., Philadelphia.

Beaven, D. W., E. A. Espiner and D. S. Hart. 1964. The suppression of cortisol secretion by steroids, and response to corticotrophin, in sheep with adrenal transplants. J. Physiol. 171:216–230.

Bellman, R., and K. L. Cooke. 1963. Differential–difference equations. Academic Press Inc., New York.

Bellman, R., J. A. Jacquez, and R. Kalaba. 1960. Some mathematical aspects of chemotherapy: I. One-organ models. Bull. Math. Biophys. 22:181-198.

Benzinger, T. H., C. Kitzinger, and A. W. Pratt. 1963. The human thermostat. pp. 637–665. In Am. Inst. Physics. Temperature—its measurement and control in science and industry. J. D. Hardy [ed.] Part 3, Biology and Medicine. Reinhold Publishing Corp., New York.

Bernard, C. 1878. Leçons sur les phénomènes de la vie communs aux animaux et aux vegetaux. tom I. pp. 67, 111–114, 123–124. Ballière, Paris.

Bernard, C. 1949. An introduction to the study of experimental medicine. [Transl. from French by H. C. Greene] Henry Schuman Inc., New York.

Bogdanove, E. M. 1963. Direct gonad-pituitary feedback: an analysis of effects of intracranial estrogenic depots on gonadotrophin secretion. Endocrinology 73:696–712.

Bonner, D. M. [ed.] 1961. Control mechanisms in cellular processes. The Ronald Press Co., New York.

Bourne, G. [ed.] 1960. The structure and function of muscle. Academic Press Inc., New York.

Brobeck, J. R. 1960. Food and temperature. Recent Progr. Hormone Res. 16:439–459.

Brobeck, J. R., J. Tepperman, and C. N. H. Long. 1943. Experimental hypothalamic hyperphagia in the albino rat. Yale J. Biol. Med. 15:831–853.

Cain, D. F., A. A. Infante, and R. E. Davies. 1962. Chemistry of muscle contraction. Adenosine triphosphate and phosphoryl creatine as energy supplies for single contractions of working muscle. Nature 196:214–217.

Campbell, H. J., G. Feuer, and G. W. Harris. 1964. The effect of intrapituitary infusion of median eminence and other brain extracts on anterior pituitary gonadotrophic secretion. J. Physiol. 170:474–486.

Cannon, W. B. 1929. Organization for physiological homeostasis. Physiol. Rev. 9:399–431.

Cannon, W. B. 1939. The wisdom of the body. Rev. ed. W. W. Norton and Co., Inc., New York.

Cannon, W. B., and A. L. Washburn. 1912. An explanation of hunger. Amer. J. Physiol. 29:441–454.

Carlson, A. J. 1916. The control of hunger in health and disease. University of Chicago Press, Chicago.

Carlson, F. D., D. J. Hardy, and D. R. Wilkie. 1963. Total energy production and phosphocreatine hydrolysis in the isotonic twitch. J. Gen. Physiol. 46:851–882.

Chalmers, T. M., A. Kekwick and G. L. S. Pawan. 1958. On the fat-mobilising activity of human urine. Lancet 1:866–869.

Clark, W. M. 1952. Topics in physical chemistry. 2nd ed. The Williams and Wilkins Co., Baltimore.

Cohn, C. and D. Joseph. 1962. Influence of body weight and body fat on appetite of "normal," lean and obese rats. Yale J. Biol. Med. 34:598–607.

Cunningham, D. J. C., and B. B. Lloyd [eds.] 1963. The regulation of human respiration. F. A. Davis Co., Philadelphia.

Davis, R. A., and F. P. Brooks. 1962. Variability of gastric secretory response to insulin hypoglycemia in fistulous beagle dogs. Amer. J. Physiol. 202:1070–1072.

Davis, R. A., and F. P. Brooks. 1963. Gastric secretion, continuously recorded blood sugar, and plasma steroids after insulin. Amer. J. Physiol. 204:143–146.

Davson, H. 1964. The mechanism of contraction of muscle. pp. 895–991. In H. Davson, A textbook of general physiology. Little, Brown and Co., Inc., Boston.

Defares, J. G. 1963. On the use of mathematical models in the analysis of the respiratory control system. pp. 319–329. In D. J. C. Cunningham and B. B. Lloyd [eds.] The regulation of human respiration. F. A. Davis Co., Philadelphia.

Dhariwal, A. P. S., R. Nallar, M. Batt, and S. M. McCann. 1965. Separation of follicle-stimulating hormone-releasing factor from luteinizing hormone-releasing factor. Endocrinology 76:290–294.

Edsall, J. T., and J. Wyman. 1958. Biophysical chemistry. Vol. 1. Academic Press Inc., New York.

Epstein, A. N., and P. Teitelbaum. 1964. Severe and persistent deficits in thirst produced by lateral hypothalamic damage. pp. 395–410. In M. J. Wayner [ed.] Thirst. First Florida State University Symposium, Tallahassee. 1963. Pergamon Press, Ltd. Oxford.

Estep, H. L., D. P. Island, R. L. Ney, and G. W. Liddle. 1963. Pituitary-adrenal dynamics during surgical stress. J. Clin. Endocrinol. 23:419–425.

Euler, C. von. 1961. Physiology and pharmacology of temperature regulation. Pharmacol. Rev. 13:361–398.

Evans, J. R. [ed.] 1964. Structure and function of heart muscle. American Heart Association Monograph number nine. Circulation Res. Suppl. 2, Vol. 15.

Everett, J. W. 1964. Central neural control of reproductive functions of the adenohypophysis. Physiol. Rev. 44:373–431.

Fenn, W. O., and H. Rahn [eds.] 1964. Handbook of physiology. Sec. 3, Vol. I. Amer. Physiol. Soc., Washington.

Fisher, C., W. R. Ingram, and S. W. Ranson. 1938. Diabetes insipidus and the neurohormonal control of water balance: A Contribution to the structure and function of the hypothalamico-hypophyseal system. Edwards Bros. Inc., Ann Arbor.

Fox, R. H., and S. M. Hilton. 1958. Bradykinin formation in human skin as a factor in heat vasodilation. J. Physiol. 142:219-232.

Gagge, A. P. 1964. Body perspiration on exposure to thermal radiation. Fed. Proc. 23:2223 (abstr.).

Ganong, W. F. 1963. The central nervous system and the synthesis and release of adrenocorticotrophic hormone, pp. 92–149. In A. V. Nalbandov [ed.] Advances in neuroendocrinology. University of Illinois Press, Urbana.

Gasnier, A., and A. Mayer. 1939. Recherches sur la régulation de la nutrition. Ann. Physiol. Physicochim. Biol. 15:157–214.

Gergely, J. 1964. The biochemistry of muscle contraction. Little, Brown and Co. Inc., Boston.

Goodwin, B. C. 1963. Temporal organization in cells. Academic Press Inc., New York.

Gray, J. S. 1950. Pulmonary ventilation and its physiological regulation. Publication 63. American Lecture Series. Charles C Thomas, Springfield.

Greep, R. O. 1961. Physiology of the anterior hypophysis in relation to reproduction. pp. 240–301. In W. C. Young [ed.] Sex and internal secretion. 3rd. ed., Vol. I. The Williams and Wilkins Co., Baltimore.

Gregersen, M. I., and L. J. Cizek. 1961. Total water balance: thirst, fluid deficits and excesses. pp. 317–331. In P. Bard [ed.] Medical physiology. 11th ed. The C. V. Mosby Co., St. Louis.

Gregg, D. E. 1964. Coronary blood supply and oxygen usage of the myocardium. pp. 325-337. In F. Dickenson and E. Neil [eds.] Symposium on oxygen in the animal organism. Pergamon Press, Ltd., Oxford.

Gregory, H., P. M. Hardy, D. S. Jones, G. W. Kenner, and R. C. Sheppard. 1964. The antral hormone gastrin-structure of gastrin. Nature 204:931–933.

Gregory, R. A. 1962. Secretory mechanisms of the gastro-intestinal tract. Edward Arnold Ltd., London.

Gregory, R. A., and H. J. Tracy. 1964. The constitution and properties of two gastrins extracted from hog antral mucosa. Gut 5:103–117.

Grodins, F. S. 1963. Control theory and biological systems. Columbia University Press, New York.

Grossman, M. I. 1963. Integration of neural and hormonal control of gastric secretion. Physiologist 6:349–357.

Grossman, S. P. 1960. Eating or drinking elicited by direct adrenergic or cholinergic stimulation of hypothalamus. Science 132:301–302.

Grossman, S. P. 1964. A neuropharmacological analysis of the role of limbic and reticular mechanisms in motivation and learning. Bol. Inst. Estud. Med. Biol. (Mex.) 22:115–128.

Grosvenor, C. E., and C. W. Turner. 1957. Release and restoration of pituitary lactogen in response to nursing stimuli in lactating rats. Proc. Soc. Exp. Biol. Med. 96:723–725.

Grosvenor, C. E. and C. W. Turner. 1958. Assay of lactogenic hormone. Endocrinology 63:530–534.

Guillemin, R. 1964. Hypothalamic factors releasing pituitary hormones. Recent Progr. Hormone Res. 20:89–130.

Halasz, B., L. Pupp, and S. Uhlarik. 1962. Hypophysiotrophic area in the hypothalamus. J. Endocr. 25:147–154.

Hamilton, C. L., and J. R. Brobeck. 1964. Food intake and temperature regulation in rats with rostral hypothalamic lesions. Amer. J. Physiol. 207:291–297.

Hammel, H. T., D. C. Jackson, J. A. J. Stolwijk, J. D. Hardy, and S. B. Strømme. 1963. Temperature regulation by hypothalamic proportional control with an adjustable set point. J. Appl. Physiol. 18:1146–1154.

Hammel, H. T., S. Strømme, and R. W. Cornew. 1963. Proportionality constant for hypothalamic proportional control of metabolism in unanesthetized dog. Life Sci. 2:933–947.

Hardy, J. D. 1961. Physiology of temperature regulation. Physiol. Rev. 41:521–606.

Hardy, J. D., and H. T. Hammel. 1963. Control system in physiological temperature regulation. pp. 613–625. In Am. Inst. Physics. Temperature—its measurement and control in science and industry. Vol. III. Reinhold Publishing Corp., New York.

Hardy, J. D., R. F. Hellon, and K. Sutherland. 1964. Temperature-sensitive neurones in the dog's hypothalamus. J. Physiol. 175:242–253.

Harris, G. W. 1960. Central control of pituitary secretion. pp. 1007–1038. In J. Field [ed.] Handbook of physiology. Sec. 1, Vol. II. Amer. Physiol. Soc., Washington.

Harris, G. W. 1964a. Sex hormones, brain development and brain function. Endocrinology 75:627–648.

Harris, G. W. 1964b. The central nervous system and the endocrine glands. Triangle 6:242–251.

Harris, G. W., and D. Jacobsohn. 1952. Functional grafts of the anterior pituitary gland. Roy. Soc. (London), Proc., B. 139:263–276.

Hasselbach, W., and M. Makinose. 1961. Die Calcium-Pumpe der "Erschlaffungsgrana" des Muskels und ihre Abhängigkeit von der ATP-Spaltung. Biochem. Z. 333:518–528.

Hensel, H. 1952. Physiologie der Thermoreception. Ergebn. Physiol. 47:166–368.

Hodges, J. R., and M. T. Jones. 1963. The effect of injected corticosterone on the release of adrenocorticotrophic hormone in rats exposed to acute stress. J. Physiol. 167:30–37.

Hodgkin, A. L. 1958. Ionic movements and electrical activity in giant nerve fibers. Roy. Soc. (London), Proc., B. 148:1–37.

Hodgkin, A. L., and P. Horowicz. 1960. Potassium contractures in single muscle fibres. J. Physiol. 153:386–403.

Houssay, B. A., A. Biasotti, and R. Sammartino. 1935. Modifications fonctionnelles de l'hypophyse après les lésions infundibulo-tubériennes chez le crapaud. Compt. Rend. Soc. Biol. (Paris) 120:725–727.

Huxley, A. F. 1957. Muscle structure and theories of contraction. Progr. Biophys. 7:255–318.

Huxley, A. F. 1959. Local activation of muscle. Ann. N. Y. Acad. Sci. 81:446–452.

Huxley, A. F., and H. E. Huxley [eds.] 1964. A discussion on the physical and chemical basis of muscular contraction. Roy. Soc. (London), Proc., B. 160:433–547.

Huxley, H. E. 1960. Muscle cells. pp. 365–481. In J. Brachet and A. Mirsky [eds.] The cell. Vol. IV. Academic Press Inc., New York.

Igarashi, M., and S. M. McCann. 1964. A hypothalamic follicle stimulating hormone-releasing factor. Endocrinology 74:446–452.

Inchina, V. I., and Ya. D. Finkinshtein. 1965. Osmoreceptors and baroreceptors of the pancreas. Fed. Proc. 24:T189–191.

Jackson, D. C., and H. T. Hammel. 1963. Reduced set point temperature in exercising dog. USAF AMRL-TDR-63-93.

James, H., N. Nichols, and R. Phillips [eds.] 1947. Theory of servomechanisms. Radiation laboratory series. Vol. 25. McGraw-Hill Book Co., Inc. New York.

Jöbsis, F. 1964. Basic processes in cellular respiration. pp. 63–124. In W. O. Fenn and H. Rahn [eds.] Handbook of physiology. Sec. 3, Vol. I. Amer. Physiol. Soc., Washington.

Jones, A. W., E. O. Feigl, and L. H. Peterson. 1964. Water and electrolyte content of normal and hypertensive arteries in dogs. Circ. Res. 15:386–392.

Kanematsu, S. and C. H. Sawyer. 1964. Effects of hypothalamic and hypophysial estrogen implants on pituitary and plasma LH in ovariectomized rabbits. Endocrinology 75:579–585.

Katz, B. 1962. The transmission of impulses from nerve to muscle, and the subcellular unit of synaptic action. Roy. Soc. (London), Proc., B. 155:455–477.

Katzenelson, J., and L. Gould. 1962. The design of non-linear filters and control systems: Part I. Information and Control 5:108–143.

Katzenelson, J., and L. Gould. 1964. The design of non-linear filters and control systems: Part II. Information and Control 7:117–145.

Keller, A. D., and E. B. McClaskey. 1964. Localization, by the brain slicing method, of the level or levels of the cephalic brainstem upon which effective heat dissipation is dependent. Amer. J. Phys. Med. 43:181–213.

Kennedy, G. C. 1953. The role of depot fat in the hypothalamic control of food intake in the rat. Roy. Soc. (London), Proc., B. 140:578–592.

Kety, S. S., and C. F. Schmidt. 1948. The nitrous oxide method for the quantitative determination of cerebral blood flow in man: theory, procedure and normal values. J. Clin. Invest. 27:476–483.

Kleiber, M. 1961. The fire of life. John Wiley and Sons, Inc., New York.

Krogh, A. 1929. The anatomy and physiology of capillaries. Rev. ed. Yale University Press, New Haven.

La Barre, J., and C. de Cespédès. 1931. Rôle du système nerveux central dans l'hyper-sécrétion gastrique consécutive a l'administration d'insuline. Compt. Rend. Soc. Biol. (Paris) 106:1249–1251.

LaJoy, M. H. 1958. Industrial automatic controls. 3rd ed. Prentice-Hall Inc., Englewood.

Landis, E. M. 1927. Micro-injection studies of capillary permeability. Amer. J. Physiol. 82:217–238.

Langlois, K. J., R. K. S. Lim, G. Rosiere, D. I. Stewart, and D. L. Stumpff. 1952. Unconditioned orogastric secretory reflex. Fed. Proc. 11:88–89.

Lee, Y. W. 1960. Statistical theory of communication. John Wiley and Sons, Inc., New York.

Linden, R. J. 1963. The control of output of the heart. pp. 330–381. In R. Creese [ed.] Recent advances in physiology. 8th ed. Little, Brown and Co., Inc., Boston.

Lisk, R. D. 1960. Estrogen-sensitive centers in the hypothalamus of the rat. J. Exp. Zool. 145:197–208.

Long, J. F., and F. P. Brooks. 1965. The comparative gastric acid and pepsin response to histamine, insulin hypoglycemia and feeding in vagally innervated pouch dogs. Quart. J. Exp. Physiol. (in press).

Lotka, A. J. 1956. Elements of mathematical biology. Dover Publications Inc., New York.

McCann, S. M., and J. R. Brobeck. 1954. Evidence for a role of the supraopticohypophyseal system in regulation of adrenocorticotrophin secretion. Proc. Soc. Exp. Biol. Med. 87:318–324.

McCann, S. M., and V. D. Ramirez. 1964. The neuroendocrine regulation of hypophyseal luteinizing hormone secretion. Recent Progr. Hormone Res. 20:131–170.

McCubbin, J. W., J. H. Green, and I. H. Page. 1956. Baroceptor function in chronic renal hypertension. Circ. Res. 4:205-210.

McHugh, P. R., W. C. Black, and J. W. Mason. 1963. Some hormonal effects of electrical "self stimulation" in the lateral preoptic region. Physiologist 6:232 (abstr.).

McHugh, P. R., and G. P. Smith. 1964. CNS control of 17-hydroxycorticosteroid secretion. In Medical aspects of stress in a military climate. Symposium at Walter Reed Army Institute of Research. Washington. (in press).

Magnus, R., and E. A. Schäfer. 1901. The action of pituitary extracts upon the kidney. J. Physiol. (Proc.). 27:ix, x.

Masland, W. S., and W. S. Yamamoto. 1962. Abolition of ventilatory response to inhaled CO_2 by neurological lesions. Amer. J. Physiol. 203:789–795.

Mayer, A. 1900. Variations de la tension osmotique du sang chez les animaux privés de liquides. Compt. Rend. Soc. Biol. 52:153–155.

Mayer, J. 1953. Genetic, traumatic and environmental factors in the etiology of obesity. Physiol. Rev. 33:472–508.

Mayer, J. 1955. Regulation of energy intake and the body weight: the glucostatic theory and the lipostatic hypothesis. Ann. N. Y. Acad. Sci. 63:15–43.

Mayerson, H. S., C. G. Wolfram, H. H. Shirley, Jr., and K. Wasserman. 1960. Regional differences in capillary permeability. Amer. J. Physiol. 198:155–160.

Minard, D. 1963. Sweat rate during work and rest at elevated internal temperatures. Fed. Proc. 22:177 (abstr.).

Misher, A., and F. P. Brooks. 1964. Inhibition of gastric secretion during electrical stimulation of the ventromedial hypothalamus. Physiologist. 7:207 (abstr.).

Misher, A., and F. P. Brooks. 1965. Augmented interdigestive gastric secretion during electrical stimulation of the lateral hypothalamus. Fed. Proc. 24:406 (abstr.).

Mitchell, R. A., H. H. Loeschcke, W. H. Massion, and J. W. Severinghaus. 1963. Respiratory responses mediated through superficial chemosensitive areas on the medulla. J. Appl. Physiol. 18:523–533.

Montemurro, D. G., and J. A. F. Stevenson. 1957. Adipsia produced by hypothalamic lesions in the rat. Can. J. Biochem. 35:31–37.

Nahas, G. G. [ed.] 1963. Regulation of respiration. Ann. N. Y. Acad. Sci. 109:411–948.

Nakayama, T., H. T. Hammel, J. D. Hardy, and J. S. Eisenman. 1963. Thermal stimulation of electrical activity of single units of the preoptic region. Amer. J. Physiol. 204:1122–1126.

Nikitovitch-Winer, M. B. 1962. Induction of ovulation in rats by direct intrapituitary infusion of median eminence extracts. Endocrinology 70:350–358.

Nikitovitch-Winer, M., and J. W. Everett. 1958. Functional restitution of pituitary grafts retransplanted from kidney to median eminence. Endocrinology 63:916–930.

Papez, J. W. 1937. A proposed mechanism of emotion. Arch. Neurol. Psychiat. 38:725–743.

Pappenheimer, J. R. 1953. Passage of molecules through capillary walls. Physiol. Rev. 33:387–423.

Parlow, A. F. 1961. Bioassay of pituitary luteinizing hormone by depletion of ovarian ascorbic acid. pp. 300–310. In A. Albert [ed.] Human pituitary gonadotrophins. Charles C Thomas, Springfield.

Parlow, A. F. 1964a. Differential action of small doses of estradiol on gonadotrophins in the rat. Endocrinology 75:1–8.

Parlow, A. F. 1964b. Discussion of paper by McCann and Ramirez. Recent Progr. Hormone Res. 20:171–172.

Peters, J. P., and D. D. Van Slyke. 1931. Quantitative clinical chemistry. Interpretations. Vol. I. The Williams and Wilkins Co., Baltimore.

Peterson, L. H. 1962a. Properties and behavior of living vascular wall. Physiol. Rev. 42 (Suppl. 5):309–327.

Peterson, L. H. 1962b. Some studies of the regulation of cardiovascular functions: Arch. Int. Pharmacodyn. 140:281–290.

Peterson, L. H. 1964. Vessel wall stress-strain relationship. pp. 263–274. In E. O. Attinger [ed.] Pulsatile blood flow. McGraw-Hill Book Co., New York.

Pitts, R. F. 1963. Physiology of the kidney and body fluids. Year Book Medical Publishers, Inc., Chicago.

President's Commission on Heart Disease, Cancer and Stroke. 1964. Vol. 1. U. S. Government Printing Office, Washington.

Raub, W. F. 1965. Time series analysis of the blood-brain potential difference with reference to the regulation of respiration. (Doctoral dissertation.) University of Pennsylvania.

Renkin, E. M. 1964. Transport of large molecules across capillary walls. Eighth Bowditch Lecture. Physiologist 1:13–28.

Ridley, P. T., and F. P. Brooks. 1965. Alterations in gastric secretion following hypothalamic lesions producing hyperphagia. Amer. J. Physiol. (in press).

Rodahl, K., and B. Issekutz, Jr. [eds.] 1964. Fat as a tissue. McGraw-Hill Book Co., New York.

Rose, S., and J. F. Nelson. 1957. The direct effect of oestradiol on the pars distalis. Austr. J. Exp. Biol. Med. Sci. 35:605–610.

Rothchild, I. 1962. Relation of central nervous system, pituitary gonadotrophins, and ovarian hormone secretion. Fertil. and Steril. 13:246–258.

Roughton, F. J. W. 1954. Respiratory functions of blood. pp. 51–102. In W. M. Boothly [ed.] Handbook of respiration. Respiratory physiology in aviation. Air University, WSAF School of Aviation Medicine, Randolph Field, Texas.

Sayers, G., and M. A. Sayers. 1948. The pituitary-adrenal system. Recent Progr. Hormone Res. 2:81–115.

Schally, A. V., and C. Y. Bowers. 1964. Purification of luteinizing hormone-releasing factor from bovine hypothalamus. Endocrinology 75:608–614.

Schmidt-Nielsen, K. 1964. Terrestrial animals in dry heat: desert rodents. pp. 493–507. In D. B. Dill, E. F. Adolph, and C. G. Wilber [eds.] Handbook of physiology. Sec. 4. Adaptation to the environment. Amer. Physiol. Soc., Washington.

Schoenheimer, R. 1942. The dynamic state of body constituents. Harvard University Press, Cambridge.

Seed, J. C., F. S. Acton, and A. J. Stunkard. 1962. A model for the appraisal of glucose metabolism. Clin. Pharmacol. Ther. 3:191–215.

Sherrington, C. S. 1947. The integrative action of the nervous system. Cambridge University Press, London.

Siggard-Andersen, O. 1964. The acid-base status of the blood, 2nd ed. The Williams and Wilkins Co., Baltimore.

Smelik, P. G. 1963a. Failure to inhibit corticotrophin secretion by experimentally induced increases in corticoid levels. Acta Endocr. 44:36–46.

Smelik, P. G. 1963b. Relation between blood level of corticoids and their inhibiting effect on the hypophyseal stress response. Proc. Soc. Exp. Biol. Med. 113:616–619.

Smith, G. P., J. J. Boren, and P. R. McHugh. 1964. Gastric secretory response to acute environmental stress. In Medical aspects of stress in a military climate. Symposium at Walter Reed Army Institute of Research. Washington (in press).

Smith, G. P., and P. R. McHugh. 1964. Gastric secretory response to amygdaloid and hypothalamic stimulation in conscious *Macaca mulatta*. Physiologist 7:259 (abstr.).

Starling, E. H. 1894. The influence of mechanical factors on lymph production. J. Physiol. 16:224–267.

Starling, E. H. 1896. On the absorption of fluids from the connective tissue spaces. J. Physiol. 19:312–326.

Starling, E. H. 1918. Linacre lecture on the law of the heart. Longmans, Green and Co., London [See also: 1918, Brit. Med. J. 1:122.]

Steelman, S. L., and F. M. Pohley. 1953. Assay of the follicle stimulating hormone based on the augmentation with human chorionic gonadotrophin. Endocrinology 53:604–616.

Stevenson, J. A. F. 1964. Current reassessment of the relative functions of various hypothalamic mechanisms in the regulation of water intake. pp. 553–567. In M. J. Wayner [ed.] Thirst. First Florida State University Symposium, Tallahassee. 1963. Pergamon Press, Ltd., Oxford.

Stevenson, J. A. F., B. M. Box, and A. J. Szlavko. 1964. A fat mobilizing and anorectic substance in the urine of fasting rats. Proc. Soc. Exp. Biol. Med. 115:424–429.

Sydnor, K. L., and G. Sayers. 1954. Blood and pituitary ACTH in intact and adrenalectomized rats after stress. Endocrinology 55:621–636.

Szent-Györgyi, A. 1953. Chemical physiology of contraction in body and heart muscle. Academic Press Inc., New York.

Szentágothai, J. 1964. The parvicellular neurosecretory system. pp. 135–146. In W. Bargmann and J. P. Schadé [eds.] Lectures on the diencephalon. Progress in brain research. Vol. 5. Elsevier Publishing Co., New York.

Taleisnik, S., and S. M. McCann. 1961. Effects of hypothalamic lesions on the secretion and storage of hypophysial luteinizing hormone. Endocrinology 68:263–272.

Talwalker, P. K., A. Ratner, and J. Meites. 1963. *In vitro* inhibition of pituitary prolactin synthesis and release by hypothalamic extract. Amer. J. Physiol. 205:213–218.

Teitelbaum, P. 1961. Disturbances in feeding and drinking behavior after hypothalamic lesions. pp. 39–69. In M. Jones [ed.] Nebraska symposium on motivation. University of Nebraska Press, Lincoln.

Tepperman, J. 1962. Metabolic and endocrine physiology. The Year Book Publishers Inc., Chicago.

Tepperman, J., and J. R. Brobeck [eds.] 1960. Symposium on energy balance. Amer. J. Clin. Nutr. 8:527–774.

Tou, Julius T. 1959. Digital and sampled data control systems. McGraw-Hill Book Co., New York.

Truxal, J. G. 1955. Automatic feedback control system synthesis. McGraw-Hill Book Co., New York.

Tschirgi, R. D., and J. L. Taylor. 1958. Slowly changing bioelectric potentials associated with the blood-brain barrier. Amer. J. Physiol. 195:7–22.

Uvnas, B. 1942. The part played by the pyloric region in the cephalic phase of gastric secretion. Acta Physiol. Scand. 4 (Suppl. 13) :1–86.

Verney, E. B. 1947. The antidiuretic hormone and factors which determine its release. Roy. Soc. (London), Proc., B. 135:25–106.

Vernikos-Danellis, J. 1965. Effect of stress, adrenalectomy, hypophysectomy and hydrocortisone on the corticotropin releasing activity of rat median eminence. Endocrinology 76:122–126.

Waddell, W. J., and T. C. Butler. 1959. Calculation of intracellular pH from the distribution of 5, 5-dimethyl-2, 4-oxazolidinedione (DMO). Application to skeletal muscle of the dog. J. Clin. Invest. 38:720–729.

Weber, A., and R. Herz. 1963. The binding of calcium to actomysin systems in relation to their biological activity. J. Biol. Chem. 238:599–605.

Weber, A., R. Herz, and I. Reiss. 1963. On the mechanism of the relaxing effect of fragmented sarcoplasmic reticulum. J. Gen. Physiol. 46:679–702.

Weiss, B., and V. G. Laties. 1961. Behavioral thermoregulation. Science 133:1338–1344.

Welt, L. G. [ed.] 1957. Essays in metabolism. pp. 175–375. In The John Punnett Peters Number, Yale J. Biol. Med. 29:174–379.

Welt, L. G. 1959. Clinical disorders of hydration and acid-base equilibrium. 2nd ed. Little, Brown and Co., Inc., Boston.

Wettendorff, H. 1901. Modifications du sang sous l'influence de la privation d'eau: contribution a l'étude de la soif. Instituts Solvay 4:353–484.

Wiener, N. 1961. Cybernetics. 2nd ed. John Wiley and Sons, Inc., New York.

Winegrad, S. 1965. Autoradiographic studies of intracellular calcium in frog skeletal muscle. J. Gen. Physiol. 48:455–479.

Wolf, A. V. 1958. Thirst: physiology of the urge to drink and problems of water lack. Charles C Thomas, Springfield.

Worthington, W. C., Jr. 1960. Vascular responses in the pituitary stalk. Endocrinology 66:19–31.

Wyrwicka, W., and C. Dobrzecka. 1960. Relationship between feeding and satiation centers of the hypothalamus. Science 132:805–806.

Yamamoto, W. S. 1960. Mathematical analysis of the time course of alveolar CO_2. J. Appl. Physiol. 15:215–219.

Yamamoto, W. S. 1962. Transmission of information by the arterial blood stream with particular reference to carbon dioxide. Biophys. J. 2:143–159.

Yamamoto, W. S., and McI. W. Edwards, Jr. 1960. Homeostasis of carbon dioxide during intravenous infusion of carbon dioxide. J. Appl. Physiol. 15:807–818.

Yates, F. E., and J. Urquhart. 1962. Control of plasma concentrations of adrenocortical hormones. Physiol. Rev. 42:359–443.

Yates, F. E., S. E. Leeman, D. W. Glenister, and M. F. Dallman. 1961. Interaction between plasma corticosterone concentration and adrenocorticotropin-releasing stimuli in the rat: evidence for the reset of an endocrine feedback control. Endocrinology 69:67–80.

Index

345